Analysis of Marine Samples in Search of Bioactive Compounds

Comprehensive Analytical Chemistry

Volume 65

Analysis of Marine Samples in Search of Bioactive Compounds

Comprehensive Analytical Chemistry

Volume 65

Edited by

Teresa Rocha-Santos
ISEIT/Viseu, Instituto Piaget, Estrada do Alto do Gaio, Galifonge, Viseu, Portugal; CESAM & Department of Chemistry, University of Aveiro, Aveiro, Portugal

Armando C. Duarte
CESAM & Department of Chemistry, University of Aveiro, Aveiro, Portugal

ELSEVIER

AMSTERDAM • BOSTON • HEIDELBERG • LONDON • NEW YORK
OXFORD • PARIS • SAN DIEGO • SAN FRANCISCO • SINGAPORE
SYDNEY • TOKYO

Elsevier
Radarweg 29, PO Box 211, 1000 AE Amsterdam, Netherlands
The Boulevard, Langford Lane, Kidlington, Oxford OX5 1GB, UK
225 Wyman Street, Waltham, MA 02451, USA

British Library Cataloguing in Publication Data
A catalogue record for this book is available from the British Library

Library of Congress Cataloging-in-Publication Data
A catalog record for this book is availabe from the Library of Congress

ISBN: 978-0-444-63359-0

For information on all Elsevier publications
visit our website at http://store.elsevier.com/

Contents

Contributors to Volume 65

Armando C. Duarte, Department of Chemistry & CESAM, University of Aveiro, Campus de Santiago, Aveiro, Portugal

K. Duarte, CESAM & Department of Chemistry, University of Aveiro, Aveiro, Portugal

Regina M.B.O. Duarte, Department of Chemistry & CESAM, University of Aveiro, Portugal

A.C. Freitas, CESAM & Department of Chemistry, University of Aveiro, Aveiro, Portugal; ISEIT/Viseu – Instituto Piaget, Estrada do Alto do Gaio, Galifonge, Lordosa, Viseu, Portugal

A.M. Gomes, CBQF/Escola Superior de Biotecnologia, Catholic University, Rua Dr. António Bernardino de Almeida, Porto, Portugal

Shafiul Haque, Gene Expression Laboratory, Department of Biosciences, Jamia Millia Islamia (A Central University), New Delhi, India

C.I.L. Justino, Department of Chemistry & CESAM, University of Aveiro, Campus de Santiago, Aveiro, Portugal

Marco F.L. Lemos, ESTM - GIRM, Polytechnic Institute of Leiria, Peniche, Portugal

Sara Leston, CFE – Centre for Functional Ecology, Department of Life Sciences, University of Coimbra, Apartado Coimbra, Portugal; CEF – Center for Pharmaceutical Studies, Health Sciences Campus, Pharmacy Faculty, Coimbra University, Azinhaga de Santa Comba, Coimbra, Portugal

Mahesh Majik, Bioorganic Chemistry Laboratory, CSIR-National Institute of Oceanography, Dona Paula, Goa, India

Sofia Isabel G.H.M. Montalvão, Centre for Drug Research, Faculty of Pharmacy, Viikki Biocenter 2, P.O. Box 56 (Viikinkaari 5 E), University of Helsinki, Finland

Margarida Nunes, CFE – Centre for Functional Ecology, Department of Life Sciences, University of Coimbra, Apartado, Coimbra, Portugal; CEF – Center for Pharmaceutical Studies, Health Sciences Campus, Pharmacy Faculty, Coimbra University, Azinhaga de Santa Comba, Coimbra, Portugal

Miguel A. Pardal, CFE – Centre for Functional Ecology, Department of Life Sciences, University of Coimbra, Apartado, Coimbra, Portugal

Fernando Ramos, CEF – Center for Pharmaceutical Studies, Health Sciences Campus, Pharmacy Faculty, Coimbra University, Azinhaga de Santa Comba, Coimbra, Portugal; CNC – Centre for Neurosciences and Cell Biology, Health Sciences Campus, Pharmacy Faculty, Coimbra University, Azinhaga de Santa Comba, Coimbra, Portugal

Teresa Rocha-Santos, ISEIT/Viseu, Instituto Piaget, Estrada do Alto do Gaio, Galifonge, Lordosa, Viseu, Portugal; Department of Chemistry & CESAM, University of Aveiro, Campus de Santiago, Aveiro, Portugal

João Rosa, CFE – Centre for Functional Ecology, Department of Life Sciences, University of Coimbra, Apartado, Coimbra, Portugal; CEF – Center for Pharmaceutical Studies, Health Sciences Campus, Pharmacy Faculty, Coimbra University, Azinhaga de Santa Comba, Coimbra, Portugal

Clementina M.M. Santos, Department of Vegetal Production and Technology, School of Agriculture, Polytechnic Institute of Bragança, Bragança, Portugal

Artur M.S. Silva, Department of Chemistry & QOPNA, University of Aveiro, Aveiro, Portugal

Keisham S. Singh, Bioorganic Chemistry Laboratory, CSIR-National Institute of Oceanography, Dona Paula Goa, India

Vineeta Singh, Microbiology Division, CSIR-Central Drug Research Institute, Sector 10, Jankipuram Extension, Sitapur Road, Lucknow, Uttar Pradesh, India

Supriya Tilvi, Bioorganic Chemistry Laboratory, CSIR-National Institute of Oceanography, Dona Paula Goa, India

Series Editor's Preface

It is my pleasure to introduce this new volume on the *Analysis of Marine Samples in Search of Bioactive Compounds* edited by Teresa Rocha-Santos and Armando C. Duarte. The reason for having such a book in the Comprehensive Analytical Chemistry series is quite obvious: there has been considerable growth in the discovery of bioactive compounds from marine samples in the last few years as a consequence of their proven potential as antibacterial, antioxidant, antifungal, anticancer and antiviral agents among other properties. Marine bioactive compounds have been incorporated as food ingredients as well. So, overall marine bioactive compounds will play a key role in our society during the coming years. This book, the first of the CAC series covering aspects of the marine environment, should help to better characterise and analyse marine bioactive compounds.

This book contains ten chapters that cover sample preparation and analysis of marine bioactive compounds. The first four chapters are devoted to an introduction to the topic as well as the collection, preservation and preparation of samples. For obvious reasons sampling in the marine environment requires tailor-made methodologies to avoid cross-contamination. The following six chapters cover a broad range of analytical techniques, from rapid screening methods such as bioassays to more conventional instrumental methods using nuclear magnetic resonance, vibrational spectroscopy, mass spectrometry and chromatography using different detectors. In the last chapter novel techniques involving on-line bioassays combined with chemical and structural characterisation of the new compounds are reported.

Overall this book will be useful to a broad spectrum of professionals, varying from marine researchers, to analytical and food chemists as well as medical doctors interested in this topic. The book can be used as well for teaching purposes either during the current academic year or at specialized summer schoolson marine chemistry.

Finally I would like to thank the volume editors and all contributors to this book for their time and efforts in preparing this excellent and useful book on the analysis of bioactive compounds in marine samples.

June 12, 2014

Prof. D. Barceló
Dept. Environmental Chemistry
IDAEA-CSIC Barcelona
ICRA, Girona

Preface

Seas and oceans, from shallow coastal waters to the deep ocean, offer a wide diversity of marine organisms and a wide range of temperature, pressure, light, and chemical conditions. Such an availability of resources can be used to obtain new products and to develop services, and in turn these will help to bring solutions to some of the challenges that affect our planet, such as a sustainable supply of food and energy and the need for new industrial materials, and processes, by providing new bioactive compounds, and enabling the development of new potential health treatments. The use of such resources can also contribute to the development of greener and smarter economies. Marine compounds have been identified that have proven antibacterial, anticoagulant, antifungal, antimalarial, anticancer, antiprotozoal, antituberculosis, antioxidant, and antiviral activities. Since the discovery of their bioactive properties, among other applications they have been used in the production of anticancer drugs and other medicines, as well as incorporated as food ingredients.

The discovery of bioactive compounds from marine samples is a hot topic considering the current need of sustainable use of marine resources and this book introduces a comprehensive overview and future perspectives of analytical methodologies for the discovery and the characterization of bioactive compounds (from all classes of compounds) isolated from (all possible) marine sources. This overview includes the assessment of sampling and preparation of extracts, separation and isolation of bioactive compounds, structural characterization and the application of bioassays for the discovery of bioactive compounds.

Chapter 1 highlights the importance of bioactive compounds, marine sources/samples including marine by-products and fisheries waste streams, the current status in classes of bioactive compounds and their bioactivity, and also the general approaches for screening bioactive compounds. Chapter 2 describes the different sampling strategies that should to be taken into account to ensure the appropriate collection of micro and macroorganisms from natural environments. Chapter 3 and chapter 4 overview and discuss the state-of-the art of classical and green analytical methodologies, respectively, for sample treatment and also the state-of-the-art of direct analysis of samples (avoiding a sample treatment step). Chapter 5 introduces the different strategies that can be followed for bioactivity screening. Chapters 6 and 7 provide an overview and discuss the analytical methodologies associated with vibrational spectroscopy and nuclear magnetic resonance spectroscopy for structural characterization of bioactive compounds. Chapter 8 describes the state-of-the-art of mass spectrometry for the recognition of bioactive compounds while Chapter 9 describes the techniques of one- and two-dimensional chromatography, focusing on the advantages and limitations of such analytical methodologies when applied to the separation of bioactive compounds. Finally, Chapter 10 focuses on the online combination of bioassays with chemical and structural characterization that enable fast screening and identification of individual bioactive compounds with several bioactivities without prior purification and bioactivity assays.

The volume editors would like to thank the Comprehensive Analytical Chemistry (CAC) series editor, Prof. Damiá Barceló, for giving us the opportunity to coordinate such an interesting thematic work on "Analysis of Marine Samples in Search of Bioactive Compounds". All the help and advice from Susan Dennis, the acquisitions editor, and Derek Coleman, the senior editorial project manager, are greatly acknowledged. Thanks are also due to the authors who helped us to assemble a close set of chapters into a book targeting a broad spectrum of readers ranging from the researchers in the field to undergraduate and graduate students interested in such an overview and a primary source of information for pursuing further studies.

Teresa Rocha-Santos and
Armando C. Duarte

Chapter 1

Introduction to the Analysis of Bioactive Compounds in Marine Samples

Teresa Rocha-Santos[1] and Armando C. Duarte[2]
[1]*ISEIT/Viseu, Instituto Piaget, Estrada do Alto do Gaio, Galifonge, Lordosa, Viseu, Portugal; Department of Chemistry & CESAM, University of Aveiro, Campus de Santiago, Aveiro, Portugal*
[2]*Department of Chemistry & CESAM, University of Aveiro, Campus de Santiago, Aveiro, Portugal*

Contents

1.1 IMPORTANCE OF BIOACTIVE COMPOUNDS

The marine environment occupies approximately three quarters of the surface of the Earth and offers a wide range of conditions of temperature, pressure, light, and chemistry, as wells as a wide diversity of marine organisms, from shallow coastal waters to the deep ocean [1]. These resources can be used to obtain new products and to develop new services, presenting solutions regarding the challenges that affect our planet. Potential solutions include a sustainable supply of food and energy, new industrial materials and processes, new bioactive compounds, and new health treatments. The use of these resources not only creates jobs and improves health, but it can also contribute to the development of greener and smarter economies.

Marine bioactive compounds are natural compounds, derived and isolated from biological sources, that have biological activity and are isolated from

Analysis of Marine Samples in Search of Bioactive Compounds, Vol. 65. DOI: 10.1016/B978-0-444-63359-0.00001-X

marine sources. The natural products also refer to secondary metabolites; that is, small molecules with molecular weight (MW) less than 2 kDa, produced by an organism that is not strictly essential for its survival [2]. The first discovery of a bioactive marine natural product was reported in the late 1950s by Bergmann [3]. The arabino and ribo-pentosyl nucleosides extracted from marine sponges were the first demonstration that naturally occurring nucleosides could contain sugars besides ribose and desoxyribose. Chemical synthesis allowed the development of two derivatives (vidarabine and cytarabine) that are two nucleosides with antiviral activity. These two nucleosides have been applied in clinical treatments for decades [3].

Concerning their applications, the marine bioactive compounds have been used mainly due to their antibacterial, anticoagulant, antifungal, anti-inflammatory, antimalarial, antiprotozoal, antituberculosis, antioxidant, antidiabetic, anticancer, and antiviral activities [4]. Additionally, marine compounds have been found to affect the immune and nervous systems [4, 5]. Schumacher *et al.* [5] reviewed the marine compounds acting on the six hallmarks of cancer; that is, the compounds that abolish self-sufficiency in growth signals, compounds that reestablish sensitivity to growth-inhibitory signals, compounds that lead to apoptosis, antiangiogenic compounds, compounds that reduce the replicative potential, and compounds that prevent tissue invasion and metastasis, demonstrating that the marine environment is rich with organisms useful for the isolation of compounds with anticancer activities.

Mayer *et al.* [4] reviewed the marine compounds used in pharmacology accordingly with their origins (organisms) and chemical classes. They found that the studies about the molecular mechanisms of action of discovered bioactive compounds are lacking, remaining undetermined in the major works at the time of publication. Since the mechanism of action is unknown, and due to the loss of the biological activity when tested *in vivo*, the majority of bioactive compounds isolated from marine organisms do not attain the clinical trials. According to Sawadogo *et al.* [6], concerning the anticancer compounds from marine origin, among the 83% of the compounds tested *in vitro*, the biological mechanisms of action of about 45% are unknown, with only 2% in clinical trial and 14% already tested *in vivo*. Thus, such specific biochemical interaction through which a drug substance produces its pharmacological effect is often unclear for the newly discovered compounds, which represents a challenging task.

1.2 SOURCES OF BIOACTIVE COMPOUNDS

Marine organisms that constitute approximately one half of the total global biodiversity are rich reservoirs of structurally diverse biofunctional components. So far, more than 22,000 compounds have been isolated from marine organisms [7] such as sponges, coelenterates, microorganisms, algae, echinoderms, mollusks, bryozoans, and from other sources such as marine processing wastes. Figure 1.1 gives an overview of marine derived sources used from 2010 to 2013

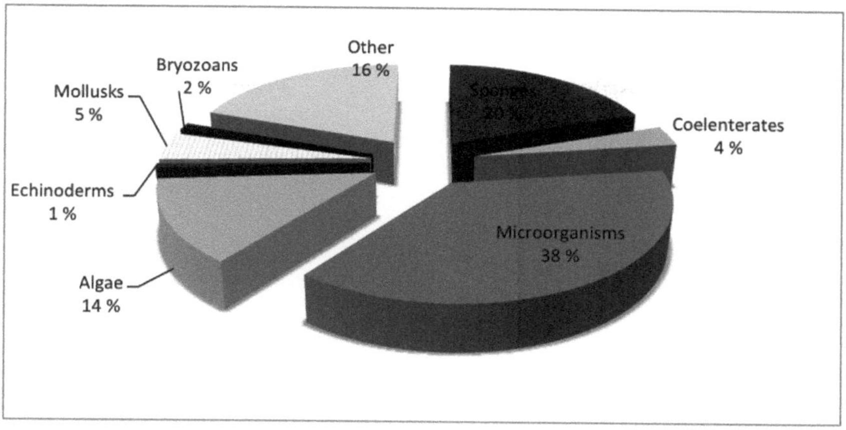

FIGURE 1.1 Marine-derived sources used on the search of new bioactive compounds.

for the discovery of new bioactive compounds. Roughly more than one-third of all new bioactive compounds reported from marine sources are derived from microorganisms. About two-fifths of new marine bioactive compounds are derived from microorganisms, including bacteria, microalgae, and fungi, and about one-fifth of the new bioactive compounds are from other marine sources such as fish processing waste and ascidian (tunicates).

Table 1.1 shows a selection of examples [8–16] of bioactive natural products from sources such as bacteria, fungi, sponge, microalgae, starfish, algae, ascidian and fish waste. There are several reviews covering the marine natural products from differences sources [7, 17–21].

The marine sponges are multicellular invertebrates attached to solid substrates in benthic habitats, and they have been isolated as sources of sterols, such as steroids, terpenoids, alkaloids, cyclic peptides, and unsaturated fatty acids [18].

The marine coelenterates include the sea anemones, the true jellyfish, and the hydroids [22]. Coelenterates have been a source of several bioactive compounds such as steroids, terpenes, and glycosides. These compounds showed antitumor activity against cancer cell lines.

The microorganisms group, and in particular the marine bacteria, have been used to isolate bioactive secondary metabolites with pharmacodynamic properties such as antitumor, antivirus, and enzyme inhibitor properties, which are important in the area of the drug discovery [23]. More than 600 marine bioactive compounds have been isolated from marine bacteria, mostly from cyanobacteria, and actinobacteria due to their particular metabolic and physiological capabilities [24]. The marine cyanobacteria, which are photosynthetic prokaryotes, are considered as excellent sources of antineoplastic, antibacterial, antiviral, and antifungal compounds such as toxins. Recently, Pagliara and Caroppo [25] have investigated the bioactive properties of cyanobacterial strains that belong

TABLE 1.1 Selected Examples of Bioactive Marine Natural Products from Different Sources

Compound	Source	Bioactivity	Reference
Veromycin E	Bacteria/*Streptomyces sp*	Antibacterial Antifungal Anticancer	[8]
Penicitide A	Fungus/*Penicillium chrysogenum*	Cytotoxicity	[9]
Simplextones A	Sponge/*Plakortis simplex*	Cytotoxicity	[10]
Bouillonamide	Microalgae/*Moorea bouillonii*	Cytotoxicity	[11]
Astrosteriosides A	Starfish/*Astropecten monacanthus*	Anti-inflammatory	[12]
Fucoxanthin	Algae/*Sargassum siliquastrum*	Antioxidant	[13]
S. plicata dermatan sulfate	Ascidian/*Styela plicata*	Anti-inflammatory	[14]
Tyr-Asn	Fish waste/clam/*Meretrix lusoria*	ACE inhibitory	[15]
Eusynstyelamide B	Bryozoan/*Tegella cf. spitzbergensis*	Antimicrobial	[16]

to the *Leptolyngbya* and *Synechococcus* genera, isolated from the Mediterranean sponge *Petrosia ficiformis.* The bioassays performed with the obtained aqueous extracts showed that such cyanobacteria exhibited citolytic effect on human erythrocytes and toxic activity against the crustacean *Artemia salina nauplii.* The marine actinobacteria, mainly including the genera Streptomyces, Actinomyces, Arthrobacter, and Micrococcus, are also responsible to the production of pharmaceutically bioactive compounds with disparate chemical structures in order to combat resistant pathogens, as recently reviewed by Manivasagan *et al.* [26]. The marine actinomycetes have been investigated for their bioactive metabolites such as salinosporamides, useful for therapeutic applications, and are the group of the streptomycetes most economically viable due to the possibility of extracting a variety of antimicrobial metabolites and antibiotics such as vancomycin, penicillin, cephalosporines, and streptomycin, as reviewed by Subramani and Aalbersberg [27].

The marine fungi have also contributed to the discovery of bioactive compounds, especially those isolated from other marine organisms such as sponges and algae [28]. Other bioactive compounds isolated from fungi have been recently reported due to their anticancer, antibacterial, and antiviral properties, among others [29]. Marine-derived fungi were the source of more than 690 new bioactive compounds from 2006 to the middle of 2010 [17].

The potential of macroalgae as a source of bioactive nitrogenous compounds has been reviewed elsewhere [30]. The marine algae are rich in sulphated polysaccharides such as fucoidans (mainly in brown algae), carrageenans (mainly in red algae), and ulvans (mainly in green algae), which have anticoagulant, anti-allergic, antiviral, antioxidative, and anticancer activities with great potential in nutraceutical, pharmaceutical, and cosmeceutical areas [31, 32]. The marine algae must adapt rapidly to new environmental conditions contributing to their survival, producing a wide variety of secondary bioactive metabolites. According to Raposo *et al.* [33], among the various marine algae, the microalgae (unicellular microscopic organisms) have been recently highlighted as the new source of marine bioactive compounds, such as polyunsaturated fatty acids, sterols, proteins, enzymes, and vitamins, due to their anti-inflammatory, antiallergic, and analgesic properties, useful for health applications.

Tunicates (ascidians) and mollusks were the source of new bioactive peptides with pharmacological properties such as cytotoxic and antimicrobial properties [34]. Recently, the marine mussels have been studied for their bioactive compounds with potential use in human health. The three major chemical classes of primary metabolites of mussel are proteins, lipids, and carbohydrates, with lipids being the most requested due to their use in the prevention and treatment of rheumatoid arthritis [21].

Marine organisms such as fish and shellfish are rich sources of several bioactive nitrogenous compounds, which show significant potential as functional food ingredients. Bioactivities such as antihypertensive, antioxidant, antimicrobial, anticoagulant, antidiabetic, anticancer, immunostimulatory, calcium-binding, hypocholesteremic, and appetite suppression have been reported [35]. Large quantities of marine processing by-products, including substandard muscles, viscera, heads, skins, fins, frames, trimmings, shellfish and crustacean shell waste, are generated each year, and represents a source for bioactive compounds mining. These by-products generally are either discarded or used as animal feed or fertilizers. Conversion of marine waste/discard/by-products into high value functional ingredients may provide a solution for dealing with the legal restrictions, high cost, and environmental problems associated with disposal of such waste material [35].

1.3 CLASSES OF BIOACTIVE COMPOUNDS

Many classes of marine-sourced compounds have been reviewed including briarane-type diterpenoids; cyclic polypeptides containing b-amino acid fragments; pyrroloiminoquinone alkaloids; antitumor peptides; cyclic imines including spirolides and gymnodimines; kahalalides; guanidine-containing alkaloids; ascidian-derived alkaloids; 2-aminoimidazole alkaloids; other assorted alkaloids; carotenoids; algal bromophenols; volatile algal halogenated compounds; brominated compounds from Aplysina sponges; furanocembranoids and norcembranoids; a-conotoxins, cladiellins, asbestinins, and briarellins; and eleutherobins, fuscosides, and pseudopterosins [19]. Mayer *et al.* [4, 36] used a modification of Schmitz's chemical classification to assign marine natural product structures to six major chemical classes: polyketides, terpenes, peptides, alkaloids, shikimates, and sugars (Table 1.2).

1.3.1 Polyketides

Polyketides constitute a highly diverse class of compounds both in terms of structure and biological activities. Compounds of this class are often highly oxygenated; they include macrolides, polyethers, polyols, and aromatics. Their biological activities are varied, and include antibiotic, anticancer, antifungal, antiparasitic, immunosuppressive, and neurotoxic activities [40, 41].

The structure of Reveromycin E (Table 1.2), which has antifungal, anticancer, and antibacterial activities, was isolated by Fremlin *et al.* [8] from the marine-derived *Streptomyces* spp. Penicitide A, a polyketide isolated from the fungus *Penicillium chrysogenum,* which exhibits moderate citotoxicity against human hepatocellular liver carcinoma cell line [9]. Bouillonamide, a new

TABLE 1.2 Selected Examples of Bioactive Marine Natural Product Structures to Six Major Chemical Classes: Polyketides, Terpenes, Peptides, Alkaloids, Shikimates, and Sugars

Class/ Structural Biogenic Type	Stucture/Compound Name	Reference
Polyketide	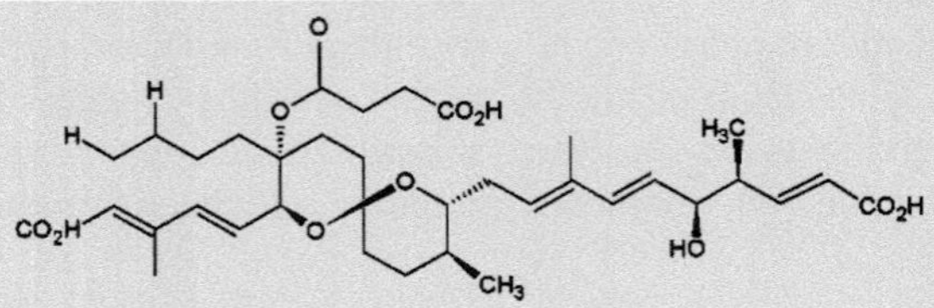 Reveromycin E	[8]
Terpene	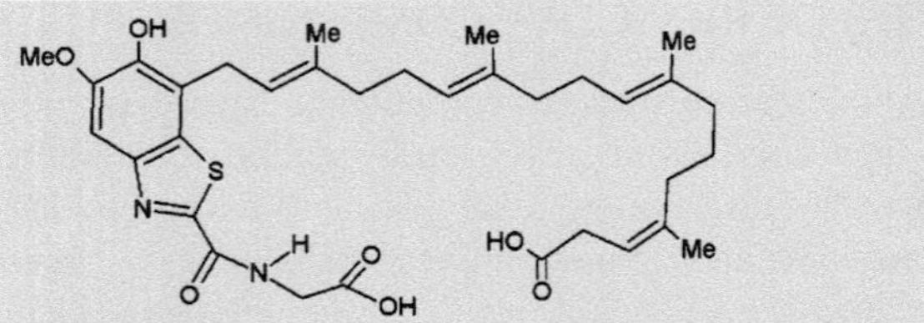 Erythrazole B	[37]

TABLE 1.2 Selected Examples of Bioactive Marine Natural Product Structures to Six Major Chemical Classes: Polyketides, Terpenes, Peptides, Alkaloids, Shikimates, and Sugars *(cont.)*

Class/ Structural Biogenic Type	Stucture/Compound Name	Reference
Peptide	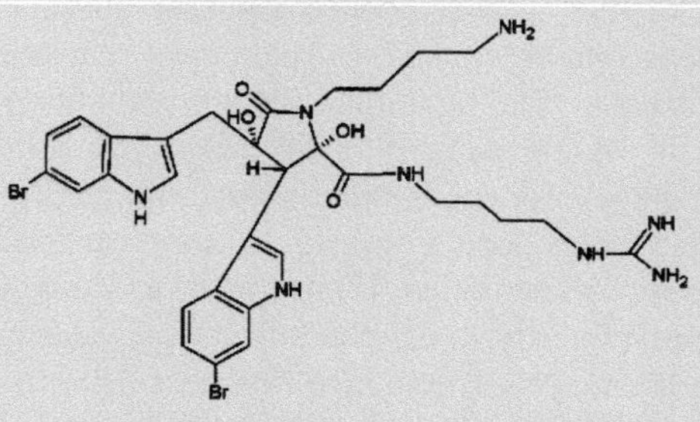 Eusynstyelamide B	[16]
Alkaloide	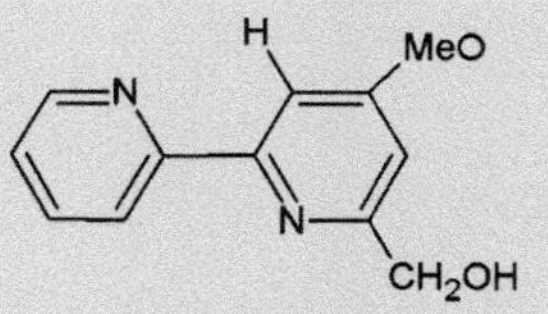 Caerulomycin F	[38]
Shikimate	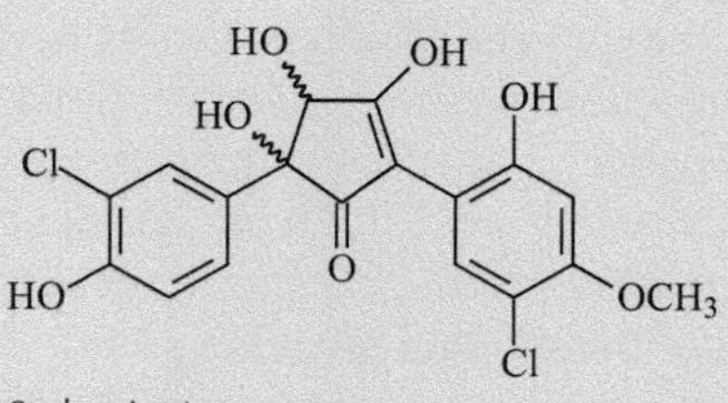 Sydowin A	[17]
Sugar	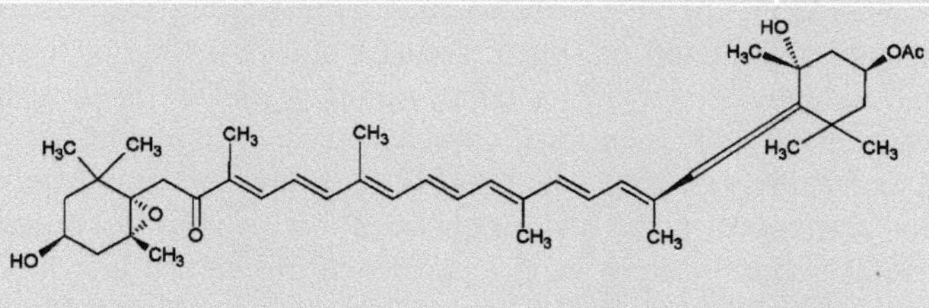 Fucoxanthin	[39]

mixed polyketide cytotoxin, isolated from the marine cyanobacterium *Moorea bouillonii*, showed a mild toxicity against neuro-2a mouse neuroblastoma cells [11].

1.3.2 Terpenes

Terpenes are primary and secondary metabolites produced by marine organisms to ensure their survival. These compounds are derived from a five-carbon isoprene structure, and depending on the combination of units, can be subdivided into biogenetic classes such as monoterpenes, sesquiterpenes, diterpenes, sesterpenes, triterpenes (steroids), and tertraterpenes (carotenoids) [42, 43]. Marine terpenes can have several functional groups, including isonitrile, isocyanate, dichloroimine, and halogenated [43]. Several groups of marine organisms produce terpenes to compete for space and reproduction, thus these compounds present different bioactivities such as cytotoxic, antiproliferative, antifouling, antifungal, and antimicrobial activities. Therefore the exploitation of such compounds has been of great interest to the pharmaceutical industry [42]. Hu *et al.* [37] found diterpene erythrazole B in active extracts of *Erythrobacter* sp., and this compound showed citotoxic properties against non-small-cell lung cancer (NSCLC) cell lines. Tsukada *et al.* [44] isolated the decarboxyhydroxycitrinone, a terpene, from the marine fungus *Arthrinium sacchari*, with antiangiogenic activity.

1.3.3 Peptides

Bioactive peptides are specific protein fragments that are inactive within parent protein, but once released impart positive functions or benefits to human health [45]. They act as sources of nitrogen and amino acids and have numerous potential physiological functions within the body such as opioid, immunomodulatory, antibacterial, antithrombotic, and antihypertensive activity. In general, biofunctional peptides range in size from 2 to 20 amino acid residues and are encrypted within the sequence of the parent protein and may be released during gastrointestinal digestion and/or during food processing [46].

In the last decade, studies focused on marine bioactive peptides have been increased [35, 46]. Due to their structural properties, sequence, and amino acid residues, marine bioactive peptides have shown several biological functions of pharmaceutical interest, including opioid agonistic, mineral binding, immunomodulatory, antimicrobial, antioxidant, antithrombotic, hypocholesterolemic, and antihypertensive functions [47]. Besides extraction from marine sources, bioactive peptides can also be produced by *in vitro* enzymatic hydrolysis of several marine sources using appropriate proteolytic enzymes. According to Harnedy and Fitzgerald [35], proteolytic enzymes from fish and aquatic invertebrates can be used for the hydrolysis process of marine products to develop bioactive peptides, as applied in the food industry. Tadesse *et al.* [16] discovered from the Arctic bryozoan *Tegella cf. spitzbergensis* the Eusynstelamide B

(Table 1.2) that is a brominated tryptophan-derived compound with antimicrobial activity against *Staphylococcus aureus.*

1.3.4 Alkaloids

Alkaloids were defined by Pelletier *et al.* [48] as "cyclic organic compounds containing nitrogen in a negative oxidation state which is of limited distribution among living organism." These metabolites have been subdivided by Kumar *et al.* [49] into seven subclass: pyridoacrine alkaloids (tetracyclic, pentacyclic, hexacyclic, heptacyclic, and octacyclic alkaloids), indole alkaloid (bisindole, indolocarbazole, ergoline, peptidoindole, β–carboline, and trisindole alkaloids), pyrrole alkaloid (bromopyrroles, pyrroloquinones, pryrrodoquinoline alkaloids, pyrrodoacridines), isoquinoline alkaloids, gadinine alkaloids, aminoimidazole alkaloids, and sterol alkaloids. Alkaloids produced by endophytic fungi and algae were reviewed by Zhang *et al.* [50] and Güven *et al.* [51] and have several bioactivities such as antiviral, antibacterial, anti-inflammatory, antimalarial, antioxidant, and anticancer. The caerulomycin F (Table 1.2) has been extracted by Fu *et al.* [38] from the bacteria *Actinoalloteichus cyanogriseus*, and it has shown cytotoxic activity. As another example, Lee *et al.* [52] isolated a new diketopiperazine-type alkaloid, Protubonines A, from the marine-derived *Aspergillus* sp. SF-5044. This compound was tested to a growth inhibitory activity against cell lines, including human acute promyelocytic leukemia cells (HL-60), human breast cancer adenocarcinoma cells (MDA-MB-231), hepatocellular carcinoma cells (Hep3B), and chronic myelogenous leukemia cells.

1.3.5 Shikimates

Shikimates are metabolite derivatives from the shikimate pathway, produced by bacteria, fungi, plants, and some protozoaroans for the biosynthesis of some aromatic amino acids, such as L-phenylalanine, L-tyrosine, and L-tryptophan. This pathway is not present in animals, and the shikimate-derived aromatic amino acids have been obtained through diet [53]. Chrysophaentin A was isolated by Plaza *et al.* [54] from the marine chrysophyte alga *Chrysophaeum taylori* and tested against several multiresistant bacteria, including methicilin-resistant *Staphylococcus aureus*, multidrug resistant *Staphylococcus aureus*, and vancomycin-resistant *Enteroccocus faecium.*

1.3.6 Sugars

The major class of carbohydrates is polysaccharides, which can be isolated from marine organisms, mostly from seaweeds, since they produce large quantities of agar, alginates, agarose, carrageenans, and fucoidans. Polysaccharides can be used for several applications in the biotechnology, pharmacy, and chemical industries, such as gel production, drug delivery systems, wound healing, tissue engineering, and blood dialysis membranes. In addition, these metabolites display bioactive properties such as anticytotoxic, antimutagenic, antitumorgenic, hypocholosterolemic, and anticoagulant activities. Zhang *et al.* [55] evaluated antioxidant and anticoagulant activities of a sulfated polysaccharide, a porphyran, extracted from a red algae *Porphyra haitanensis*.

1.4 GENERAL APPROACHES FOR SCREENING BIOACTIVE COMPOUNDS

There are two main approaches used in the discovery of bioactive compounds from extracts: bioassay-guided fractionation and pure compound screening [56]. Figure 1.2 provides a schematic comparison of the essential steps in the search of bioactive compounds by both approaches.

The bioassay-guided fractionation displayed in Figure 1.2A is the most used procedure, and it involves the following steps: assessment of the potential bioactivity using a batch biological assay; extraction by the use of different combinations of solvents followed by bioactivity assessment; repeated fractionation of bioactive extracts and fractions until the successful isolation of the bioactive compounds; structural characterization of the bioactive components using several spectroscopic techniques; and pharmacological and toxicological testing. In the bioassay-guided fractionation procedure the extracts can be excluded

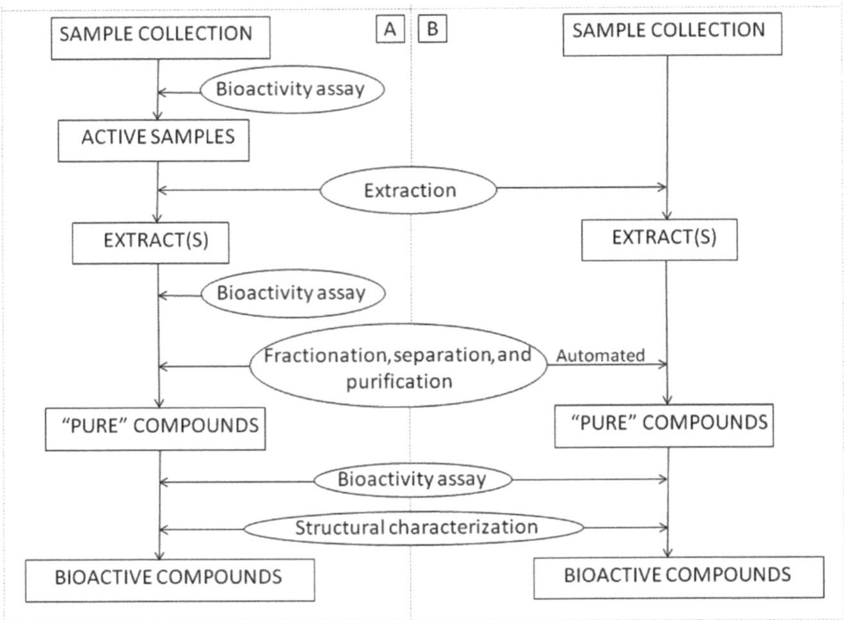

FIGURE 1.2 Comparative scheme of the essential steps in the search of bioactive compounds by (A) bioassay-guided fractionation and (B) pure compound screening reprinted from Duarte *et al.* [29]. Copyright (2012), with permission from Elsevier.

from further work, when they do not show bioactivity. Kjer *et al.* [57] observed that it takes at least three months from sampling fungi in the marine environment until completion of the structure elucidation and the bioactivity screening. In this procedure, the steps that consume the most time and resources are the repeated fractionation and the structural characterization, which many times lead to isolated compounds that are either already known or uninteresting. The main advantage of this procedure is the possible exclusion of extracts and fractions from further work, while the main disadvantage is the extensive usage of expensive resources such as time and solvents.

The procedure displayed in Figure 1.2B, pure compound screening, is a procedure less frequently used, and usually has the following steps: first, it starts off with an automated isolation and structure elucidation of the compounds in the extract; second, it assesses whether the purified and structurally elucidated compounds have the desired bioactivity using a biological assay; and third, it tests the pharmacological and toxicological activity. In this case it is necessary to select extracts containing compounds that are not present in the available libraries of pure compounds. The pure compounds screening is labor-intensive and time-consuming and can easily lead to compounds without bioactivity, since bioactivity is checked only after isolation and structure elucidation.

Other less common approaches are the online combination of bioassays and chemical and structural characterization [58]. These approaches have the advantages of screening and identifying bioactive compounds without the need for cumbersome purification and subsequent screening.

Both procedures should identify as many bioactive fractions and compounds in the extracts as possible at the earliest stage, in order to avoid redundant work and waste of resources.

REFERENCES

[1] Green Paper, From Challenges to Opportunities: Towards a Common Strategic Framework for EU Research and Innovation funding, European Commission, 2011, available at: http://ec.europa.eu/research/csfri/pdf/com_2011_0048_csf_green_paper_en.pdf#view=fit&pagemode=none. Last accessed on 09/01/2014.

[2] V. Seidel, In: S.D. Sarker, Z. Latif, A.I. Gray (Eds.), Methods in biotechnology, Natural Products Isolation, Vol. 20, Humana Press Inc, 2006, pp. 27–46.

[3] J.F. Imhoff, A. Labes, J. Wiese, Biotechnol. Adv. 29 (2011) 468–482.

[4] A.M.S. Mayer, A.D. Rodríguez, O. Taglialatela-Scafati, N. Fusetani, Mar. Drugs 11 (2013) 2510–2573.

[5] M. Schumacher, M. Kelkel, M. Dicato, M. Diederich, Biotechnol. Adv. 29 (2011) 531–547.

[6] W.R. Sawadogo, M. Schumacher, M.-H. Teiten, C. Cerella, M. Dicato, M. Diederich, Molecules 18 (2013) 3641–3673.

[7] J.W. Blunt, B.R. Copp, M.H. Munro, P.T. Northcote, M.R. Prinsep, Nat. Prod. Rep. 28 (2011) 196–268.

[8] L. Fremlin, M. Farrugia, A.M. Piggott, Z. Khalil, E. Lacey, R.J. Capon, Org. Biomol. Chem. 9 (2011) 1201–1211.

[9] S.-S. Gao, X.-M. Li, F.-Y. Du, C.-S. Li, P. Proksch, B.-G. Wang, Mar. Drugs 9 (2011) 59–70.
[10] X.-F. Liu, Y.-L. Song, H.-J. Zhang, F. Yang, H.-B. Yu, W.-H. Jiao, S.-J. Piao, W.-S. Chen, H.-W. Lin, Org. Let. 13 (2011) 3154–3157.
[11] L.T. Tan, T. Okino, W.H. Gerwick, Mar. Drugs 11 (2013) 3015–3024.
[12] N.P. Thao, N.X. Cuong, B.T.T. Luyen, N.V. Thanh, N.X. Nhiem, Y.-S. Koh, B.M. Ly, N.H. Nam, P.V. Kiem, C.V. Minh, Y.H. Kim, J. Nat. Prod. 76 (2013) 1764–1770.
[13] S.-J. Heo, Y.-J. Jeon, J. Photochem. Photobiol. 95 (2009) 101–107.
[14] C.L.R. Belmiro, M.T.L. Castelo-Branco, L.M.C. Melim, A. Schanaider, C. Elia, K. Madi, M.S.G. Pavão, H.S.P. de Souza, J. Biol. Chem. 284 (2009) 11267–11278.
[15] J.-S. Tsai, J.-L. Chen, B.S. Pan, Process Biochem. 43 (2008) 743–747.
[16] M. Tadesse, J.N. Tabudravu, M. Jaspars, M.B. Strøm, E. Hansen, J.H. Andersen, P.E. Kristiansen, T. Haug, J. Nat. Prod. 74 (2011) 837–841.
[17] M.E. Rateb, R. Ebel, Nat. Prod. Rep. 28 (2011) 290–344.
[18] T.R.A. Thomas, D.P. Kavlekar, P.A. LokaBharathi, Mar. Drugs 8 (2010) 1417–1468.
[19] J.W. Blunt, B.R. Copp, R.A. Keyzers, M.H.G. Munroa, M.R. Prinsep, Nat. Prod. Rep. 30 (2013) 237–323.
[20] J.W. Blunt, B.R. Copp, R.A. Keyzers, M.H.G. Munroa, M.R. Prinsep, Nat. Prod. Rep. 29 (2012) 144–222.
[21] U. Grienke, J. Silke, D. Tasdemir, Food Chem. 142 (2014) 48–60.
[22] G.A. Lewbart, M. Stoskopf, Coelenterates, in Invertebrate Medicine, Blackwell Publishing Professional, (2008).
[23] C. Nikapitiya, Adv. Food Nutr. Res. 65 (2012) 363–387.
[24] P.G. Williams, Trends Biotechnol. 27 (2009) 45–52.
[25] P. Pagliara, C. Caroppo, Toxicon. 57 (2011) 889–896.
[26] P. Manivasagan, J. Venkatesan, K. Sivakumar and S.-K. Kim, Microbiol. Res., *in* press.
[27] R. Subramani, W. Aalbersberg, Microbiol. Res. 167 (2012) 571–580.
[28] Y. Ohkawa, K. Miki, T. Suzuki, K. Nishio, T. Sugita, K. Kinoshita, K. Takahashi, K. Koyama, J. Nat. Prod. 73 (2010) 579–582.
[29] K. Duarte, T.A.P. Rocha-Santos, A.C. Freitas, A.C. Duarte, Trends Anal. Chem. 34 (2012) 97–110.
[30] P.A. Harnedy, R.J. Fitzgerald, J. Phycol. 47 (2011) 218–232.
[31] I. Wijesekara, R. Pangestuti, S.-K. Kim, Carbohydr. Polym. 84 (2011) 14–21.
[32] A. Jiménez-Escrig, E. Gómez-Ordóñez, P. Rupérez, Adv. Food Nutr. Res. 64 (2011) 325–337.
[33] M.F.J. Raposo, R.M.S.C. Morais, A.M.M.B. Morais, Life Sci. 93 (2013) 479–486.
[34] A. Aneiros, A. Garateix, J. Chromatogr. B 803 (2004) 41–53.
[35] P.A. Harnedy, R.J. Fitzgerald, J. Funct. Food 4 (2012) 6–24.
[36] A.M.S. Mayer, A.D. Rodríguez, R.G.S. Berlinck, N. Fusetani, Comp. Biochem. and Physiol. C 153 (2011) 191–222.
[37] Y. Hu, J.B. MacMillan, Org. Lett. 13 (2011) 6580–6583.
[38] P. Fu, S. Wang, K. Hong, X. Li, P. Liu, Y. Wang, W. Zhu, J. Nat. Prod. 74 (2011) 1751.
[39] J. Staunton, K.J. Weissman, Nat. Prod. Rep. 18 (2001) 380–416.
[40] C. Hertweck, Angew. Chem. Int. Ed. 48 (2009) 4688–4716.
[41] H. Gross, G.M. König, Phytochem. Rev. 5 (2006) 115–141.
[42] R. Ebel, Mar. Drugs 8 (2010) 2340–2368.
[43] M. Tsukada, M. Fukai, K. Miki, T. Shiraishi, T. Suzuki, K. Nishio, T. Sugita, M. Ishino, K. Kinoshita, K. Takahashi, M. Shiro, K. Koyama, J. Nat. Prod. 74 (2011) 1645–1649.
[44] A.C. Freitas, J.C. Andrade, F.M. Silva, T.A.P. Rocha-Santos, A.C. Duarte, A.M. Gomes, Curr. Med Chem. 20 (2013) 4575–4594.

[45] D.-H. Ngo, T.-S. Vo, D.-N. Ngo, I. Wijesekara, S.-K. Kim, Int. J. Biol. Macromol. 51 (2012) 378–383.
[46] S. Kim, I. Wijesekara, J. Funct. Foods 2 (2010) 1–9.
[47] S. W. Pelletier, S. W. Pelletier (Eds.), The nature and definition of an alkaloid, New York, 1983.
[48] D. Kumar, D. S. Rawat, V. K. Tiwari, Mishra B. B. (Eds.), Opportunity, Challenge and Scope of Natural Products in Medicinal Chemistry, New Delhi, 2011.
[49] Y. Zhang, T. Han, Q. Ming, L. Wu, K. Rahman, L. Qin, Nat Prod Commun. 7 (2012) 963–968.
[50] K.C. Güven, A. Percot, E. Sezik, Mar. Drugs 8 (2010) 269–284.
[51] S.U. Lee, Y. Asami, D. Lee, J.-H. Jang, J.S. Ahn, H. Oh, J. Nat. Prod. 74 (2011) 1284–1287.
[52] P. Dewick Medicinal Natural Products: A Biosynthetic Approach, Second Edition, New York, 2002.
[53] A. Plaza, J.L. Keffer, G. Bifulco, J.R. Lloyd, C.A. Bewley, J. Am. Chem. Soc. 132 (2010) 9069–9077.
[54] Z. Zhang, Q. Zhang, J. Wang, H. Song, H. Zhang, X. Niu, Carbohyd. Polym. 79 (2010) 1124–1129.
[55] D. Wolf, K. Siems, Chimia 69 (2007) 339–345.
[56] J. Kjer, A. Debbab, A.H. Aly, P. Proksch, Nat. Protoc. 5 (2010) 479–490.
[57] S.-Y. Shi, H.-H. Zhou, Y.-P. Zhang, X.-Y. Jiang, X.-Q. Chen, K.-L. Huang, Trends Anal. Chem. 28 (2009) 865–877.

Chapter 2

Prospection, Collection, and Preservation of Marine Samples

Sara Leston[1,2], Margarida Nunes[1,2], João Rosa[1,2], Marco F.L. Lemos[3], Fernando Ramos[2,4] and Miguel A. Pardal[1]

[1]*CFE – Centre for Functional Ecology, Department of Life Sciences, University of Coimbra, Apartado Coimbra, Portugal*

[2]*CEF – Center for Pharmaceutical Studies, Health Sciences Campus, Pharmacy Faculty, Coimbra University, Azinhaga de Santa Comba, Coimbra, Portugal*

[3]*ESTM, GIRM, Polytechnic Institute of Leiria, Peniche, Portugal*

[4]*CNC – Centre for Neurosciences and Cell Biology, Health Sciences Campus, Pharmacy Faculty, Coimbra University, Azinhaga de Santa Comba, Coimbra, Portugal*

Contents

2.1 INTRODUCTION

Although there is a lack of uniformity in the definition of bioprospecting, it can be defined as the collection of biological material and the analysis of its material properties, or its molecular, biochemical, or genetic content for the purpose of developing a commercial product (as defined by the New Zealand government).

Therefore, bioprospecting can be considered as a tool for biodiversity conservation if conducted in an environmentally responsible way. In fact, the International Society of Chemical Ecology has proposed the Biorational Approach to "narrow the field of search to manageable proportions and increase the likelihood of discovering a useful natural product, bridging bioprospecting with biodiversity conservation" [1–3]. There is a global interest in

Analysis of Marine Samples in Search of Bioactive Compounds, Vol. 65. DOI: 10.1016/B978-0-444-63359-0.00002-1

conservation, with nondestructive use of natural resources to avoid losses due to overexploitation and even species extinction accompanying habitat destruction. Thus, bioprospecting will contribute greatly to increase scientific knowledge on natural marine ecosystems, regardless of whether the target compound will lead to financial returns or not. Furthermore, from an economical point of view, these actions can promote innovation, scientific research increase, and potentiate aquaculture production of target organisms, which will in turn, generate wealth. However, special care must be given to ethical issues in bioprospecting and conservation policies: research must respect host institutes and organizations regarding new discoveries, including new localities of rare and endangered species. Also, prospecting raises issues of environmental disturbance during the extractive process. For instance, the movement of vessels can cause pollutants to be released or lead to a disruption of currents that may cause unwanted movement of organisms. Moreover, discarding scientific samples can act as a source of biological contamination [3,4]. Therefore, mitigation measures to reduce the impact to a minimum must be undertaken. Also, special care must be given in protecting traditional knowledge on natural products regarding the rights of indigenous people and local communities, safeguarded by the Rio Declaration and the United Nations Convention on Biological Diversity [4,5].

Traditionally, the greatest contributors of bioactive compounds have come from terrestrial habitats and include plants, bacteria, and fungi [6,7]. With the search extending into the marine environment, new groups have surfaced as sources of bioactive substances, including algae and invertebrates.

The definition of marine microorganisms refers to all microscopic species found in salt water, which is in fact an ecological designation rather than a taxonomic classification [8]. The vast biological diversity of marine microorganisms includes archaea, bacteria, viruses, diatoms, yeasts, and fungi [9], but has been explored only to a very limited extent. Nonetheless, the diversity of secondary metabolites produced by these communities is expected to be equally diverse and abundant [7,2,10]. Current estimates point to hundreds of millions of marine species representing over 90% of total marine biomass [8]. For fungi alone, despite the fact that only about 465 species are described, the most realistic approach estimates the existence of 1.5 million [7,10]. Marine bacteria represent about 10% and are found mainly in sediments, but also in endophytic/symbiotic associations with other marine organisms [7,11]. It is assumed that all marine organisms are associated with microorganisms, which can constitute in some cases up to 40% of total biomass. In fact, in some cases, bioactive compounds primarily attributed to invertebrates were later found to be produced by bacteria or fungal species.

As mentioned before there is a long-lasting tradition of exploring plants and plant extracts in the search for bioactive compounds. Even now the estimates point to 25% of all currently administered drugs having originated from plants. As it happens with marine microorganisms, marine plants include various taxonomic categories and again are an ecologically based classification composed of microalgae, macroalgae (seaweeds), and flowering plants. Also, when compared to terrestrial species, research on bioactive compounds has yet to reach the same level [7].

Finally, macroorganisms constitute another important source of bioactive compounds. This is also a diverse taxonomic group representing several trophic levels of the food web and includes both invertebrate and vertebrate animal species. Among the ascidians, sponges and mollusks have gained importance due to the number of substances with bioactivity being discovered [12]. Research has focused on sessile invertebrates, because they possess no means of physical defense but have evolved powerful chemical substances as defense mechanisms for themselves and to protect egg masses against predation, among other effects [12–14]. Also, common to all the chemical compounds released in the aquatic environment where dilution would constitute a problem, these substances are usually very potent.

2.2 BIOPROSPECTING IN EXTREME ENVIRONMENTS

Marine organisms living in the deep sea, polar regions, and other extreme environments are exposed to severe conditions in terms of temperature, light, pressure, oxygen, salinity, and even chemical toxicity [15–17]. Such extremophile organisms developed unique biochemical properties in order to survive under these circumstances, and therefore they are promising sources of unique bioactive compounds [18–20]. Nevertheless, shallow temperate and tropical habitats have been the most explored so far [21,22].

Regarding deep sea, significant efforts have been made to prospect around hydrothermal vents in particular, because of the unexploited biodiversity and unknown adaptations present in these hostile environments [20]. Hydrothermal vents are considered one of the most extreme and dynamic environments on Earth, with temperatures ranging from 4 to 400 °C, high pressure conditions, a complete absence of light, and abrupt chemical and pH gradients [23]. Although deep sea hydrothermal vents are usually characterized by low macroorganism biodiversity, they support one of the highest microbial diversity on the planet [24,25]. The variety of endemic organisms living in these ecosystems relies on normally toxic sulfides and hydrocarbons for primary production. The capacity of these unusual organisms to utilize, detoxify, or degrade toxic compounds has led to the current intense interest from the industry involved in bioactive compounds. The hyperthermophilic bacteria and Archae that colonize hydrothermal vents are also able to produce proteins that remain stable at temperatures close to or higher than the water boiling point [26], and thus, can be of great use in the biotechnology industry. On the other hand, deep-sea barophilic bacteria can produce enzymes more functional under high pressure than at atmospheric pressure, which can be used in biotechnological applications such as bioreactor systems [27]. In addition, bacteria isolated from hydrothermal vents have also exhibited the ability to produce unusual extracellular polymers, and may have novel applications such as viscosifiers, gelling agents, emulsifiers, stabilizers, and texture enhancers [28].

The cold marine environment has already proved to be a valuable source of compounds with interesting biological properties [29], and the industry has been paying attention to Arctic and Antarctica as never before. In particular, much effort has been focused on microorganisms able to survive and grow in the polar regions. Bioprospecting for organisms living in permanently low temperatures is centered on finding cold-adapted enzymes that may be employed in a range of industrial applications [30–32], and polyunsaturated fatty acids (PUFAs) [33,34].

Hydrothermal vents and polar habitats constitute a source of new biologically active natural compounds, however research missions in these remote areas with difficult access demand complex logistics, and consequently, high costs [21]. While most shallow waters in tropical and temperate environments can be accessed with relative ease by SCUBA diving or snorkeling, the collection of samples from deeper and cold environments needs to be accomplished using closed circuit rebreather diving, dredging and trawling, or submersibles, whether remotely operated vehicles (ROVs) or autonomous underwater vehicles (AUVs; Figure 2.1).

Another challenge to bioprospecting in these locations, as in all marine ecosystems, is to have access to a sufficient amount of biological material and subsequently to pure bioactive compounds [35,36]. In addition, due to the remoteness of some of these ecosystems and the logistics and financial support required, it may be difficult to resample an already visited location to collect

FIGURE 2.1 Sample collection for bioactive compound prospection using SCUBA apparatus (A); snorkeling sample collection (B); deployment of a remotely operated vehicle (ROV) (C); ROV collecting samples in Azores, Portugal deep sea (D).

more organisms if the initial results are promising and more material is necessary [37]. Furthermore, seasonal variation may also influence the production of secondary metabolites [36,38]. Thus, it is necessary to collect as much as possible while sampling, letting the freezer capacity on the vessel to be the only limiting factor. Such an approach raises questions regarding sustainability and the environmental impacts of bioprospecting [39]. Generally, these impacts are thought to be relatively small at early stages of collection, but if a given species shows biotechnology potential, repeated collection may involve larger quantities of biological material, raising the possibility of environmental impact. Nevertheless, the culture of organisms and/or the synthesis of the bioactive compounds of interest could eliminate the need for repeated collection, and limit the impacts of bioprospecting [40].

2.3 COLLECTION OF MARINE SAMPLES

In order to have a full and unequivocal natural compound evaluation and in that sense, to assure the properties discovered can in fact be attributed to the organism collected, several factors must be considered. The safety and quality of a given compound is influenced by both intrinsic and extrinsic factors (see Table 2.1) [14,41]. As we will discuss further, specimens obtained from local suppliers will be more easily accessible to bioactive compounds, especially if larger amounts are required. However this percentage is insignificant, taking into account the ocean's biodiversity. A summary of bioactive compounds are presented in Table 2.1, with the collection method detailed whenever available. Only compounds with tested antimicrobial activity are present, since some compounds are detected in marine organisms, but further tests were not performed in order to detect their activity. Other works isolated the bioactive compounds from specimens' libraries (e.g., [42–44]) and for that reason they are also not included.

Genetics is the main intrinsic factor that will dictate the properties of a bioactive compound and thus it becomes essential, when possible, to survey several populations of a given organism and to build background knowledge on its ecology, distribution, and abundance patterns [41]. Variations in secondary metabolites production are common between different samples of the same species as a result of genotypic diversity [41]. Other important intrinsic factors are diurnal and seasonal variations that affect the chemical production and accumulation of compounds. For example, plants will produce some substances at different rates according to the stages of growth but usually will peak at the time of flowering with subsequent decline [41] (Figure 2.2).

Regarding the environment, it is unlikely that the prospector may have an influence on its quality and the effects it will induce, but rather a selection of the most suitable habitat is possible to increase safety and quality of the product. Also, variations on bioactive compounds from samples of the same species taken in different habitats will occur, and therefore, knowledge on environmental

TABLE 2.1 Overview of the Natural Products from Marine Specimens, Including Sampling Procedures

Organism	Specie		Collection Methodology	Compound	Reference
Actinobacteria from Sponge	*Actinomadura sp. SBMs009 from Suberites japonicus*	Tissue	n/a	Steroids: Bendigole D-F and 3-keto sterol	[82]
Tunicate	*Styela clava*	Hemocytes	Freshly harvested from a local supplier	Antimicrobial peptides: clavanins	[83]
	Styela clava	Pharyngeal tissues	Freshly harvested	Antibacterial peptide: clavaspirin	[84]
	Halocynthia aurantium	Hemocytes	Purchased at a local sea market	Antimicrobial peptides: dicynthaurin	[85]
	Halocynthia papillosa	Hemocytes	Collected in Mediterranean Sea along the Catalan coast	Antimicrobial peptides: halocyntin, papillosin	[86]
Green sea urchin	*Strongylocentrotus droebachiensis*	Coelomic fluid	Collected off the coast, Norway; maintained in fresh flowing seawater until sample collection	Antimicrobial peptides: strongylocins	[87]
	Strongylocentrotus droebachiensis	Coelomycetes	Collected off the coast, Norway; maintained in fresh flowing seawater until sample collection	Antimicrobial peptides: centrocins 1 and 2	[88]
Polychaeta	*Arenicola marina*	Coelomycetes	Collected at a coastal zone, Russia, in the middle of August when reproduction season is over	Antimicrobial peptides: arenicins	[89]

Scyphoid jellyfish	*Aurelia aurita*	Mesogela	Collected at a coastal zone, Russia	Antimicrobial peptides: aurelins	[90]
Sea hare	*Dolabella auricularia*	Body wall—skin and mucus	Collected at Kominato, Japan	Antimicrobial peptide: dolabellanin B2	[91]
Bay scallop	*Argopecten irradians*	Haemolymph	Collected from a scallop farm	Antimicrobial peptide: big defensin AiBD	[92]
Mussel	*Mytilus galloprovincialis*	Hemocytes and plasma	Obtained from a commercial shellfish farm	Antimicrobial peptide: myticin	[93]
	Mytilus galloprovincialis *Mytilus edulis*	Haemolymph	Obtained from an aquaculture farm	Antimicrobial peptides: MytilinB and MGD2	[51]
	Mytilus edulis	Blood	Collected from a sea natural population, Russia	Antimicrobial pepitdes: Mytilin and insect defensin	[94]
Chinese shrimp	*Fenneropenaeus chinensis*	Haemolymph	Purchased from a local shrimp farm	Antimicrobial protein: crustin-like, CruFc and CshFc	[62]
Tiger shrimp	*Penaeus monodon*	Whole body homogenization	Obtained from a shrimp farm	Chelonianin	[63]
	Penaeus monodon	Hemocytes	N/A	Antimicrobial protein: crustin-like, Crus-likePm	[95]
Pacific blue shrimp	*Litopenaeus stylirostris*	Hemocytes	N/A	Antimicrobial peptide: Ls-Stylicin1	[96]
brachyuran crab	*6 species*	Haemolymph and hemocytes	Collected from neritic and oyster zone environment in Vellar estuary	Antimicrobial lipids	[97]

(Continued)

TABLE 2.1 Overview of the Natural Products from Marine Specimens, Including Sampling Procedures ***(cont.)***

Organism	Specie		Collection Methodology	Compound	Reference
Horseshoe crab	*Tachypleus tridentatus*	Haemolymph (hemocytes or amebocytes)	N/A	Clotting factors, proteinase inhibitors, lectins, antimicrobial peptides, humoral factors	[115]
	Tachypleus tridentatus	Hemocyte debris	N/A	Antimicrobial protein: Tachycitin	[98]
	Tachypleus tridentatus	Hemocyte debris	N/A	Antimicrobial protein: Tachystatin A, B, C	[99]
	Tachypleus tridentatus	Haemolymph plasma and hemocytes	Collected in Fukuoka, Japan	Antibiotic peptide: big defensin	[100]
	Tachypleus tridentatus	Hemocytes	N/A	Antimicrobial peptides: tachyplesin II, polyphemusins I, II	[101]
Spider crab	*Hyas araneus*	Hemocytes	Obtained by trawling at a coastal zone, Norway	Antimicrobial peptides: arasin 1	[102]
	Hyas araneus	Hemocytes	Collected at a coastal zone, Norway	Antimicrobial peptides: hyastatin	[103]

Shore crab	*Carcinus maenas*	Hemocytes	Collected in baited creels, Scotland	Antibacterial peptide: proline-rich	[104]
	Carcinus maenas	Granular hemocytes	Collected in baited baskets, Scotland	Antibacterial protein: cysteine-rich	[105]
	Carcinus maenas	Hemocytes	Collected in St. Andrews Bay, Scotland	Antibacterial protein: carcinin	[106]
Mud crab	*Scylla serrata*	Seminal plasma	Purchased from a local farm	Antibacterial protein: Scygonadin	[107]
	Scylla serrata	Hemocytes	Purchased from the local fish market	Antimicrobial peptides: SSAP	[108]
	Scylla partamamosan	Hemocytes	Purchased from local fish markets	Antimicrobial peptides: arasin-likeSp, GRPSp	[109]
European lobster	*Homarus gammarus*	Haemolymph	Obtained from East Neuk Shellfish company	Antibacterial peptide: crustin-like	[110]
American lobster	*Homarus americanus*	Haemolymph	Obtained from local suppliers	Antibacterial peptide: crustin-like	[111]
	Homarus americanus	Hemocytes	N/A	Antimicrobial peptides: CAP-1 and CAP-2	[112]
Marine annelid	*Nereis diversicolor*	Coelomocytes	Collected at two different points, France	Antimicrobial peptide: hedistin	[113]
Marine clamworm	*Perinereis aibuhitensis*	Whole body homogenization	N/A	Antimicrobial peptide: perinerin	[114]

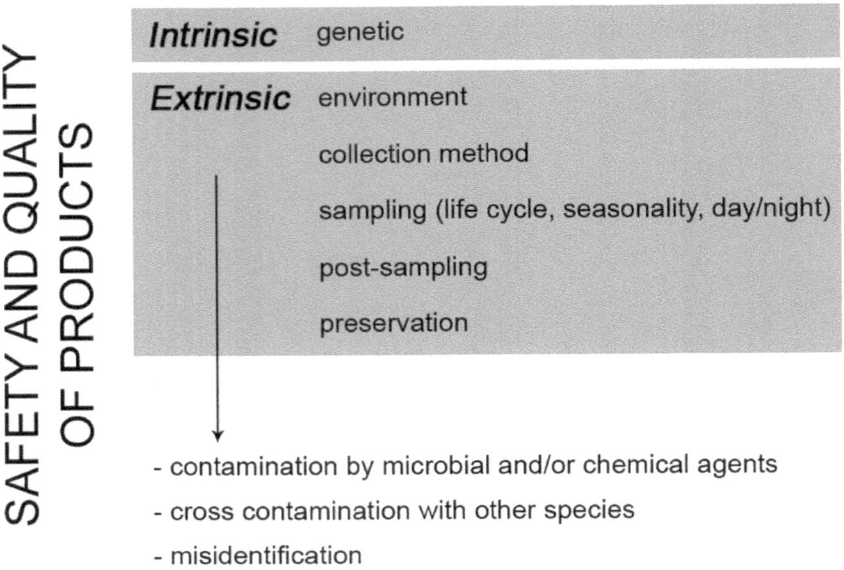

FIGURE 2.2 Factors conditioning safety and quality of marine bioactive compounds.

variations is very useful, particularly on abiotic and biotic conditions [2,45] (Figure 2.2).

Another very important aspect is the collection method. Before entering into detail, it is important to remember that biodiversity conservation should always be kept in mind, mitigating the impacts as much as possible. Collection of organisms must be restricted to minimum amounts and care with rare or endangered species must be taken into account, together with care in preserving the surrounding individuals and communities.

The first step that must be assured when collecting specimens from wild populations during the prospection process is that the organisms are accurately identified (at least to the nearest taxonomic level), that the location of collection is accurate (preferably GPS coordinates), and also that they are in perfect condition to be fully evaluated, either by chemical screening or through bioassays [2,45]. Identification often requires morphological and histological examinations (macroorganisms) based on preserved specimens, making *in situ* documentation a very important tool, which includes photography (before and after collection) and the gathering of precise information such as microhabitat, substrate, depth, and time, among others [45]. These measures are especially important in case recollections are necessary since it is unrealistic that only samples of organisms available in large quantities are collected. It is necessary to assure that the identity and location are correct if further biological material is required, especially if there is a successful drug discovery, which will need sufficient quantities for follow-up research and for compound production

FIGURE 2.3 Marine microorganism prospection: collection of marine waters with a water sampler (A); water sample processing, by filtration, for microorganism culture (B); and a laboratory fermentor for high-throughput microorganism culture (C).

[2,45]. Therefore, there must be a balance between small amounts initially collected and subsequent recollections, and large initial quantities with the possible restriction of the diversity of samples taken [2,14] (Figure 2.2).

Possibly the most determinant factor for a successful discovery is the sampling methodology. After securing the steps for an accurate identification, the target organisms will be collected. Depending on the type of habitat and the species ecology, different techniques are necessary.

Microorganisms (Figure 2.3) populate distinct areas of the marine environment: (1) in free suspension in the water column; (2) in flocculated particulates in the water column; (3) in the sediment; (4) on surfaces of both living and nonliving bodies; and (5) in endophytic/symbiotic associations [46]. The most serious risks regarding the sampling of these organisms are contamination and cross-contamination. Therefore, the most suitable methods rely on sampling the material together with supporting materials, which include water, sediment, portions of plants and other macroorganisms, or other substracts [46]. Water samples must be taken with sterilized glass containers to avoid contamination and proliferation of cultures. In the case of sediments from intertidal areas, sampling can be performed by removing core samples while in subtidal regions, and when suitable, sediment grabs (like Ponar and Van Veen grabs) may be used. However, in more hostile environments such as the deep sea, more sophisticated techniques are required. Parkes and colleagues (2009) developed a system (DeepIsoBUG) for high-pressure systems, which allows taking cores and slices of the sediment, with each slice being transferred to a low pressure vessel, thus reducing the probabilities of contamination and maintaining pressure and temperature [46,47]. After sampling is concluded, preservation must be performed as quickly as possible to avoid genetic and phenotypic alterations of the cultures [46].

Both plants and animals may be sampled by similar techniques. Again, in intertidal areas sampling can be conducted during low-tide where sessile and sedentary species can be carefully detached from substrate to maintain the

FIGURE 2.4 Organism collection at the intertidal zone during a low tide: organism collection in tidal pools (A); macroalgae collection in a rocky beach exposed during low tide (B and C); and animal collection (limpets) and immediate freezing in liquid nitrogen (D).

organism intact (Figure 2.4). In the case of mobile and buried animal species (usually attached to small stones or shells), sampling can be performed by capturing the individuals directly or with nets, and if necessary, using shovels, trowels, and sieves. Each individual must be independently placed in properly labeled containers and transported under refrigerated conditions [45]. In shallow inshore waters, snorkel and SCUBA diving can also be applied for sampling (Figures 2.1A and 2.1B). In subtidal areas and deep sea, including extreme environments, ship-based collections are conducted with bottles, nets, grabs (Ponar or Van Veen grabs), benthic dredges, sleds, and commercial trawls, especially in the case of vertebrate species. Recently, collection via ROVs and AUVs has been used, especially in extreme environments [48] (Figures 2.1C and 2.1D). The development of such robotic systems has launched a revolution in oceanographic science, but mainly in marine environment exploration and discovery [49,50]. Despite the high cost, such equipment can be a real game changer in the discovery of new marine natural products, since 45 to 50 years ago the collection of marine organisms was limited to those obtained by skin diving [48]. Depths up to 6000 meters can become routinely achievable; together with improved battery technology, dive times beyond 24h and the possibility of carrying a vast array of acoustic, optical, and chemical sensors, make the marine environment increasingly explored [49].

There are undeniable advances in the discovery of natural products from marine organisms, but the lack of information on how samples are collected is an actual issue, as well as cautions necessary in order to preserve the integrity of bioactive compounds (e.g., horseshoe crab, *Tachypleus tridentatus*, was used in five distinct studies presented in Table 2.1, with only one referring to the location of sample collection). There is very limited information on the effects of temperature stress in antimicrobial activity. Nonetheless, in marine mussels haemocytes, a heat shock treatment of 33 °C was sufficient to cause changes at the gene expression level [51]. Therefore, care should be taken in order to protect marine species from heat during collection and transport to laboratory.

2.4 SUSTAINABLE PRODUCTION OF BIOACTIVE COMPOUNDS

With the intensification of oceans exploration, a notable and growing number of bioactive natural products are being isolated from a variety of marine organisms. However, despite the great potential of such bioactive compounds, their steady production is a key limiting factor that may hold back the development of commercial processes [52–54]. Bioprospecting of some marine compounds usually depends on wild specimens harvesting. Wild harvest only satisfies the demand partially and emerges as an unsuitable production route [55–58], since it can involve a few major issues concerning sustainability and replicability. Sustainability issues may be related to the fact that bioactive compounds of marine origin are present in small quantities in the producer organism [59], with larger amounts of fresh material being required for the complete biological and chemical evaluation, and eventual production of the bioactive compounds. Concentrations of such compounds in marine invertebrates can represent less than 1-6% of the wet weight, dropping to lower values with traditional extraction procedures [60,61]. Using large amounts of biomass may have an impact on the number of specimens, and for rare species, it might not even be possible to collect enough organisms for the research pipeline. So, independently of how attractive a compound might be, unless an adequate supply can be assured, the compound will remain only of novelty value [13]. Replicability might be an important issue as a result of environmental and population variability that might result in site-specific or even time-frame-specific target metabolite production—"loss of the source" [59]. In fact, a major part of the antimicrobial peptides discovered have been isolated not only from common and abundant invertebrates, but from aquaculture species, being easier to re-collect if more material is needed [58]. In a study performed by Lee and colleagues (1997), the hemocytes extracted from the solitary tunicate *Styela clava*, which were freshly harvested from a local supplier (Long Beach, US), were used to obtain a family of four α-helical antimicrobial peptides, named clavanins. The possibility of obtaining such organisms within local suppliers makes the access to bioactive compounds easier. Similarly, Mitta and coworkers (2000) collected adult mussels (*Mytilus galloprovincialis* and *Mytilus edulis*) from an aquaculture farm to study the gene structure and expression of two antimicrobial peptides of those marine mussels. Shrimp farms were also used to purchase specimens in order to discover novel natural products from shrimp [62,63] as well as other organisms as presented in Table 2.1.

Thus, unless more realistic alternatives for harvesting are introduced, many of these target molecules will remain unknown [54]. Consequently, the focus will have to be directed to the development of more efficient culture techniques [64]. Target species aquaculture, whether seaweeds, invertebrates, or vertebrates, may overcome these bottlenecks because biomass can be continuously produced (sustainable year-round supply) using standardized conditions—a realization that has prompted much research on aquaculture for bioactive compound production to tackle one of the major concerns regarding the commonly reported "supply issue" [65] (Figures 2.5 and 2.6).

The selection of the most appropriate culture technique will depend on the nature of the compound and its concentration on the organism. Therefore, if the substance is in high concentration, the cultivation of adult specimens would be the best choice, while *in vitro* cell cultures may constitute a more suitable method for products found in low concentrations [64,66]. Nonetheless, *in vitro* systems have presented difficulties in long-term production [67,68]. When

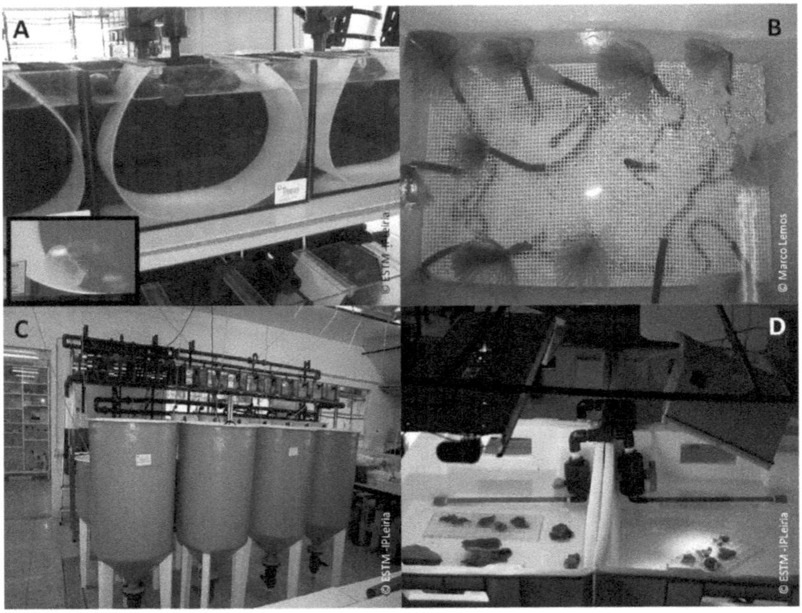

FIGURE 2.5 Aquaculture of marine organisms for biotechnological use purposes: aquaculture of jellyfish in planktonkreisel tanks (A); growth of filter feeding polychaetes (B); cylindro-conical tanks for copepod biomass production (C); and different species of corals under different light sources to enhance growth and bioactive compound production (D).

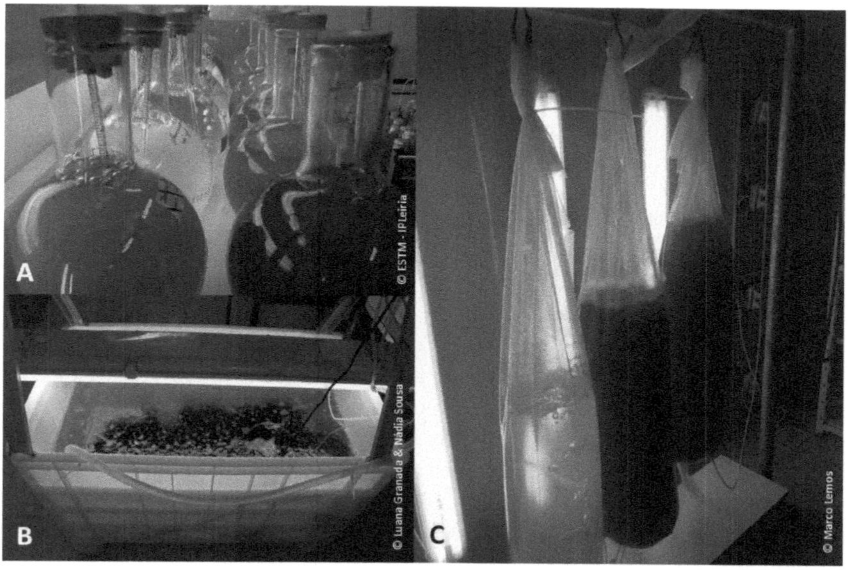

FIGURE 2.6 Aquaculture of marine algae: diverse microalgae growth in round bottom flasks (A): red macroalgae growth under different conditions for agar extraction (B); and microalgae growth in plastic sleeves (C).

looking at sponges production, aquaculture has been extensively proposed as a provider of sponge materials for biotechnological purposes in sea-based cultures [13,57,69,70]. The disadvantages of this alternative are the numerous risks to which sponges are exposed, as well as diseases or adverse weather conditions [64,71], which in turn translate into highly fluctuant survival rates that strongly depend on species ecology, location, seasonality, and technique [56,72,73–77]. There are several compounds already obtained by aquaculture of their producers, like the antitumor halichondrins, detected in *Lissodendoryx*, a New Zealand deep water sponge that was successfully maintained [13,58]. Similarly, *Bugula neritina* and *Ecteinascidia* turbinate aquacultures already exist in California. This bryozoan and ascidian are the source of the antitumor compounds bryostatin and ecteinascidin, respectively [58]. The supplies of marine metabolites being tested in the clinic also include total and semisynthesis, as well as fermentation of microbes that produce these bioactive compounds [13]. Fermentation has been the most viable method for bioactive compounds production when it comes to microorganisms, especially if the compounds are needed on a commercial scale [78] (Figure 2.3).

When a natural product, either marine or not, successfully presents interesting activity, techniques for large-scale commercial supply need to be employed. Furthermore, bioactive compounds can only be realistically obtained by harvesting from natural origins, aquaculture, fermentation, or by synthesis [13].

2.5 PRESERVATION OF MARINE SAMPLES

Preservation of organism samples is a prerequisite for field studies in remote areas or when there are impediments to return the samples to the laboratory in due time. The collection of marine organisms for the study of bioactive compounds is many times performed in open sea, where immediate tests and analyses are often impossible [79]. One problem arising in the case of microorganisms as producers for bioactive compounds is that host macroorganisms obtained during collection, in some cases, have to be stored on ships for some time. Therefore, upon the arrival at the lab, the microorganisms used for the culture might not be in pristine condition [80].

As most marine specimens are quite susceptible to fast degradation, the samples should be quickly frozen with dry ice and stored frozen at –20 °C until processing in the shortest gap of time possible [81]. Several preservation methods are discussed among the literature, most of them depending on the type of bioactive compound that is going to be extracted. Storage of organism sections in 5% dimethyl sulfoxide at –70 °C and storage of macerates in sintered glass beads, and 5% trehalose at –20 °C was the best method for storage of marine specimens, out of five different methodologies, for recovering a higher diversity of bacteria [80]. When the preservation of marine tissues for DNA analyses is the objective, Dawson and colleagues (1998) demonstrated a solution of dimethylsulfoxide and sodium chloride to best preserve a wide range of marine tissues. These authors state, however, that it is better to test several alternative methods prior to field collection, or use more than one method to preserve the samples, since the length and temperature of the storage also influence the success of sample preservation. Also, various methods of preservation and/or fixation have advantages and disadvantages. Although dry or ethanol preservation, fixation by formalin, and glutaraldehyde can be used for general anatomy or taxa study of organisms, when it comes to the study of natural products, freezing with liquid nitrogen and later storing at (–80 °C), and then freeze-drying are the only alternatives according to the Center for Systematic Biology & Evolution (Philadelphia, US). Using ethanol is rated as good; messenger RNA prep technique and dehydration is not recommended; and formalin, paraformaldehyde, glutaraldehyde, and Davison's solution is incompatible.

2.6 FINAL CONSIDERATIONS

With more than 70% of the world's surface and with the least known habitats and species, the ocean plays a critical role in the search for new solutions to ancient problems related to human health, as it is an important source of bioactive compounds with enormous potential for fighting disease. Bioprospecting and biodiversity conservation have progressed closely in this quest for new substances. Collection and preservation techniques play a crucial role in guaranteeing the discovery of new products while attempting to maintain natural diversity. Exploration of marine organisms for their contribution for the resolution of problems is vital for future generations, and with the search for new culture techniques, seems to be the route toward sustainability and high biomass production with the necessary stability concerning the previously discovered bioactivity.

ACKNOWLEDGEMENTS

The authors are grateful to the EMEPC – Estrutura de Missão para a Extensão da Plataforma Continental, Escola Superior de Turismo e Tecnologia do Mar – Polytechnic Institute of Leiria, André Horta, Celso Alves, Luana Granada, Marco Lemos, Nádia Sousa, Nuno Vasco Rodrigues and Sílvia Jesus for kindly providing the photos used in this chapter.

REFERENCES

[1] T. Eisner, J. Meinwald, Chemoecol. 1 (1990) 38.
[2] H.G. Wildman, IUPAC, 1999.
[3] L. Slobodian, R. Kinna, A. Kambu, L. Ognibene, United Nations Environment Programme (http://www.unep.org/delc/Portals/119/Biosprecting-Issuepaper.pdf)
[4] A.J. Beattie, M. Hay, B. Magnusson, E. de Nys, J. Smeathers, J.F.V. Vincent, Austral Ecol. 36 (2011) 341–356.
[5] A.L. Harvey, Drug Discov. Today 13 (2008) 894–901.
[6] T.P. Anand, A.W. Bhat, Y.S. Shouche, R. Roy, J. Siddharth, S.P. Sarma, Microbiol. Res 161 (2006) 252–262.
[7] M. Kulka (Ed.), Using Old Solutions to New Problems - Natural Drug Discovery in the 21st Century, InTech, Rijeka, 2013.
[8] P. Dewapriya, S.-K. Kim, Food Res. Int. 56 (2014) 115–125.
[9] C. Bowler, D.M. Karl, R.R. Colwell, Nature 459 (2009) 180–184.
[10] B. Haefner, DDT 8 (2003) 536–544.
[11] A. Horta, S. Pinteus, C. Alves, N. Fino, J. Silva, S. Fernandez, A. Rodrigues, R. Pedrosa, Mar. Drugs 12 (2014) 1676–1689.
[12] A. Aneiros, A. Garateix, J. Chromatogr. B 803 (2004) 41–53.
[13] M.H.G. Munro, J.W. Blunt, E.J. Dumdei, S.J.H. Hickford, R.E. Lill, S. Li, C.N. Battershill, A.R. Duckworth, J. Biotechnol. 70 (1999) 15–25.
[14] K. Benkendorff, Bioactive Molluscan Resources and their Conservation: Biological and Chemical Studies on the Egg Masses of Marine Molluscs, PhD thesis, University of Wollongong, 1999 (http://ro.uow.edu.au/theses/278).
[15] A. McMinn, A. Pankowski, T. Delfatti, J. Phyc. 41 (2005) 732–741.
[16] M. Synnes, Clean Techn. Environ. Policy 9 (2007) 53–59.
[17] M.F. Lemos, I.S. Lima, K.L. da Fonseca, H.R. Monteiro, A.C. Esteves, Curr. Opin. Biotechnol. 22 (2011) S80.
[18] C. Matz, J.S. Webb, P.J. Schupp, S.Y. Phang, A. Penesyan, S. Egan, P. Steinberg, S. Kjelleberg, PLoS One 3 (2008) 1–7.
[19] A.M.S. Mayer, K.B. Glaser, C. Cuevas, R.S. Jacobs, W. Kem, R.D. Little, J.M. McIntosh, D.J. Newman, B.C. Potts, D.E. Shuster, Trends Pharmacol. Sci. 31 (2010) 255–265.
[20] N. Jobstvogt, N. Hanley, S. Hynes, J. Kenter, U. Witte, Ecol. Econ. 97 (2014) 10–19.
[21] M.C. Leal, J. Pulga, J. Serôdio, N.C.N. Gomes, R. Calado, PLoS One 7 (2012) 1–15.

[22] A. Postec, F. Lesongeur, P. Pignet, B. Ollivier, J. Querellou, A. Godfroy, Extremophiles 11 (2007) 747–757.
[23] A. Martins, T. Tenreiro, G. Andrade, M. Gadanho, S. Chaves, M. Abrantes, P. Calado, R. Tenreiro, H. Vieira, Mar. Drugs 11 (2013) 1506–1523.
[24] P.J. Butler, J.A. Koslow, P.V.R. Snelgrove and S.K. Juniper, CSIRO Marine Research, Australia, 2001.
[25] J. Guezennec, J. Ind. Microbiol. Biotechnol. 29 (2002) 204–208.
[26] E. Mombelli, E. Shehi, P. Fusi, P. Tortora, Biochem. Biophys. Acta 1595 (2002) 392–396.
[27] K. Horikoshi, Curr. Opin. Microbiol. 1 (1998) 291–295.
[28] C.A. Mancuso Nichols, J. Guezennec, J.P. Bowman, Mar. Biotechnol. 7 (2005) 253–271.
[29] C. Avila, S. Taboada, L. Núnez-Pons, Mar. Ecol. 29 (2008) 1–71.
[30] M. Rina, C. Pozidis, K. Mavromatis, M. Tzanodaskalaki, M. Kokkinidis, V. Bouriotis, Eur. J. Biochem. 267 (2000) 1230–1238.
[31] N. Zhang, W.C. Suen, W. Windsor, L. Xiao, V. Madison, A. Zaks, Protein Eng. Des. Sel. 16 (2003) 599–605.
[32] J.H. Jeon, J.T. Kim, S.G. Kang, J.H. Lee, S.J. Kim, Mar. Biotechnol. 11 (2009) 307–316.
[33] D.S. Nichols, J.P. Bowman, K. Sanderson, C.M. Nichols, T.E. Lewis, T.A. McMeekin, P.D. Nichols, Curr. Opin. Biotechnol. 10 (1999) 240–246.
[34] V.V. Jadhav, M.M. Jamle, P.D. Pawar, M.N. Devare, R.K. Bhadekar, Ann. Microbiol. 60 (2010) 693–699.
[35] P. Proksch, R.A. Edrada, R. Ebel, Appl. Microbial. Biotechnol. 59 (2002) 125–134.
[36] T.F. Molinski, D.S. Dalisay, S.L. Lievens, J.P. Saludes, Nat. Rev. Drug Discov. 8 (2009) 69–85.
[37] J. Svenson, Phytochem. Rev. 12 (2013) 567–578.
[38] M. Tadesse, M.B. Strom, J. Svenson, M. Jaspars, B.F. Milne, V. Torfoss, J.H. Andersen, E. Hansen, K. Stensvag, T. Haug, Org. Lett. 12 (2010) 4752–4755.
[39] L. Bohlin, U. Goransson, C. Alsmark, C. Weden, A. Backlund, Phytochem. Rev. 9 (2010) 279–301.
[40] PP. Dando and S.K. Juniper, Report from an InterRidge Workshop, Institute of Ocean Sciences, Sidney (Victoria), B.C., 2001.
[41] I. Ahmad, F. Aqil, M. Owais (Eds.), Modern Phytomedicine. Turning Medicinal Plants into Drugs, WILEY-VCH Verlag GmbH & Co. KGaA, Weinheim, 2006.
[42] T.C. Bartlett, B.J. Cuthbertson, E.F. Shepard, R.W. Chapman, P.S. Gross, G.W. Warr, Mar. Biotechnol. 4 (2002) 278–293.
[43] F. Jiménez-Vega, G. Yepiz-Plascencia, K. Söderhäll, F. Vargas-Albores, Biochem. Biophys. Res. Commun. 314 (2004) 681–687.
[44] F. Vargas-Albores, G. Yepiz-Plascencia, F. Jiménez-Vega, A. Ávila-Villa, Comp. Biochem. Physiol. B, Biochem. Mol. Biol. 138 (2004) 415–422.
[45] V. Häussermann, J. Mar. Bio. Assoc. UK 84 (2004) 931–936.
[46] K. Duarte, T.A.P. Rocha-Santos, A.C. Freitas, A.C. Duarte, Trends Anal. Chem. 34 (2012) 97–110.
[47] R.J. Parkes, G. Sellek, G. Webster, D. Martin, E. Anders, A.J. Weightman, H. Sass, Environ. Microbiol. 11 (2009) 3140–3153.
[48] G.M. Cragg, D.J. Newman, Biochim. Biophys. Acta 1830 (2013) 3670–3675.
[49] C. Roman, R. Mather, Proc. Inst. Mech. Eng. M J. Eng. Marit. Environ. 224 (2010) 327–340.
[50] J. Cao, C. Liut, B. Yao, L. Lian, Oceans 1–6 (2013).
[51] G. Mitta, F. Hubert, E.A. Dyrynda, P. Boudry, P. Roch, Dev. Comp. Immunol. 24 (2001) 381–393.
[52] J.F. Imhoff, A. Labes, J. Wiese, Biotechnol. Adv. 29 (2011) 468–482.

[53] C.E. Salomon, N.A. Magarvey, D.H. Sherman, Nat. Prod. Rep. 21 (2004) 105–121.
[54] P.M. Murray, S. Moane, C. Collins, T. Beletskaya, O.P. Thomas, A.W.F. Duarte, New Biotechnol. 30 (2013) 839–850.
[55] R. Montaser, H. Luesch, Future Medicinal Chemistry 3 (2011) 1475–1489.
[56] O. Bergman, B. Mayzel, M.A. Anderson, M. Shpigel, R.T. Hill, Mar. Drugs 9 (2011) 2201–2219.
[57] R. Osinga, J. Tramper, R.H. Wijffels, Mar. Biotechnol. 1 (1999) 509–532.
[58] S.A. Pomponi, Oceanogr 14 (2001) 78–87.
[59] G.M. Cragg, D.J. Newman, K.M. Snader, J. Nat. Prod. 60 (1997) 52–60.
[60] S.V. Sperstad, T. Haug, H.-M. Blencke, O.B. Styrvold, C. Li, K. Stensvåg, Biotechnol. Adv. 29 (2011) 519–530.
[61] M.E. Hay, J. Exp. Mar. Biol. Ecol. 200 (1996) 103–134.
[62] J. Zhang, F. Li, Z. Wang, J. Xiang, J. Biotechnol. 127 (2007) 605.
[63] J.Y. Chen, H. Chuang, C.-Y. Pan, C.-M. Kuo, Fish & Shellfish Immunol. 18 (2005) 179–183.
[64] K.J. Schippers, D. Sipkema, R. Osinga, H. Smidt, S.A. Pomponi, D.E. Martens, R.H. Wijffels, , Advances in marine biology62Academic Press, (2012) 273-337.
[65] M.C. Leal, R. Calado, C. Sheridan, A. Alimonti, R. Osinga, Trends Biotech. 31 (2013) 555–561.
[66] D. Sipkema, R. Osinga, W. Schatton, D. Mendola, J. Tramper, R.H. Wijffles, Biotechnol. Bioeng. 90 (2005) 201–222.
[67] W.G. Müller, V. Grebenjuk, G. Le Pennec, H.C. Schröder, F. Brümmer, U. Hentschel, I.M. Müller, H. Breter, Mar Biotechnol 6 (2004) 105–117.
[68] B. Rinkevich, J. Biotechnol. 70 (1999) 133–153.
[69] A. Duckworth, Mar. Biotechnol. 11 (2009) 669–670.
[70] R. Pronzato, R. Manconi, Mar. Ecol. 29 (2008) 146–166.
[71] N.S. Webster, A.P. Negri, R.I. Webb, R.T. Hill, Mar. Ecol. Prog. Ser. 232 (2002) 305–309.
[72] D. Louden, S. Whalan, E. Evans-Illidge, C. Wlf, R. de Nys, Aquaculture 270 (2007) 57–67.
[73] O. Bergman, M. Haber, B. Mayzel, M. Anderson, M. Shpigel, R. Hill, Mar. Biotechnol. 13 (2011) 1169–1182.
[74] S. de Caralt, H. Otjens, M. Uriz, R. Wijffels, Mar. Biotechnol. 9 (2007) 592–605.
[75] S. de Caralt, J. Sánchez-Fontenla, M.J. Uriz, R.H. Wijffels, Mar. Drugs 8 (2010) 1731–1742.
[76] N.J. de Voogd, J. Mar. Biol. Assoc. UK 87 (2007) 1777–1784.
[77] R. Osinga, M. Sidri, E. Cerig, S.Z. Gokalp, M. Gokalp, Open Mar. Biol. J. 4 (2010) 74–81.
[78] A.L. Demain, Biotechnol. Adv. 18 (2000) 499–514.
[79] M.N. Dawson, K.A. Raskoff, D.K. Jacobs, Mol. Marine Biol. Biotechnol. 7 (1998) 145–152.
[80] K. Siebert, M. Busl, I. Asmus, J. Freund, A. Muscholl-Silberhorn, R. Wirth, Appl. Environ. Microbiol. 70 (2004) 5912–5915.
[81] B. Dhorajiya, M. Malani, B. Dholakiya, Chemical Sciences Journal, 2012.
[82] L. Simmons, K. Kaufmann, R. Garcia, G. Schwär, V. Huch, R. Müller, Bioorg. Med. Chem. 19 (2011) 6570–6575.
[83] I.H. Lee, C. Zhao, Y. Cho, S.S. Harwig, E.L. Cooper, R.I. Lehrer, FEBS Lett. 400 (1997) 158–162.
[84] I.H. Lee, C. Zhao, T. Nguyen, L. Menzel, A.J. Waring, M.A. Sherman, R.I. Lehrer, J. Pept. Res. 58 (2001) 445–456.
[85] I.H. Lee, Y.S. Lee, C.H. Kim, C.R. Kim, T. Hong, L. Menzel, L.M. Boo, J. Pohl, M.A. Sherman, A. Waring, R.I. Lehrer, Biochem. Biophys. Acta 1527 (2001) 141–148.
[86] R. Galinier, E. Roger, P.-E. Sautiere, A. Aumelas, B. Banaigs, G. Mitta, J. Pept. Sci. 15 (2009) 48–55.

[87] C. Li, T. Haug, O.B. Styrvold, T.Ø. Jørgensen, K. Stensvåg, Dev. Comp. Immunol. 32 (2008) 1430–1440.
[88] C. Li, T. Haug, M.K. Moe, O.B. Styrvold, K. Stensvåg, Dev. Comp. Immunol. 34 (2010) 959–968.
[89] T.V. Ovchinnikova, G.M. Aleshina, S.V. Balandin, A.D. Krasnosdembskaya, M.L. Markelov, E.I. Frolova, Y.F. Leonova, A.A. Tagaev, E.G. Krasnodembsky, V.N. Kokryakov, FEBS Lett. 577 (2004) 209–214.
[90] T.V. Ovchinnikova, S.V. Balandin, G.M. Aleshina, A.A. Tagaev, Y.F. Leonova, E.D. Krasnodembsky, A.V. Men'shenin, V.N. Kokryakov, Biochem. Biophys. Res. Commun. 348 (2006) 514–523.
[91] R. Iijima, J. Kisugi, M. Yamazaki, Dev. Comp. Immunol. 27 (2003) 305–311.
[92] J. Zhao, L. Song, C. Li, D. Ni, L. Wu, L. Zhu, H. Wang, W. Xu, Mol. Immunol. 44 (2007) 360–368.
[93] G. Mitta, F. Hubert, T. Noël, P. Roch, Eur. J. Biochem. 265 (1999) 71–78.
[94] M. Charlet, S. Chernysh, H. Philippe, C. Hetru, J.A. Hoffmann, P. Bulet, J. Biol. Chem. 271 (1996) 21808–21813.
[95] P. Amparyup, H. Kondo, I. Hirono, T. Aoki, A. Tassanakajon, Mol. Immunol. 45 (2008) 1085–1093.
[96] J.L. Rolland, M. Abdelouahab, J. Dupont, F. Lefevre, E. Bachère, B. Romestand, Mol. Immunol. 47 (2010) 1269–1277.
[97] S. Ravichandran, S. Wahidulla, L. D'Souza, G. Rameshkumar, Appl. Biochem. Biotechnol. 162 (2010) 1039–1051.
[98] S. Kawabata, R. Nagayama, M. Hirata, T. Shigenaga, K.L. Agarwala, T. Saito, J. Cho, H. Nakajima, T. Takagi, S. Iwanaga, J. Biochem. 120 (1996) 1253–1260.
[99] T. Osaki, M. Omotezako, R. Nagayama, M. Hirata, S. Iwanaga, J. Kasahara, J. Hattori, I. Ito, H. Sugiyama, S. Kawabata, J. Biol. Chem. 274 (1999) 26172.
[100] T. Saito, S. Kawabata, T. Shigenaga, Y. Takayenoki, J. Cho, H. Nakajima, M. Hirata, S. Iwanaga, J. Biochem. 117 (1995) 1131–1137.
[101] T. Miyata, F. Tokunaga, T. Yoneya, K. Yoshikawa, S. Iwanaga, M. Niwa, T. Takao, Y. Shimonishi, J. Biochem. 106 (1989) 663–668.
[102] K. Stensvåg, T. Haug, S.V. Sperstad, Ø. Rekdal, B. Indrevoll, O.B. Styrvold, Dev. Comp. Immunol. 32 (2008) 275–285.
[103] S.V. Sperstad, T. Haug, T. Vasskog, K. Stensvåg, Mol. Immunol. 46 (2009) 2604–2612.
[104] D. Schnapp, G.D. Kemp, V.J. Smith, Eur. J. Biochem. 240 (1996) 532–539.
[105] J.M. Relf, J.R. Chisholm, G.D. Kemp, V.J. Smith, Eur. J. Biochem. 264 (1999) 350–357.
[106] V. Brockton, J.A. Hammond, V.J. Smith, Mol. Immunol. 44 (2007) 943–949.
[107] W.S. Huang, K.J. Wang, M. Yang, J.J. Cai, S.J. Li, G.Z. Wang, J. Exp. Mar. Biol. Ecol. 339 (2006) 37–42.
[108] R.D. Yedery, K.V.R. Reddy, Acta Biochim. Pol. 56 (2009) 71–82.
[109] C. Imjongjirak, P. Amparyup, A. Tassanakajon, Fish & Shellfish Immunol. 30 (2011) 706–712.
[110] C. Hauton, V. Brockton, V.J. Smith, Mol. Immunol. 43 (2006) 1490–1496.
[111] A.E. Christie, S. Rus, C.C. Goiney, C.M. Smith, D.W. Towle, P.S. Dickinson, Mol. Immunol. 44 (2007) 3333–3337.
[112] A.L. Battison, R. Summerfield, A. Patrzykat, Fish & Shellfish Immunol. 25 (2008) 181–187.
[113] A. Tasiemski, D. Schikorski, F. Le Marrec-Croq, C. Pontoire-Van Camp, C. Boidin-Wichlacz, P.E. Sautiere, Dev. Comp. Immunol. 31 (2007) 749–762.
[114] W. Pan, X. Liu, F. Ge, J. Han, T. Zheng, J. Biochem. 135 (2004) 297–304.
[115] S. Iwanaga, Curr. Opin. Immunol. 14 (2002) 87.

Chapter 3

Classical Methodologies for Preparation of Extracts and Fractions

C.I.L. Justino[1], K. Duarte[1], A.C. Freitas[2,3], Armando C. Duarte[4] and Teresa Rocha-Santos[5,6]

[1]*CESAM & Department of Chemistry, University of Aveiro, Aveiro, Portugal*
[2]*CESAM & Department of Chemistry, University of Aveiro, Aveiro, Portugal*
[3]*ISEIT/Viseu – Instituto Piaget, Estrada do Alto do Gaio, Galifonge, Lordosa, Viseu, Portugal*
[4]*CESAM & Department of Chemistry, University of Aveiro, Aveiro, Portugal*
[5]*CESAM & Department of Chemistry, University of Aveiro, Aveiro, Portugal*
[6]*ISEIT/Viseu – Instituto Piaget, Estrada do Alto do Gaio, Galifonge, Lordosa, Viseu, Portugal*

Contents

3.1 SAMPLE PREPARATION OF BIOACTIVE COMPOUNDS FROM THE MARINE ENVIRONMENT

In general, it is more difficult to obtain large quantities of bioactive compounds from marine organisms than terrestrial species since some marine organisms produce only trace amounts of bioactive compounds [1]. However, in the latest years, the bioactive compounds from the marine environment have gained scientific attention since a broad range of their biological effects, such as antimicrobial, anti-inflammatory, antiviral, and antifungal activities, have been acknowledged. For example, the cyanobacterium *Lyngbya* has been studied as a source of various new bioactive compounds due to its high secondary metabolite production [2]. Matthew *et al.* [3] have isolated three new cyclodesipeptides, named tiglicamides, from *Lyngbya confervoides*, and Tan *et al.* [4] have discovered 12 new secondary metabolites from the class of polyketide-polypeptide in

Analysis of Marine Samples in Search of Bioactive Compounds, Vol. 65. DOI: 10.1016/B978-0-444-63359-0.00003-3

Lyngbya majuscule with antifouling properties, such as dolastatin 16 A, hantupeptin C, majusculamide A, and isomalyngamide.

In order to be available for practical applications, the bioactive compounds should be isolated from marine organisms following a procedure with several steps. First, the marine organisms are sampled from the environment and grown at the laboratory in a nutritional media. In this step, special attention should be taken since the culture conditions, such as temperature, aeration, pH of the media, incubation time, and media composition, can affect the production of the desired bioactive compound (discussed by Penesyan *et al.* [5]). The next step involves screening individual isolates for biological activity; for example in the case of antimicrobials, the inhibition of growth of microorganisms contained in the test organism is performed in order to identify the biological activities of the target compounds from the test organism [5]. One of the final steps is the extraction and fractionation of bioactive compounds, which could be obtained from the whole extract and fractions of the marine organisms, followed by the identification of the chemical structures of the bioactive compounds.

The analytical techniques used for the extraction and fractionation of marine fungi, as a source of bioactive compounds, have been recently reviewed by Duarte *et al.* [6]. Nevertheless, the marine environment has a wide variety of other organisms such as bacteria, algae, and invertebrates such as sponges, corals, and tunicates, which could be used as sources of bioactive compounds with pharmaceutical and therapeutical interest.

This chapter presents state of the art strategies based on classical methodologies for the isolation of the bioactive compounds found in the marine environment organisms, including the preparation of extracts and fractions. Practical examples are also reported corresponding to recent literature, between 2010 and 2013, identifying the advantages and limitations of the classical methodologies employed, and Tables 3.1 and 3.2 show the methods of extraction and fractionation used in recent works for the isolation of bioactive compounds from marine organisms such as algae, bacteria, and fungi (Table 3.1) as well as sponges, plants, and mollusks (Table 3.2).

As shown in Tables 3.1 and 3.2, there is a plethora of marine organisms that could be useful as a source of bioactive compounds. In some specific situations, there are marine organisms that should be isolated from another marine organism due to the existence of symbiotic relationships. Commonly, symbiotic bacteria are associated with corals, invertebrates, algae, and sponges. As shown in Table 3.1, the bacteria *Bacillus pumilus* and *Bacillus licheniformis* SAB1 should be isolated from the mucus of the black coral *Antipathes* sp. [19] and from a sponge (*Halichondria* sp.) [15], respectively, in order to obtain bioactive compounds. Abdelmohsen *et al.* [42] isolated 90 actinomycetes, associated with different species of marine sponges, due to their anti-infective activities against clinically relevant Gram-positive (*Enterococcus faecalis* and *Staphylococcus aureus*) and Gram-negative (*Escherichia coli*, *Pseudomonas aeruginosa*) bacteria, fungi (*Candida albicans*), and human parasites (*Leishmania major* and *Trypanosoma brucei*).

TABLE 3.1 Methods of Extraction and Fractionation for the Isolation of Bioactive Compounds from Marine Organisms such as Algae, Bacteria, and Fungi

Marine Organism	Origin	Bioactive Compounds	Solvent Extraction	Fractionation and Purification	Reference
			ALGAE		
Brown alga *Lobophora variegata*	Mexico	Antiprotozoal compounds: sulfoquinovosyl-diacylglycerols	Dichloromethane: methanol (7:3)	Solvent partition into methanol-water (9:1), hexane, chloroform, ethyl acetate, and n-butanol. The chloroform fraction (best bioactivity against protozoa) was subjected to column chromatography on Sephadex LH-20, eluted with hexane-chloroform-methanol (3:2:1).	[7]
Brown algae *Turbinaria ornata*	Hawaii	Carotenoids such as fucoxanthin	Methanol	RP-HPLC (C-18 column, with gradient elution from 5% methanol:water to 100% methanol).	[8]
Brown algae *Myagropsis myagroides*	Korea	Fucoxanthin	Methanol 80%	Solvent partition into chloroform. Fractionation by silica column chromatography with stepwise elution of chloroform:methanol (from 100:1 to 1:1) and then Sephadex LH-20 column chromatography with 100% methanol. HPLC (C-18 column by stepwise elution with methanol:water gradient) was used to purify the resulting fractions.	[9]
Macroalgae such as *Chaetomarpha linum* and *Enteromorpha compressa*	India	Antibacterial soluble compounds	Soxhlet extraction with chloroform and ethyl acetate	—	[10]

(Continued)

TABLE 3.1 Methods of Extraction and Fractionation for the Isolation of Bioactive Compounds from Marine Organisms such as Algae, Bacteria, and Fungi *(cont.)*

Marine Organism	Origin	Bioactive Compounds	Solvent Extraction	Fractionation and Purification	Reference
Macroalgae *Sargassum wightii*	South East coast of India	Phenolic compounds	Soxhlet extraction with methanol, chloroform, and diethyl ether	—	[11]
	India	Steroids and phenolic compounds	Methanol, chloroform, and benzene	TLC on silica gel plates with chloroform:methanol (9:1) as mobile phase for phenolic compounds, and benzene:methanol (9:1) for steroids. HPLC (C-18 column with mobile phase of 0.1% (v/v) methanol (solvent A) and water (solvent B)).	[12]
Macroalgae *Kappaphycus alvarezii*	India	Phenolic compounds, terpenes, and tannins	Soxhlet extraction with ethanol, methanol, and acetone	—	[13]
Macroalgae *Bryothamnion triquetrum*	Brazil	Bioactive compounds with antinociceptive and anti-inflammatory activities	Soxhlet extraction with methanol	—	[14]

BACTERIA					
Bacillus licheniformis SAB1 isolated from a sponge (*Halichondria* sp.)	Indo-Pacific region	Antimicrobial compounds: indole, 3-phenylpropionic acid, and 4,4′-oxybis (3-phenylpropionic acid)	Methanol	Silica gel column chromatography (60–120 mesh) with increasing concentrations of ethyl acetate in petroleum ether as eluent.	[15]
Pigmented bacteria	Norway	Carotenoids	Methanol	HPLC with methanol and dichloromethane as mobile phases.	[16]
Streptomyces albidoflavus	North Sea, West Pacific	Antifouling compounds with 2-furanone ring: α,β-unsaturated lactones	Ethyl acetate	Macroporous resin column chromatography using a gradient solvent system from water to acetone. The fraction with better bioactivity was purified on ODS RP-HPLC with water-methanol solvent system.	[17]
Streptomyces VITSVK5 sp.	Southern India	Larvicidal compound: 5-(2,4-dimethylbenzyl) pyrrolidin-2-one	N-butanol	Silica gel column chromatography with chloroform and methanol (increasing solvent concentrations between 10:0 and 7:3). The fraction with better bioactivity was separated by TLC on silica gel using chloroform and methanol (8:2) as solvent system.	[18]
Bacillus pumilus isolated from the coral *Antipathes* sp.	Panama	Indole alkaloids	Ethyl acetate	C-18 SPE cartridges eluted with methanol in water (stepwise elution gradient between 20 and 100%), followed by RP-HPLC (C-18 column with elution of methanol:water (85:15)).	[19]

(Continued)

TABLE 3.1 Methods of Extraction and Fractionation for the Isolation of Bioactive Compounds from Marine Organisms such as Algae, Bacteria, and Fungi *(cont.)*

Marine Organism	Origin	Bioactive Compounds	Solvent Extraction	Fractionation and Purification	Reference
Streptomyces spp.	Canada	New bioactive compounds analogues to novobiocin	Ethyl acetate	Solvent partition with ethyl acetate and water. Column chromatography with Sephadex-LH20 with methanol:dichloromethane (4:1) as eluent. RP-HPLC (C-18 column with acetonitrile: 0.05% aqueous trifluoroacetic acid, 1:1) was used to isolate the pure compounds.	[20]
Lyngbya majuscula	Singapore	Bioactive compounds from polyketides-polypeptide class	Dichloromethane: methanol (1:1, v/v)	Column chromatography on normal phase Si using a combination of hexanes, ethyl acetate, and methanol of increasing polarity. HPLC (with Sphereclone 5 μm ODS, and 8:2 methanol:water) was used to purify the resulting polar fractions.	[4]
			FUNGI		
Hypocrea vinosa	Japan	Antiangiogenic metabolites: hypochromins A and B	Chloroform and ethyl acetate	Silica gel column chromatography (stepwise gradient solvent system of 0–100% chloroform-methanol).	[21]
Aspergillus versicolor isolated from sponge *Petrosia* sp.	Korea	Aromatic polyketide derivative, xanthones, and anthraquinones	Ethyl acetate	Solvent partition into n-hexane and methanol 90%. The partition with better bioactivity was purified with RP-HPLC (Shodex C8-5E with 55% methanol).	[22]

Trichoderma koningii	South China Sea	Polyketide derivatives	Ethyl acetate:methanol: acetic acid (80:15:5)	Solvent partition with ethyl acetate. Silica gel column chromatography (200–300 mesh) using a gradient of chloroform in methanol and Sephadex LH-20 column chromatography with petroleum ether:chloroform:methanol mixture (5:5:1). RP-HPLC (C-18 column) was further used for purification of compounds.	[23]
Ascomycetous fungus	Germany	Macrolide	Ethyl acetate	Silica gel column chromatography (0.063–0.200 µm with various solvent systems[1] and 0.015–0.040 µm with various solvent systems[2]). TLC (silica gel). RP-HPLC with a gradient solvent system of 10–100% methanol.	[24]
Fungus isolated from macroalgae *Kappaphycus alvarezii*	Indonesia	New bioactive compound: $C_{12}H_{10}O_4$	Soxhlet extraction with ethyl acetate	Column chromatography with Sephadex LH-20 using methanol as mobile phase. The fraction with better bioactivity was separated by TLC (silica gel).	[25]
Aspergillus fumigatus	Japan	New indole alkaloids: 2-(3,3-dimethylprop-1-ene)-costaclavine and 2-(3,3-dimethylprop-1-ene)-epicostaclavine	Acetone:methanol (1:1, v/v)	Column chromatography with silica gel (n-hexane and ethyl acetate). TLC with dichloromethane:methanol (10:1). HPLC was also used for final purification of compounds (C-18 column with 50–100% methanol in water).	[26]

(Continued)

TABLE 3.1 Methods of Extraction and Fractionation for the Isolation of Bioactive Compounds from Marine Organisms such as Algae, Bacteria, and Fungi *(cont.)*

Marine Organism	Origin	Bioactive Compounds	Solvent Extraction	Fractionation and Purification	Reference
Myceliophthora lutea		New bioactive compounds: isoacremine D and acremine A	Ethyl acetate and dichloromethane	Column chromatography with silica gel (40/100 μm) and Sephadex LH-20. TLC was used for further fractionation (silica gel).	[27]
Fasciatispora nypae, *Caryosporella rhizophorae*, *Melaspilea mangrovei*, and *Leptosphaeria* sp.	Malaysia	2,2,7-trimethyl-2H-chromen-5-ol	Ethyl acetate and dichloromethane	Column chromatography with silica gel using a stepwise elution with mixtures of dichloromethane and methanol (between 100:0 and 0:100). TLC was used for further fractionation (silica gel).	[28]

RP-HPLC: Reverse-phase high performance liquid chromatography; TLC: Thin-layer chromatography; SPE: Solid Phase Extraction.
[1]*ethanol:hexane:methanol (65:35:5), ethanol:methanol (95:5), ethanol:methanol (50:50), and methanol*
[2]*dichloromethane:ethanol (75:25) and ethanol*

TABLE 3.2 Methods of Extraction and Fractionation for the Isolation of Bioactive Compounds from Marine Organisms such as Sponges, Plants, and Mollusks

Marine Organism	Origin	Bioactive Compounds	Solvent Extraction	Fractionation and Purification	Reference
			SPONGES		
Xestospongia testudinaria	South Pacific	Halenaquinone-type polyketides	Dichloromethane and dichlorometane: methanol (1:1)	Silica gel column chromatography (230–400 mesh) with 20% methanol in dichloromethane as eluent. TLC (silica gel) used for further fractionation with dichloromethane:methanol (8:2).	[29]
Dysidea fragilis	China	Terpenes	Acetone	Solvent partition into diethyl ether and water. Silica gel (200–300 mesh) and Sephadex LH-20 column chromatographies. RP-HPLC (methanol:water as mobile phase, 80:20) was used for further purification of compounds.	[30]
Haliclona oculata	India	Alkaloids	Methanol	Solvent partitions into hexane, chloroform, and n-butanol. The chloroform fraction (better bioactivity) was purified with column chromatography.	[31]
Petromica citrina	Rio de Janeiro, Brazil	Sterol: halistanol trisulphate	Methanol	Solvent partition into hexane, chloroform, and ethyl acetate, as well as into aqueous residue. The aqueous residue fraction (better bioactivity) was purified with Sephadex LH-20 column chromatography (eluted with a system of solvents of different polarity from water to methanol).	[32]

(Continued)

TABLE 3.2 Methods of Extraction and Fractionation for the Isolation of Bioactive Compounds from Marine Organisms such as Sponges, Plants, and Mollusks *(cont.)*

Marine Organism	Origin	Bioactive Compounds	Solvent Extraction	Fractionation and Purification	Reference
Lanthella cf. *flabelliformis*	Australia	Alkaloids: sesquiterpenes and indole alkaloids	Ethanol	Solvent partitions into hexane, dichloromethane, and methanol. Fractions were purified with HPLC (C-8 column with gradient elution of 10–100% acetonitrile-water).	[33]
Haliclona exigua	Gulf of Mannar	Fatty acids	Soxhlet extraction with ethyl acetate	—	[34]
Dendrilla nigra					[35]
Aurora globostellata	India	3-hydroxy tetradecanoic acid	Ethyl acetate	Column chromatography	[36]
Hyrtios spp.	Red Sea	New alkaloids: hyrtioerectines	Methanol: dichloromethane (1:1)	Solvent partition into n-hexane. Column chromatography with silica gel (70–230 mesh) and Sephadex LH20 with methanol as mobile phase. HPLC (C-18 column) was further used for purification of compounds using 20% of acetonitrile in water.	[37]
PLANTS and MOLLUSKS					
Sea fennel (plant) *Crithmum maritimum L.*	France	Polyacetylene: falcarindiol	Chloroform	Solvent partition into hexane. Purification with Sephadex LH-20 column chromatography (elution with dichloromethane and acetone).	[38]

Mollusk *Scaphander lignarius*	Norway	Polyunsaturated fatty acids	Dichloromethane: methanol (1:1, v/v)	Solvent partition into hexane, acetonitrile, and purified water. Final purification of compounds with HPLC (C-18 column with a gradient solvent system from 20 to 100% of acetonitrile in water).	[39]
Gastropod *Cellana radiata*	East Coast of South India	Anticancer bioactive compound	Soxhlet extraction with diethyl ether	—	[40]
Mollusk *Amussium pleuronectus*		Anti-coagulant compounds: glycosaminoglycans	Methanol (85%, v/v)	Ion-exchange column chromatography with DEAE-cellulose (eluted with distilled water and NaCl) and Amberlite IRA-900. Final purification of compounds with column chromatography with Sephadex G-100.	[41]

RP-HPLC: Reverse-phase high performance liquid chromatography

The bacteria associated with marine algae have also been frequently investigated due to their anticancer and antibiotic activities, for example with the specie *Leucobacter* sp. [43]. On the other hand, the cyanobacteria, which are photosynthetic prokaryotes, can live not only free but also in association with other organisms, such as corals, ascidians, and sponges, and although these associations are not actually well understood, such symbionts are responsible for the production of bioactive compounds as a response to ecological pressures [44,45].

Besides the production of secondary metabolites associated with the sponge–bacteria relationships, the symbiotic functions are responsible for the nutrient acquisition, the stabilization of sponge skeleton, and the processing of metabolic wastes, as reviewed by Thomas *et al.* [44]. However, other marine organisms could be isolated from sponges, such as fungi. Lee *et al.* [22] have isolated the fungi *Aspergillus versicolor* from a *Petrosia* sp. sponge in order to obtain bioactive compounds such as xanthones and anthraquinones, which have exhibited a significant cytotoxicity against human solid tumor cell lines, as well as antibacterial activity against Gram-positive strains.

Before designing the steps of the isolation procedure, the nature of the target compound should be considered—that is, solubility, acid-base properties, charge, stability, and molecular size—and Sarker *et al.* [46] reviewed the various experiments that could be easily performed for such study: the choice of the best methodology for further separation leads to fastest isolation procedure.

3.1.1 Extraction of Bioactive Compounds

The choice of the extraction method is dependent on the bioactive compound to be isolated and the nature of the organism from where the extract will be obtained. The objective of the extraction should be clear since the bioactive compounds of interest to be isolated could be unknown, known, or structurally similar to a group of known compounds [46]. On the other hand, it should be taken into account that the success of an extraction process is affected by the content of bioactive compounds in the marine organisms. For example, it is known that the protein content of marine algae varies with the algae species; that is, high levels of proteins (maximum of 47% (w/w) dry weight) could be found in red macroalgae and low levels (3–15% (w/w) dry weight) could be found in brown algae [47]. In the same way, the seasonal variation could also affect the content of algae proteins, mainly due to the nutrient supply, water temperature, available light and salinity, types of proteins present, and fluctuations in carbohydrate level [47]. According to Barbarino *et al.* [48], the extraction of algal proteins is influenced by the chemical composition of the algae species, its morphological and structural characteristics, and even the content of the algal proteins is dependent on the extraction procedures.

A specific and precise protocol should be followed when the marine organism of interest is associated with other marine organisms, as in the case of the bacteria associated with algae or sponges. Before the extraction of the bioactive compounds and biological assays, first the bacteria should be isolated and purified up to the third generation in order to assure the integrity of pure colonies, and second, they should be characterized in terms of macroscopic morphology and Gram strain, as well as submitted to phylogenetic analysis [43].

As shown in Tables 3.1 and 3.2, the marine bioactive compounds are mainly obtained by solvent extraction and Soxhlet extraction, but other methodologies of extraction could be employed such as aqueous, acid, and alkaline extractions, as discussed in the following subsections.

3.1.1.1 Extraction by Solvents

The extraction of bioactive compounds consists mostly of the use of solvents with different polarities accordingly to the nature of the bioactive compound of interest. Table 3.3 reports some examples of bioactive compounds and the solvents commonly used for their extraction, taking into account the review paper of Duarte *et al.* [6].

According to Bhakuni and Rawat [1], for example, the lipophilic compounds are mostly present in the hexane and chloroform fractions, and the nonpolar compounds that are extracted in hexane, benzene, and chloroform are generally esters, ethers, terpenoids, sterols, and fatty acids.

The extraction by solvents could follow the principle of either "liquid-liquid" or "solid-liquid" extractions. The liquid-liquid extraction is the classical technique in chemistry to isolate a target component from a mixture. Thus, the selective partitioning of such components of interest into one of the two

TABLE 3.3 Examples of Bioactive Compounds and the Solvents Commonly Used for Their Extraction

Type of Bioactive Compounds	Examples of Bioactive Compounds	Solvents Commonly Used
Polar organic compounds	Alkaloids Shikimates Polyketides Sugars Amino acids Polyhydroxysteroids Saponins	N-butanol Chloroform Ethyl acetate Acetone Methanol Ethanol Water
Medium-polarity compounds	Peptides	Dichloromethane Methanol Carbon tetrachloride
Low-polarity compounds	Terpenes Hydrocarbons Fatty acids	Carbon tetrachloride Hexane

immiscible phases results from the choice of the most adequate extraction solvent, as shown in Table 3.3. When the optimal conditions are not applied, low recoveries are achieved and further extraction should be made in order to find the optimal combination of extraction solvents to obtain high recovery and higher purity of the liquid–liquid extraction.

Since a marine organism is a solid matrix, the solid-liquid extraction is the extraction most applied for obtaining marine bioactive compounds, and it can be performed according to the following steps:

1. The material (marine organism) is placed in contact with a liquid solvent.
2. The solvent diffuses into the cells and solubilizes the metabolites.
3. The metabolites are diffused out of the cells into the solvent media.

The solid–liquid extraction can be performed using a Soxhlet apparatus, as discussed in Section 3.1.1.2. The yield of a chemical extraction can depend on the type of solvents with varying polarities, pH, extraction time, and temperature as well as on the chemical composition of the sample. Under the same conditions of time and temperature, the solvent and the chemical properties of the sample become the two most important factors. Earlier in 2008, solvents such as petroleum ether, ethyl acetate, methanol, dichloromethane, and butanol have commonly were used for the extraction of phenolic compounds from marine microalgae, as studied by Ganesan *et al.* [49], and the ethyl acetate fraction exhibited the higher antioxidant activity in comparison to the other solvent fractions. Yu *et al.* [30] suggest that the acetone extraction of terpenes (Table 3.2)—that is, dysifragilisins A and B, isolated from the sponge *Dysidea fragilis*—implies artifacts since the group CH_3COCH_2 present in such compounds, due to the use of acetone, were not detected in a chloroform extraction.

Recently, Almeida *et al.* [50] reviewed the bioactivities from marine algae of the genus *Gracilaria*, and verified that the ethanol is the solvent most used for the extraction of bioactive compounds, such as palmitic acid and steroids, which have antibacterial activity, both using the entire plant and the talus of the algae (*Gracilaria domingensis*), either dried or freshly collected. Makin *et al.* [51] reported that when the extraction of steroids is performed, special attention should be paid to the glassware used since many steroids can bind very tightly to glass. Thus, the extraction requires silanization of all glassware by washing with dimethyldichlorosilane (e.g., 1% v/v in toluene) and then with methanol. On the other hand, plastic should be excluded due to the occurrence of phthalates in extracts, which can interfere in the final analysis [51]. Cantillo-Ciau *et al.* [7] have shown that the chloroform fraction obtained from alga extracts (*Lobophora variegata*) showed the best antiprotozoal activity against the three protozoa *Giardia intestinalis*, *Entamoeba histolytica*, and *Trichomonas vaginalis*, which are human-infective parasites, when compared with fractions obtained from partitions of methanol-water (9:1), hexane, ethyl acetate, and n-butanol (Table 3.1).

The carotenoids, considered the most abundant pigment group, have antioxidant properties, and they are commercialized as food colorants, animal feed supplements, and nutraceuticals for cosmetic and pharmaceutical applications. They are lipophilic and hydrophobic, but soluble in organic solvents such as acetone, alcohols, ethyl ether, chloroform, and ethyl acetate [16]. Among the various organic solvents, Stafsnes *et al.* [16] have shown that methanol leads to the best overall extraction efficiency, when carotenoids were isolated from pigmented bacteria (Table 3.1). Lipids were also extracted by chloroform (1:5, w/v) by Wilson-Sanchez *et al.* [52] from the tail muscle of white shrimp (*Litopenaeus vannamei*) in order to study potential antimutagenic and antiproliferative properties.

The biochemical details involved in the extraction of bioactive compounds are also important in the selection of both the extraction method and the solvent used. For example, Moraes *et al.* [53] observed that the extraction of phycobiliproteins, which are one of the most important groups of proteins from macroalgae, occurs after the disruption of the algae cells, leading to the release of proteins.

The main advantages of the use of solvent extraction are its low processing cost and the ease of operation. On the other hand, the drawbacks are low selectivity, low extraction efficiency, production of solvent residue, and environmental pollution [54].

3.1.1.2 Extraction by Soxhlet Methodology

The Soxhlet extraction is a classical extraction methodology also used for the extraction of bioactive compounds from marine resources (Tables 3.1 and 3.2). Commonly, the Soxhlet extraction is required when the target analytes have a limited solubility in a solvent, but this methodology can also be used for soluble materials.

The principle of Soxhlet extraction is based on placing a solid material containing the target analytes inside a thimble made from filter paper, which is loaded in the main chamber of the Soxhlet extractor. Then, the extractor is placed onto a flask containing the extraction solvent. As the Soxhlet is equipped with a condenser, the solvent is heated to reflux and the target analytes are dissolved in the solvent. According to McCloud [55], the negative feature of the Soxhlet extraction is the long-term boiling in organic solvent of materials, being an important drawback of such methodology when used to extract bioactive compounds from marine organisms. Thus, the Soxhlet extraction methodology is not suitable for the extraction of thermo-sensitive compounds, since the sample is constantly heated [56]. However, Bhimba *et al.* [34] and Bhimba *et al.* [35] have applied the Soxhlet methodology to the extraction of fatty acids from sponges *Haliclona exigua* and *Dendrilla nigra*, respectively (Table 3.2).

Marine bioactive compounds such as phenolic compounds were also extracted by Soxhlet methodology from fresh marine macroalgae such as *Gracilaria edulis*, *Gracilaria vercosa*, *Acanthospora spicifera*, *Ulva lacta*, *Kappaphycus spicifera*, *Sargassum ilicifolium*, *Sargassum wightii*, and *Padina gymonospora*

[11]. Thirunavukkarasu *et al.* [11] have found that the methanolic, chloroform, diethyl ether extracts of *Sargassum wightii*, and the acetone extract of *Gracilaria vercosa* produced a maximum zone of inhibition against fish pathogenic bacteria *Vibrio alginolyticus*. Although Thirunavukkarasu *et al.* [11] have identified only phenolic compounds in the extracts of *Sargassum wightii*, such macroalgae is a rich source of phytoconstituents such as steroids, alkaloids, saponins, flavonoids, and phenolic compounds (Table 3.1); that is, constituents that exhibit phytochemical properties, as studied by Marimuthu *et al.* [12].

Halim *et al.* [57] have extracted lipids from microalgae *Chlorococcum* sp. through supercritical carbon dioxide (a green extraction methodology, discussed in Chapter 4) and Soxhlet extraction with hexane, in order to compare their extract yields. Halim *et al.* [57] have found that both extraction methodologies achieved comparable lipid yields (approximately 0.058 g of lipid extract per g of dried microalgae) with less efficiency recorded with the Soxhlet procedure. Spiric *et al.* [58] also compared the extraction of marine bioactive compounds (fatty acids and cholesterol) in carp fish muscles obtained with a green extraction methodology; that is, the pressurized liquid extraction (also discussed in Chapter 4) and the classical methodology of Soxhlet extraction. Spiric *et al.* [58] have found that the Soxhlet extraction methodology have a higher extraction yield of omega-6-fatty acid than that obtained by the green methodology.

3.1.1.3 Extraction by Other Methodologies

The aqueous, acid, and alkaline extractions could also be used for the extraction of protein fractions and sulphated polysaccharides from macroalgae [47,59]. For example, the laminarans (polysaccharides of glucose), which are water soluble, can be extracted by aqueous methodology, the fucans (sulphated polysaccharides) can be extracted with dilute hydrochloric acid, and alginates can be extracted through alkaline extraction [59].

Cheng *et al.* [60] have evaluated the impact of several extraction methods by using water, acidic, or alkaline extracting media on the antioxidant activity of polysaccharides isolated from mussels *Mytilus edulis*. Cheng *et al.* [60] have found that the antioxidant activity increased with the increasing concentrations of polysaccharides, and that the water and alkaline extracts of such biological compounds have a stronger activity that the acid ones. Martínez-Maqueda *et al.* [61] reported that the alkaline extractions are simple due to the ready availability of the reagents required but the protein quality can be affected by such extraction since undesirable reactions can occur such as racemisation of amino acids, formation of toxic compounds, loss of essential amino acids, and decrease in nutritive values.

Recently, Khaniki *et al.* [62] have optimized the extractions of carotenoids from *Penaeus semisulcatus* shrimp wastes using alkaline extraction with NaOH and enzymatic extraction with alcalase. Khaniki *et al.* [62] have found similar carotenoids extraction yields obtained with alkaline extraction (170 mg/L of carotenoids) and alcalase extraction (234 mg/L of carotenoids).

Concerning the extraction of bioactive compounds from marine fungi, Blunt *et al.* [63] state that a majority of marine fungi produce hydrophobic compounds, when isolated through organic solvent extraction. However recently, He *et al.* [64] reported that bioactive hydrophilic compounds can also be obtained from marine fungi. For example, novel hydrophilic compounds hypochromins A and B were isolated from the specie *Hypocrea vinosa*, showing the higher inhibitory activity when obtained from the ethanol extract of such fungi specie [21]. Le Ker *et al.* [65] have also found that the aqueous extraction is more effective to extract hydrophilic compounds from marine fungi than the organic extraction with methanol. However, Le Ker *et al.* [65] also stated that the aqueous extraction requires a complementary method of mechanical or enzymatic nature and a loss of the bioactivity of aqueous extracts is reported when compared to the bioactivity showed with organic extracts, constituting two disadvantages of the aqueous extraction process.

As shown in this discussion, some limitations of mentioned classical methodologies interfere in the yield of the extracts of bioactive compounds, due to the use of solvents, together with the labor-intensive procedures and the time-consuming nature of the extraction methodologies. However, it should be also highlighted that novel analytical techniques have been developed for the extraction of bioactive compounds but command a high investment due to the modern technologies required, such as the high pressure operation in the case of supercritical fluid extraction [66,67]. Another problem of the supercritical fluid extraction using carbon dioxide as the solvent is its nonpolar nature, which necessitates the use of polar modifiers or cosolvents in order to change the polarity of the supercritical fluid and to increase its solvating power toward the analyte of interest, as explained by Ibañez *et al.* [67].

3.1.2 Fractionation of Extracts to Obtain Fractions of Increasing Purity in Bioactive Compounds

Fractionation is used to separate the bioactive compounds from the extract mixture, which could contain neutral, acidic, basic, lipophilic, and amphiphilic compounds, in order to obtain fractions of increasingly pure bioactive compounds. In general, the solvent extracts are divided into water-soluble and non-water-soluble fractions, which are then submitted to biological assays.

The chromatographic and membrane separations have been the most used technologies for isolation of marine bioactive compounds. The membrane separation is mainly used for the enrichment of peptides, for example, from the protein hydrolysates of fish [68]. The ultrafiltration membranes have been used for the fractionation of fish protein hydrolysates in order to obtain fractions of bioactive peptides, increasing their biological activity [68]. As stated by Samarakoon and Jeon [69], after the enzyme-assisted extractions of the bioactive compounds such as proteins from the marine algae, the protein hydrolysates may be fractionated for different distributions of molecular weight (MW) by using ultrafiltration membranes with different pore sizes. Thus, the protein hydrolysates, which are complex mixtures of free amino acids, peptides with MW up to 7 kDa, and, in a lower proportion, lipids and sodium chloride, have biological activities depending on their MW and amino acid sequences [70,71].

The fractionation of bioactive compounds has been performed by the bioassay-guided fractionation procedures, which is the main approach to screening such compounds, together with the pure compound screening. The difference is that, in the pure compound screening, the extracts with compounds that are not already present in the library of pure compounds should be selected [72].

The bioassay-guided fractionation approach, mainly *in vitro*, involves several steps [6,72]:

1. The assessment of the potential bioactivity of the sample using a bioassay.
2. The extraction using different solvents and assessment of bioactivity.
3. The repeated fractionation of active extracts and fractions in order to obtain the successful isolation of the target bioactive compounds.
4. The structural characterization of the bioactive compounds, followed by pharmacological and toxicological testing.

The pure compound screening involves [6,72]:

1. An automated process of isolation of the compounds of the extract and elucidation of their structure.
2. The screening of the purified and structurally elucidated bioactive products.
3. The pharmacological and toxicological testing of the bioactive compound.

It is important to highlight that the fractionation is a crucial step in obtaining pure bioactive compounds but it is not always beneficial with respect to bioactivity, which is its main limitation [73]. As studied by Sarmadi and Ismail [74], in some cases, mixtures of peptides, amino acids, and sugars show higher bioactivity (e.g., antioxidant activity) than single purified peptides. On the other hand, concerning the bioactive metabolites from mussels, it is shown that although lipids have the highest potential for the commercial development of new bioactive compounds, in comparison to the other two major groups of mussel primary metabolites (proteins and carbohydrates), the increasing instability of lipids during the purification process can limit the research on single lipid components [75]. Thus, the purification procedures focus mainly on the characterization of lipid extracts or fractions rather than on pure compounds [74]. This is a limitation of the purification step when mussels are the marine organisms of interest for the extraction and purification of bioactive compounds.

Concerning proteomics, the need for fractionation and separation procedures is essential due to the presence of very complex mixtures of proteins within biological systems, as reviewed by Martínez-Maqueda *et al.* [61] and Issaq *et al.* [75]. Thus, the fractionation of the mixtures of proteins and peptides should take into account their various properties such as solubility, hydrophobicity, MW, and isoelectric point, among others [61].

3.1.2.1 Fractionation by Solvent Partition

Fractionation by solvent partition separates the active extract from the inactive, while discarding a large part of the inactive material but keeping the chemical complexity of the active fractions. Alkaloids and sterols are predominantly isolated by solvent partition (Tables 3.1 and 3.2), which involves the use of sets of two immiscible solvents in a separating funnel. The compounds are distributed in the two solvents according to their different partition coefficients [76]. The advantage of the solvent partition is the total recovery of target compounds, since it is known, for example, that fats will be present in the hexane fraction while the inorganic salts will be present in the aqueous fraction [77]. The following step is the purification of such compounds, as discussed next.

3.1.2.2 Separation and Purification by Chromatography

After the solvent partition, the separation and purification of the active fractions by chromatography should be performed in order to obtain fractions of increasing purity in bioactive compounds. If the separation is successful, the bioactivity should be concentrated in a specific fraction. The active fractions can be fractioned by column chromatography of several types such as absorption on silica gel or alumina, ion-exchange, and gel permeation, using a variety of solvent systems adapted to the polarity of the active fraction [1]. Various chromatographic processes should also be used for obtaining a final fraction with high purity. In a final phase of isolation of pure compounds, other techniques such as thin-layer chromatography, high performance liquid chromatography, or electrophoresis should be required, as shown in works presented in Tables 3.1 and 3.2.

Ion-exchange chromatography is the most versatile and efficient methodology for the isolation of amino acids, due to adsorption onto a strong acid cation-exchange resin [47], with the proteins separated according to their isoelectric point. The acidic proteins are usually fractionated by anion-exchange chromatography while the basic proteins are fractionated by cation-exchange chromatography. Saravanan and Shanmugam [41] have used the ion-exchange chromatography to fractionate polyanionic sulfated polysaccharide (i.e., glycosaminoglycans) from marine mollusk *Amussium pleuronectus* (Table 3.2).

Thin-layer chromatography (TLC) has the advantage that a large number of samples can be processed in a single chromatography run [51], but the associated difficulty is the identification of the area on the thin-layer plate corresponding to each component. Wilson-Sanchez *et al.* [52] have extracted lipids from white shrimp (*Litopenaeus vannamei*) by solvent extraction, and they have applied the thin-layer chromatography for the fractionation of shrimp extracts. A sequential fractionation of the active extracts was performed with a mixture of chloroform and acetone (9:1, v/v), obtaining lipidic fractions with antimutagenic and antiproliferative properties. As shown in Tables 3.1 and 3.2, TLC can be used before the HPLC, for example for the fractionation of phenolic compounds and steroids [12], after the column chromatography [18,25,27,29], or as an intermediate fractionation step between the column chromatography and the HPLC techniques [24,26].

According to Weller [78], liquid chromatography is one of the most used methods for offline or online separation of complex mixtures, and its main advantage is the possibility of hyphenation, for example, to autosamplers and detectors. In the field of the discovery of bioactive compounds, the main limitation of liquid chromatography is the use of organic solvents and other additives, which could be incompatible with biochemical assays [78]. For example, the RP-HPLC has the advantage of high capacity, recovery, reproducibility, and chromatographic resolution compared to most separation methods, separating proteins according to their hydrophobicity. The proteins are adsorbed on a stationary phase containing hydrophobic groups, and they are eluted with increasing concentration of an organic solvent, for example, acetonitrile [61]. Due to the low yield of extracts provided from marine microorganisms, the separation of compounds only with RP could lead to some problems such as the purity of the products, the total recovery from the extract, or the permanent loss of activity due to the instability or degradation [79]. Thus, the combination of RP with NP chromatography on silica gel (low cost) or the use of bonded phases such as polyethyleneimine or diol could become an alternative solution [79]. As shown in Tables 3.1 and 3.2, and in the majority of the works, the HPLC was used as the final step of the purification of bioactive compounds.

3.2 FINAL CONSIDERATIONS

After the extraction and fractionation of bioactive compounds, the final step of the isolation of marine bioactive compounds is the drug discovery and potential clinical use. However, the majority of bioactive compounds isolated from marine organisms do not reach the clinical trials, since their biological mechanisms of action remain unknown. Thus, such specific biochemical interaction through how a drug substance produces its pharmacological effect is often unclear for the newly discovered compounds, which represents a challenging approach. The clinical trials also cannot be attained due to the loss of the biological activity when tested *in vivo*, although interesting properties could be reported *in vitro*. According to Sawadogo *et al.* [80], and concerning the anticancer compounds from marine origin, among the 83% of the compounds tested *in vitro*, the biological mechanisms of action of about 45% are unknown, with only 2% in clinical trial and 14% already tested *in vivo*.

REFERENCES

[1] D.S. Bhakuni, D.S. Rawat (Eds.), Bioactive marine natural compounds, Springer, 2005.
[2] T. Teruya, H. Sasaki, K. Kitamura, T. Nakayama, K. Suenaga, Org. Lett. 11 (2009) 2421–2424.
[3] S. Matthew, V.J. Paul, H. Luesch, Phytochemistry 70 (2009) 2058–2063.
[4] L.T. Tan, B.P.L. Goh, A. Tripathi, M.G. Lim, G.H. Dickinson, S.S.C. Lee, S.L.M. Teo, Biofouling 26 (2010) 685–695.

[5] A. Penesyan, S. Kjelleberg, S. Egan, Mar. Drugs 8 (2010) 438–459.
[6] K. Duarte, T.A.P. Rocha-Santos, A.C. Freitas, A.C. Duarte, Trends Anal. Chem. 34 (2012) 97–110.
[7] Z. Cantillo-Ciau, R.M.-P.L. Quijano, Y. Freile-Pelegrín, Mar. Drugs 8 (2010) 1292–1304.
[8] D. Kelman, E.K. Posner, K.J. McDermid, N.K. Tabandera, P.R. Wright, A.D. Wright, Mar. Drugs 10 (2012) 403–416.
[9] S.-J. Heo, W.-J. Yoon, K.-N. Kim, G.-N. Ahn, S.-M. Kang, D.-H. Kang, A. Affan, C. Oh, W.-K. Jung, Y.-J. Jeon, Food Chem. Toxicol. 48 (2010) 2045–2051.
[10] J.K. Patra, A.P. Patra, N.K. Mahapatra, H.N. Thatoi, S. Das, R.K. Sahu, G.C. Swain, Mal. J. Microbiol. 5 (2010) 128–131.
[11] R. Thirunavukkarasu, P. Pandiyan, D. Balaraman, K. Subaramaniyan, G. Edward, G. Jothi, S. Manikkam, B. Sadaiyappan, J. Coastal Life Med. 1 (2013) 6–13.
[12] J. Marimuthu, P. Essakimuthu, J. Narayanan, B. Anantham, R. Joy, J.M. Tharmaraj, S. Arumugam, Asian Pac. J. Trop. Dis. 2 (2012) 109–113.
[13] V. Prabha, D.J. Prakash, P.N. Sudha, Int. J. Pharm. Sci. Rev. Res. 4 (2013) 306–310.
[14] L.H.A. Cavalcante-Silva, C.B. Brito da Matta, M.V. Araújo, J.M. Barbosa-Filho, D. Pereira de Lira, B.V. Oliveira Santos, G.E.C. de Miranda, M.S. Alexandre-Moreira, Mar. Drugs 10 (2012) 1977–1992.
[15] P. Devi, S. Wahidullah, C. Rodrigues, L.D. Souza, Mar. Drugs 8 (2010) 1203–1212.
[16] M.H. Stafsnes, K.D. Josefsen, G. Kildahl-Andersen, S. Valla, T.E. Ellingsen, P. Bruheim, J. Microbiol. 48 (2010) 16–23.
[17] Y. Xu, H. He, S. Schulz, X. Liu, N. Fusetani, H. Xiong, X. Xiao, P.-Y. Qian, Biores. Technol. 101 (2010) 1331–1336.
[18] K. Saurav, G. Rajakumar, K. Kannabiran, A.A. Rahuman, K. Velayutham, G. Elango, C. Kamaraj, A.A. Zahir, Parasitol. Res. 112 (2011) 215–226.
[19] S. Martínez-Luis, J.F. Gómez, C. Spadafora, H.M. Guzmán, M. Gutiérrez, Molecules 17 (2012) 11146–11155.
[20] D.S. Dalisay, D.E. Williams, X.L. Wang, R. Centko, J. Chen, R.J. Andersen, PLoS ONE 8 (2013) 1–14.
[21] Y. Ohkawa, K. Miki, T. Suzuki, K. Nishio, T. Sugita, K. Kinoshita, K. Takahashi, K. Koyama, J. Nat. Prod. 73 (2010) 579–582.
[22] Y.M. Lee, H. Li, J. Hong, H.Y. Cho, K.S. Bae, M.A. Kim, D.-K. Kim, J.H. Jung, Arch. Pharm. Res. 33 (2010) 231–235.
[23] F. Song, H. Dai, Y. Tong, B. Ren, C. Chen, N. Sun, X. Liu, J. Bian, M. Liu, H. Gao, H. Liu, X. Chen, L. Zhang, J. Nat. Prod. 73 (2010) 806–810.
[24] M.A.M. Shushni, R. Singh, R. Mentel, U. Lindequist, Mar. Drugs 9 (2011) 844–851.
[25] K. Tarman, U. Lindequist, K. Wende, A. Porzel, N. Arnold, L.A. Wessjohann, Mar. Drugs 9 (2011) 294–306.
[26] D. Zhang, M. Satake, S. Fukuzawa, K. Sugahara, A. Niitsu, T. Shirai, K. Tachibana, J. Nat. Med. 66 (2012) 222–226.
[27] O.F. Smetanina, A.N. Yurchenko, A.I. Kalinovskii, D.V. Berdyshev, A.V. Gerasimenko, M.V. Pivkin, N.N. Slinkina, P.S. Dmitrenok, N.I. Menzorova, T.A. Kuznetsova, S.S. Afiyatullov, Chem. Nat. Comp 47 (2011) 385–390.
[28] N. Zainuddin, S.A. Alias, C.W. Lee, R. Ebel, N.A. Othman, M.R. Mukhtar, K. Awang, Bot. Mar. 53 (2010) 507–513.
[29] A. Longeon, B.R. Copp, M. Roué, J. Dubois, A. Valentin, S. Petek, C. Debitus, M.-L. Bourguet-Kondracki, Bioorg. Med. Chem. 18 (2010) 6006–6011.
[30] Z.-G. Yu, J. Li, Z.-Y. Li, Y.-W. Guo, Chem. Biodiversity 6 (2010) 858–862.

[31] J. Gupta, S. Misra, S.K. Mishra, S. Srivastava, M.N. Srivastava, V. Lakshmi, S. Misra-Bhattacharya, Experim. Parasitol. 130 (2012) 449–455.
[32] P.R. Marinho, N.K. Simas, R.M. Kuster, R.S. Duarte, S.E.L. Fracalanzza, D.F. Ferreira, M.T.V. Romanos, G. Muricy, M. Giambiagi-DeMarval, M.S. Laport, J. Antimicrob. Chemother. 67 (2010) 2396–2400.
[33] W. Balansa, R. Islam, D.F. Gilbert, F. Fontaine, X. Xiao, H. Zhang, A.M. Piggott, J.W. Lynch, R.J. Capon, Bioorg. Med. Chem. 21 (2013) 4420–4425.
[34] V. Bhimba, M.C. Beulah, V. Vinod, Asian J. Pharm. Clin. Res. 6 (2013) 1–3.
[35] V. Bhimba, V. Vinod, M.C. Beulah, Asian Pac. J. Trop. Dis. 1 (2011) 299–303.
[36] K. Chairman, M. Jeyamala, S. Sankar, A. Murugan, A.J.A. Ranjit Singh, Int. J. Mar. Sci. 3 (2013) 151–157.
[37] D.T.A. Youssef, L.A. Shaala, H.Z. Asfour, Mar. Drugs 11 (2013) 1061–1070.
[38] L. Meot-Duros, S. Cérantola, H. Talarmin, C. Le Meur, G. Le Floch, C. Magné, Food Chem. Toxicol. 48 (2010) 553–557.
[39] T. Vasskog, J.H. Andersen, E. Hansen, J. Svenson, Mar. Drugs 10 (2012) 2676–2690.
[40] R. Krishnamoorthi, A. Yogamoorthi, Int. J. Biophar. Res. 2 (2013) 146–149.
[41] R. Saravanan, A. Shanmugam, Appl. Biochem. Biotechnol. 160 (2010) 791–799.
[42] U.R. Abdelmohsen, S.M. Pimentel-Elardo, A. Hanora, M. Radwan, S.H. Abou-El-Ela, S. Ahmed, U. Hentschel, Mar. Drugs 8 (2010) 399–412.
[43] I.E. Soria-Mercado, L.J. Villarreal-Gómez, G.G. Rivas, N.E.A. Sánchez, in: R.H. Sammour (Ed.), Biotechnology – molecular studies and novel applications for improved quality of human life, Intech, 2012.
[44] T.R.A. Thomas, D.P. Kavlekar, P.A. LokaBharathi, Mar. Drugs 8 (2010) 1417–1468.
[45] P. Pagliara, C. Caroppo, Toxicon. 57 (2011) 889–896.
[46] S.D. Sarker, Z. Latif, A.I. Gray, in: S.D. Sarker, Z. Latif, A.I. Gray (Eds.), Methods in biotechnology, Volume 20 - natural products isolation, Humana Press Inc., 2006.
[47] P.A. Harnedy, R.J. Fitzgerald, J. Phycol. 47 (2011) 218–232.
[48] E. Barbarino, S.O. Lourenco, J. Appl. Phycol. 17 (2005) 447–460.
[49] P. Ganesan, C.S. Kumar, N. Bhaskar, Bioresour. Technol. 99 (2008) 2717–2723.
[50] C.L.F. Almeida, H.S. Falcão, G.R.M. Lima, C.A. Montenegro, N.S. Lira, P.F. Athayde-Filho, L.C. Rodrigues, A.F.V. Souza, J.M. Barbosa-Filho, L.M. Batista, Int. J. Mol. Sci. 12 (2011) 4550–4573.
[51] H.L.J. Makin, J.W. Honour, C.H.L. Shackleton, W.J. Griffiths, in: H.L.J. Makin, D.B. Gower (Eds.), Chapter 3 – General methods for the extraction, purification, and measurement of steroids by Chromatography and Mass Spectrometry, Springer, 2010.
[52] G. Wilson-Sanchez, C. Moreno-Félix, C. Velazquez, M. Plascencia-Jatomea, A. Acosta, L. Machi-Lara, M.-L. Aldana-Madrid, J.-M. Ezquerra-Brauer, R. Robles-Zepeda, A. Burgos-Hernandez, Mar. Drugs 8 (2010) 2795–2809.
[53] C.C. Moraes, J.F.D.M. Burkert, S.J. Kalil, J. Food Biochem. 34 (2010) 133–148.
[54] L. Najafian, A.S. Babji, Peptides 33 (2012) 178–185.
[55] T.G. McCloud, Molecules 15 (2010) 4526–4563.
[56] S.P. Tan, L. O'Sullivan, M.P. Prieto, P. McLoughlin, P.G. Lawlor, H. Hughes, G.E. Gardiner, In: B. Hernández-Ledesma, M. Herrero (Eds.) Chapter 13 – Seaweed antimicrobials: isolation, characterization, and potential use in functional foods, Springer, 2014.
[57] R. Halim, B. Gladman, M.K. Danquah, P.A. Webley, Biores. Technol. 102 (2011) 178–185.
[58] A. Spiric, D. Trbovic, D. Vrabic, J. Djinovic, R. Petronijevic, V. Matekalo-Sverak, Anal. Chim. Acta 672 (2010) 66–71.
[59] A. Jiménez-Escrig, E. Gómez-Ordóñez, P. Rupérez, Adv. Food Nutr. Res. 64 (2011) 325–337.

[60] S. Cheng, X. Yu, Y. Zhang, Shipin Gongye Keji 31 (2010) 132–134.
[61] D. Martínez-Maqueda, B. Hernández-Ledesma, L. Amigo, B. Miralles, J.Á. Gómez-Ruiz, In: F. Toldrá, L.M.L. Nollet (Eds.) Chapter 2 – Extraction/fractionation techniques for proteins and peptides and protein digestion, Springer, 2013.
[62] G.J. Khaniki, P. Sadighara, R.N. Nodehi, M. Alimohammadi, N.V. Saatloo, J. Coastal Life Med. 1 (2013) 96–98.
[63] J.W. Blunt, B.R. Copp, W.P. Hu, H.G. Murray, M.H. Munro, P.T. Northcote, M.R. Prinsep, Nat. Prod. Rep. 26 (2009) 170–244.
[64] J.-Z. He, Q.-M. Ru, D.-D. Dong, P.-L. Sun, Molecules 17 (2012) 4373–4387.
[65] C. Le Ker, K.-E. Petit, J.-F. Biard, J. Fleurence, Mar. Drugs 9 (2011) 82–97.
[66] M. Fernández-Ronco, A. Lucas, J.F. Rodríguez, M.T. García, I. Gracia, J. Supercrit. Fluids 79 (2013) 345–355.
[67] E. Ibañez, M. Herrero, J.A. Mendiola, M. Castro-Puyana, in: M. Hayes (Ed.), Marine bioactive compounds: sources, characterization and applications, Springer, 2012, pp. 55–98.
[68] A. Chabeaud, L. Vandanjon, P. Bourseau, P. Jaouen, M. Chaplain-Derouiniot, F. Guerard, Sep. Purif. Technol. 66 (2009) 463–471.
[69] J.K. Lee, J.-K. Jeon, S.-K. Kim, H.G. Byun, Adv. Food Nutr. Res. 65 (2012) 47–72.
[70] D.H. Ngo, Z.J. Qian, B.M. Ryu, J.W. Prak, S.K. Kim, J. Funct. Foods 2 (2010) 107–117.
[71] K. Hsu, E. Li-Chan, C. Jao, Food Chem. 126 (2011) 617–622.
[72] D. Wolf, K. Siems, Chimia 61 (2007) 339–345.
[73] U. Grienke, J. Silke, D. Tasdemir, Food Chem. 142 (2014) 48–60.
[74] B.H. Sarmadi, A. Ismail, Peptides 31 (2010) 1949–1956.
[75] H.J. Issaq, T.P. Conrads, G.M. Janini, T.D. Veenstra, Electrophoresis 23 (2002) 3048–3061.
[76] H. Otsuka, in: S.D. Sarker, Z. Latif, A.I. Gray (Eds.), Methods in biotechnology, Volume 20 – natural products isolation, Humana Press Inc., 2006.
[77] W.E. Houssen, M. Jaspars, in: S.D. Sarker, Z. Latif, A.I. Gray (Eds.), Methods in biotechnology, Volume 20 – natural products isolation, Humana Press Inc., 2006.
[78] M.G. Weller, Sensors 12 (2012) 9181–9209.
[79] M. Mansson, R.K. Phipps, L. Gram, M.H.G. Munro, T.O. Larsen, K.F. Nielsen, J. Nat. Prot. 73 (2010) 1126–1132.
[80] W.R. Sawadogo, M. Schumacher, M.-H. Teiten, C. Cerella, M. Dicato, M. Diederich, Molecules 18 (2013) 3641–3673.

Chapter 4

Green Analytical Methodologies for Preparation of Extracts and Analysis of Bioactive Compounds

K. Duarte[1], C.I.L. Justino[1], A.M. Gomes[2], Teresa Rocha-Santos[3,4] and Armando C. Duarte[5]

[1]*Department of Chemistry & CESAM, University of Aveiro, Campus de Santiago, Aveiro, Portugal*

[2]*CBQF/Escola Superior de Biotecnologia, Catholic University, Rua Dr. António Bernardino de Almeida, Porto, Portugal*

[3]*Department of Chemistry & CESAM, University of Aveiro, Campus de Santiago, Aveiro, Portugal*

[4]*ISEIT/Viseu, Instituto Piaget, Estrada do Alto do Gaio, Galifonge, Lordosa, Viseu, Portugal*

[5]*Department of Chemistry & CESAM, University of Aveiro, Campus de Santiago, Aveiro, Portugal*

Contents

4.1 GREEN EXTRACTION TECHNIQUES TO OBTAIN BIOACTIVE COMPOUNDS

In the field of Analytical Chemistry, the green methodologies are known as clean and environmentally friendly techniques, which have been used as alternatives to the offline treatment of wastes and residues, replacing polluting

Analysis of Marine Samples in Search of Bioactive Compounds, Vol. 65. DOI: 10.1016/B978-0-444-63359-0.00004-5

The 12 Principles of Green Chemistry

1. Prevent waste.
2. Maximize atom economy.
3. Use safe solvent and reaction conditions to improve the use of water or eco-friendly solvents.
4. Design less hazardous chemical synthesis.
5. Design safer chemicals and products.
6. Increase energy efficient work practices.
7. Use renewable feedstocks.
8. Eliminate chemical derivatizations.
9. Use catalysts, not stoichiometric reagents.
10. Design chemicals and products to degrade after use.
11. Analyze in real time.
12. Minimize the potential of accidents.

FIGURE 4.1 Identification and description of the 12 principles of Green Chemistry reprinted from de la Guardia and S. Armenta [5]. Copyright 2011, with permission from Elsevier.

methodologies [1]. Since the 1990s, the concept of Green Analytical Chemistry has increasingly been recognized as crucial in the development of new techniques and in the improvement of classical methodologies, promoting a sustainable development. According to Anastas [2], the "Green Chemistry is the use of chemistry techniques and methodologies that reduce or eliminate the use and generation of feedstocks, products, by-products, solvents, and reagents, that are hazardous to human health or the environment." Thus, the reduction of both the amount and the toxicity of solvents and reagents required for the sample pretreatment and in the measurement step, and the development of alternative direct analytical methodologies not requiring either solvents or reagents, has been suggested in order to reduce the adverse environmental impact of analytical methodologies [1].

Anastas and Warner [3] have suggested the 12 principles of Green Chemistry shown in Figure 4.1 as a framework for scientists to promote the design of new processes and materials, new energy sources, alternative methodologies, and renewable sources for the research laboratories.

In 2001, Namiesnik [4] adapted the 12 principles of Green Chemistry proposed by Anastas and Warner [3] to the following four top priorities in the field of the Green Analytical Chemistry:

- Elimination, or at least significant reduction, of consumption of reagents and organic solvents for analytical procedures
- Reduction in the emission of vapors and gases, as well as in solid wastes produced in analytical laboratories
- Elimination of reagents displaying high toxicity and/or ecotoxicity from the analytical procedures
- Reduction of labor and energy consumption in analytical procedures

Taking into account the 12 principles of Green Chemistry, various strategies have been implemented in order to reduce the environmental impacts of methodologies followed in Analytical Chemistry. Figure 4.2 identifies both the

Strategies for Green Chemistry		Methodologies
• Use of alternative samples treatments	⟶	SFE; PLE; UAE; MAE; SPE; SPME
• Direct analysis of untreated samples	⟶	Development and analysis by real time sensors
• Use of alternative reactions conditions	⟶	Photochemical reactions; Microwave and ultrasound assisted organic synthesis techniques; Catalytic reactions
• Minimization of energy consumption	⟶	Automation; Simultaneous treatment of several samples; Multi-analyte determination in a single run; Use of miniaturized and portable instruments
• Use of alternative solvents	⟶	e.g., water and ionic liquids
• Use of alternative feedstocks	⟶	e.g., biomass

FIGURE 4.2 Strategies for Green Analytical Chemistry and corresponding methodologies. SFE: Supercritical Fluid Extraction; PLE: Pressurized Liquid Extraction; UAE: Ultrasound Assisted Extraction; MAE: Microwave Assisted Extraction; SPE: Solid Phase Extraction; SPME: Solid Phase Microextraction.

strategies and the methodologies that can be applied according to the principles of Green Chemistry.

The use of alternative extraction methods for sample treatment, such as Supercritical Fluid Extraction (SFE), Pressurized Liquid Extraction (PLE), Ultrasound Assisted Extraction (UAE), Microwave Assisted Extraction (MAE), Solid Phase Extraction (SPE), and Solid Phase Microextraction (SPME), contributes to solventless preparation techniques, and avoids risks to operators and environment. The aforementioned methodologies are discussed in Section 4.1.

According to de la Guardia and Armenta [5], the direct analysis using real-time sensors offers the best analytical alternative since it allows the on-site measurement of untreated samples without the production of hazardous substances and the use of reagents. Besides the two first strategies identified in Figure 4.2, the use of "greener" reaction conditions could also be an alternative for applying the principles of Green Chemistry. For instance, photochemical reactions, microwave and ultrasound assisted techniques, and catalytic reactions improve the product yield, saving energy and minimizing wastes [6]. The photochemical reaction uses the energy of light (ultraviolet and visible) as the catalyst for many reactions largely used in green chemical processes, thus replacing toxic metals [6]. The most widely used techniques in the synthesis of organic compounds in Green Chemistry are the microwave- and ultrasound-assisted techniques since they allow high product yields without solvents, and low waste production and energy requirements. The microwave-assisted synthesis is based on the irradiation of microwaves, which leads to an efficient internal heating of polar solvent molecules [7].

On the other hand, sonochemistry (i.e., the use for ultrasound for synthesis) has been widely applied in mixed phase synthesis, materials chemistry, and biomedical uses. There are three classes of sonochemical reactions: homogeneous sonochemistry of liquids, heterogeneous sonochemistry of liquid–liquid or solid–liquid systems, and sonocatalysis [6]. The catalytic reactions constitute a powerful tool in green chemistry since they allow maximize selectivity to reduce the temperature of reaction, to use natural resources for production of chemicals, to avoid solvents using heterogeneous catalysis, to minimize or even avoid waste formation, and consequently to reduce process complexity, as well as to avoid the formation of unwanted side reactions [8,9]. Finally, the following principles should also be kept in mind in order to minimize the impact of chemistry in the environment: (1) the minimization of energy consumption by the use of portable instrumentation, also considering the automation of sample measurement processes, simultaneous treatment of several samples, and multianalyte determinations in a single run; (2) the use of alternative solvents to degrade after use under high temperature; and (3) the use of alternative feedstocks, such as renewable raw materials [5]. Therefore, the conventional procedures of Analytical Chemistry should be replaced by clean and green techniques, namely in the isolation and identification of natural products and bioactive compounds from different natural matrices such as marine sources.

In the following subsections of this chapter, the green methodologies used for preparation of bioactive compounds from marine sources in terms of green extraction methodologies will be discussed.

Table 4.1 summarizes the results obtained in various studies where extraction methodologies such as SFE, PLE, UAE, MAE, and enzyme-assisted extraction (EAE) were applied for the extraction of bioactive compounds from marine sources. The green extraction techniques, such as SFE, PLE, UAE, and MAE use water or carbon dioxide as solvents. Although, in some cases, organic solvents are still used in such methods, the quantities are very low when compared to the amounts used in traditional extractions. In addition, these extractions can be performed at reduced temperature and for a short period of time, which makes them an improvement toward the greening of analytical methods. Concerning the EAE methodology, no solvents are used, which becomes an advantage of paramount importance when compared to other extraction methodologies, as discussed in subsection 4.1.5.

4.1.1 Supercritical Fluid Extraction (SFE)

The supercritical fluid extraction (SFE) is based on extracting analytes in the solid sample from marine sources by a fluid in supercritical conditions, which is any liquid at a temperature and pressure above its critical point, where distinct liquid and gas phases do not exist. SFE is a fast, efficient, and clean method that allows the increase of yield, productivity, and selectivity of extraction of bioactive compounds from marine sources and for the removal of harmful compounds, such as organic pollutants and pesticides.

TABLE 4.1 Extraction Methodologies Based on Principles of Green Chemistry and Used to Obtain Bioactive Compounds from Marine Sources

Extraction Methodology	Marine Sources	Bioactive Compounds	Solvent / Enzyme	Optimal Extraction Conditions[1]	Yield of Extraction	References
SFE	Squid viscera of *Todarodes pacificus*	Lecithin	CO_2	45 °C 25 MPa 2.5 h	0.34 g/g squid viscera	[10]
	Microalgae *Scenedesmus almeriensis*	β-carotene	CO_2	60 °C 400 bar 5 h	1.5 mg/g dry weight of microalgae	[11]
	Microalgae *Crypthecodinium cohnii*	Docosahexaenoic acid	CO_2	30.0 MPa 323 K 3 h	42.7 g extract/100 g dry microalgae	[12]
	Macroalgae *Sargassum muticum*	Phenolic compounds	CO_2 with 12% of ethanol	60 °C 15.2 MPa 90 min	3.45% of dry weight	[13]
	Macroalgae *Sargassum muticum*	Isoflavones	CO_2 with 3% of methanol:water (9:1, v/v)	40 °C 35 MPa 60 min	-	[14]

(Continued)

TABLE 4.1 Extraction Methodologies Based on Principles of Green Chemistry and Used to Obtain Bioactive Compounds from Marine Sources *(cont.)*

Extraction Methodology	Marine Sources	Bioactive Compounds	Solvent / Enzyme	Optimal Extraction Conditions[1]	Yield of Extraction	References
PLE	Microalgae *Haematococcus pluvialis*	Carotenoids	Hexane, ethanol, and water	100 °C 10.3 MPa 20 min	5.88% of dry weight	[15]
	Macroalgae *Sargassum muticum*	Phenolic compounds	Ethanol:water (75:25, v/v)	120 °C 10.3 MPa 20 min	10.18% of dry weight	[13]
	Microalgae *Chlorella vulgaris*	β-carotene	Acetone, ethanol, and water	50–200 °C 10.3 MPa 20 min	39.1 g/100 g of dry weight	[16]
	Microalgae *Arthrospira platensis*	γ-linolenic acid	Ethanol:ethyl lactate (50:50, v/v)	180 °C 20.7 MPa 15 min	20.7% (w/w)	[17]
UAE	Microalgae *Chlorella vulgaris*	β-carotene	Acetone, hexane, and ethanol	25 °C — 25–30 min	4.79 g/100 g of dry weight	[16]
	Microalgae *Spirulina maxima*	Polysaccharides	Ethanol and water	60 °C — 30 min	19.3% (w/w)	[18]
	Microalgae *Haematococcus pluvialis*	Astaxanthin	Ethanol and ethyl acetate	41.1 °C — 16 min	27.58 mg/g	[19]
	Microalgae *Chaetoceros gracilis*	Lipids	Chloroform: methanol (2:1, v/v)	25 °C —20 min	60.28% (w/w)	[20]

MAE	Macroalgae *Laminaria Ochroleuca*	Iodine	Tetramethylammonium hydroxide and water	200 °C —5 min	3703 μg/g	[21]
	Macroalgae *Caulerpa racemosa*	Phenolic compounds	Ethanol and water	50 °C —40 min	67.89 mg/100 g of dried sample	[22]
	Macroalgae *Fucus vesiculosus*	Fucoidan	Water	—120 psi 1 min	18.22%	[23]
	Microalgae *Scenedesmus obliquus*	Palmitic acid		95 °C —30 min	31.38%	[24]
EAE	Macroalgae *Porphyra columbina*	Peptides from phycobiliproteins	Trypsin and alcalase[2]	—	—	[25]
	Microalgae *Navicula incerta*	Peptides	Papain	—	—	[26]
	Microalgae *Chlorella ellipsoidea*		Pepsin	—	—	[27]
	Mackerel *Magalaspis cordyla* (viscera)		Pepsin, trypsin, and α-chymotrypsin	—	—	[28]
	Mackerel *Magalaspis cordyla* and croaker *Otolithes ruber* (skins)			—	—	[29]
	Nemipterus japonicus and *Exocoetus volitans* (backbones)		Pepsin and trypsin	—	—	[30]
	Nemipterus japonicus and *Exocoetus volitans* (muscles)			—	—	[31]

SFE: Supercritical Fluid Extraction; PLE: Pressurized Liquid Extraction; UAE: Ultrasound Assisted Extraction; MAE: Microwave Assisted Extraction; EAE: Enzyme Assisted Extraction
[1]*Temperature, pressure, and time*
[2]*With previous solvent extraction with methanol*

The CO_2 is widely used as supercritical solvent in SFE since it shows several environmental advantages such as safety, nontoxicity, and cost-effectiveness. On the other hand, it allows the reduction of the content of toxic organic solvents, and high extraction efficiency, providing a fast and selective extraction. The CO_2 is not harmful for humans and environment, and it is inexpensive, nonflammable, and easily removed by reducing the pressure, in order to obtain a solvent-free extract [5]. On the other hand, supercritical CO_2 produces high extraction efficiency from nonpolar to low polarity compounds since it is a nonpolar solvent. This apolarity of CO_2 can be remediated using cosolvents, also called modifiers. Thus, the use of modifiers increases the speed of the mass transfer from matrix to the extraction fluid. The modifiers are mainly polar organic compounds and when they are added to the supercritical fluid, it allows the collection of more polar compounds, increasing the polarity of the extracting mixture [32].

The most important advantage of SFE is the possibility of changing operational conditions, such as temperature, pressure, and cosolvent concentration, in order to facilitate the extraction of specific compounds, since the solvating power of the supercritical fluid is directly related to these parameters [5,33,34].

The SFE using CO_2, hereby identified as SC-CO_2, has been found as useful for the extraction of carotenoids from microalgae *Scenedesmus almeriensis*, fatty acids from microalgae *Crypthecodinium cohnii*, or even phenolic compounds from *Sargassum muticum*, as shown in Table 4.1 [11–13]. For example, Macías-Sánchez *et al.* [11] have applied the SFE to obtain carotenoids from the microalgae *Scenedesmus almeriensis*, and they have determined the influence of pressure and temperature on the SFE. Macías-Sánchez *et al.* [11] have found that 50% of the content of β-carotene can be extracted from the microalgae with SFE (Table 4.1), resulting in a better yield than with the reference extraction process under nonsupercritical conditions. On the other hand, regarding the application of an experimental design (2-parameter, 5-level surface response methodology), Macías-Sánchez *et al.* [11] have observed that the temperature, pressure, and interaction of these two variables do not influence the yield of lutein and β-carotene extraction in a significant way, since p value is higher than 0.05. Couto *et al.* [12] applied the SFE technique to extract docosahexaenoic acid from the heterotrophic microalgae *Crypthecodinium cohnii*. The SC-CO_2 was used to extract lipids in a temperature range between 313 and 323 K, and pressure range of 20 to 30 MPa. The highest extraction yield of docosahexaenoic acid was obtained under operational conditions of 323 K and 30 MPa, resulting in 42.7 g of extract per 100 g of dry microalgae. However, the yield of extraction obtained is lower than that obtained with the conventional lipid extraction method of Bligh and Dyer, which is commonly used. This finding suggests that the SFE of *C. cohnii* biomass is mainly controlled by the diffusion kinetics of the lipids with the microalgae cells [12]. Thus, Couto *et al.* [12] have used methods of disrupting the *C. cohnii* cell walls as a pretreatment step in order to enhance the lipid yield extraction by SFE. The SFE was also applied to the extraction of phenolic compounds such as simple phenols and flavonoids from marine sources. For example, Klejdus *et al.* [14] have extracted isoflavones, which are a component of functional foods and are also used in the pharmaceutical industry. The isoflavones were extracted from sea macroalgae such as *Sargassum muticum*, using SC-CO_2 modified with 3% (v/v) of methanol:water (9:1, v/v) at 35 MPa and 40 °C for 60 minutes, with a previous sonication step.

Wang *et al.* [35] have extracted flavonoids from *Chlorella vulgaris* with SC-CO_2 at 4500 psi and using a modifier based on 50% of aqueous ethanol. Wang *et al.* [35] have compared the SC-CO_2 extraction with the ultrasound assisted extraction (UAE), verifying that the SFE provide higher flavonoid extract content (3.18 mg per g of lyophilized extract) than UAE (0.86 mg per g of lyophilized extract).

Uddin *et al.* [10] have applied the SC-CO_2 for the extraction of lecithin, which is a phospholipid mixture, from the oil of squid viscera residues of *Todarodes pacificus* (Table 4.1). The sample was extracted at different temperatures (35, 40, and 45 °C) and pressures (15, 20, and 25 MPa). As the solvation power depends on the temperature and pressure conditions, the highest extraction yield was found at 45 °C and 25 MPa, obtaining 0.34 g of lecithin per gram of squid viscera, corresponding to the isolation of 4.25% (w/w) of lecithin from squid viscera residues. It is important to highlight that the marine rest raw materials are sometimes used for the extraction of bioactive compounds as recently shown by Uddin *et al.* [10]. After washing the wastes with cold water such materials can be used for the extraction of bioactive compounds. As reviewed by Hayes [36], the enzymes can also be used to obtain natural biopolymers, such as chitosan and chitin, from marine rest raw materials, through the principles discussed in subsection 4.1.5.

4.1.2 Pressurized Liquid Extraction (PLE)

Pressurized liquid extraction (PLE; also referred to as pressurized solvent extraction; accelerated solvent extraction; high-pressure solvent extraction; high-pressure, high-temperature solvent extraction; pressurized hot solvent extraction; subcritical solvent extraction; and enhanced solvent extraction) is one of the most promising techniques in bioactive compound extraction [5,33]. In PLE, the solvents are used near their supercritical region, where high temperatures produce high solubility and high diffusion rates of solutes in the solvent, while the high pressure, in keeping the solvent below its boiling point, enables high penetration of the solvent in the sample [1]. This technique is a solid–liquid extraction process performed at high temperatures (50–200 °C) with high pressures (50–300 psi) for short time periods (5–10 min) in solid and semisolid matrices [5]. Thus, the solvent is maintained in the liquid state during the whole extraction process. Therefore, the high temperatures increase the extraction rate through improvement of solvation power promoting mass transfer of organic compounds to the solvent, and increasing the extraction efficiency [5]. This method requires a small amount of organic solvents, in the presence of large quantities of water in sample, providing high extraction efficiency [1]. The method development in PLE is easier than in SFE, since a lower number of parameters influencing the extraction should be considered (i.e., the extraction time and the extraction temperature) [32].

Tanniou *et al.* [13] have conducted a comparative study between SFE and PLE to extract phenolic compounds from marine macroalgae *Sargassum muticum* (Table 4.1). The SC-CO_2 was performed under a temperature of 60 °C and a pressure of 15.2 MPa during 90 minutes, and using SC-CO_2 with 12% of ethanol as solvent. In the PLE, the temperature and pressure conditions were 120 °C and 10.3 MPa, respectively, for a period of 20 minutes, and using as solvent a mixture of 75% ethanol and 25% water. The authors concluded that PLE is an extraction process more efficient than SFE for the extraction of polyphenols, since the SFE does not extract the totality of active phenolic compounds.

In turn, linolenic acid was extracted from the microalgae *Arthrospira platensis* using PLE [17] at a temperature of 180 °C and a pressure of 20.7 MPa, for a period of 15 minutes (Table 4.1). The pressurized mixture of 50% ethanol and 50% ethyl acetate (v/v) allows an extraction yield of 20.7% (w/w). A response surface methodology concerning a central composite design concluded that the temperature was the main factor for the total yield of extraction and that the high temperatures caused a decrease in the solubility of linoleic acid. The pressure contributed to only a small extent on the extraction yield.

4.1.3 Ultrasound Assisted Extraction (UAE)

Ultrasound assisted extraction (UAE) is also used in the search for bioactive compounds from marine sources and it is based on the effects of acoustic cavitation. The propagation of ultrasonic waves provides a greater solvent penetration into the sample matrix, increasing the contact between the sample and the solvent (or reagent) and improving the mass transfer rates. In addition, this technique is also useful for the extraction of compounds from living organisms since it promotes the breaking of biological cell walls. This technique allows performing simultaneous extractions, the use of low quantities of solvent, the reduction of working times, and the increase in yield and quality of extract. Moreover, UAE is also inexpensive, fast, and versatile compared to traditional techniques, since it can use several solvents of different polarities. However, UAE has some drawbacks, including difficulties in combination with other instruments and automation [5,37]. Oh *et al.* [18] compared the extraction of polysaccharides from microalgae *Spirulina maxima* and found that the extraction yield of UAE is improved at least from 25 to 30% in comparison to extraction with hot water or 80% ethanol using traditional extraction processes. Oh *et al.* [18] also found that the ultrasonic frequency is more effective at improving extraction yields than extraction time, in order to obtain *S. maxima* extracts, which are most important due to their anticancer activity. In this way, UAE also provides extracts with lower cytotoxicity than conventional solvent and hydrothermal extraction methodologies.

A nutraceutical carotenoid, astaxanthin, was extracted from *Haematococcus pluvialis* using the UAE technique by Zou *et al.* [19]. The UAE was performed at 41 °C, for a period of 16 minutes using a mixture of 48% ethanol and 52% ethyl acetate as solvent (optimal conditions) and a liquid-to-solid ratio of 20:1 mL/g, under ultrasound irradiation of 200 W (Table 4.1). The extraction yield of astaxanthin under such conditions was estimated at 27.58 ± 0.40 mg/g. A response surface methodology, based on a Box-Behnken experimental design, was used to optimize the effects of several experimental parameters such as extraction solvent, extraction temperature, and extraction time, on the extraction efficiency of astaxanthin from *H. pluvialis*. Zou *et al.* [19] have found that all experimental parameters, both in linear and quadratic terms, exhibit a significant effect on the yield of astaxanthin ($p < 0.0001$), which is not observed for the interactions of the experimental parameters with a p value higher than 0.0001.

Plaza *et al.* [16] also performed a comparative study of the ability of two extraction methods, PLE and UAE, in the extraction of bioactive compounds (carotenoids) from *Chlorella vulgaris*. In the PLE experiments, the optimal conditions of temperature and pressures were, respectively, 200 °C and 10.3 MPa, for a period of 20 minutes, using water as solvent, and the UAE was carried out at 25 °C for 25 minutes with ethanol. The results obtained by Plaza *et al.* [16] suggest that PLE is more efficient than UAE, providing higher yield extraction (39.1 g per 100 g of dry weight) than UAE (extraction yield of 4.79 g per 100 g of dry weight). On the other hand, the authors suggest that PLE is more amenable to control and automation, thus becoming a faster technique.

4.1.4 Microwave Assisted Extraction (MAE)

Microwave assisted extraction (MAE) is based on heating the solvent through absorption of microwave energy by polar molecules, thus increasing the solvent penetration into the sample matrix. MAE has been applied to the extraction of organic compounds from very different types of matrix because it saves solvent and it is rapid and efficient in terms of energy use. This method allows the acceleration of energy transfer, facilitating the solvation of analytes, and also promoting the disruption of weak hydrogen bonds. The effects of microwave energy depend on several factors, such as the nature of both solvents and solid matrix, the type of target analyte to be extracted, and especially the sample and solvent dielectric constants. The dielectric constants of the system are crucial to MAE, since higher dielectric constants promote an increasing of the amount of energy absorbed. Although this method cannot be applied to extract termolabile compounds, it becomes an important tool for green analytical chemistry, mainly for the extraction of natural products from marine sources, since it reduces significantly solvent volume and extraction times. Recently, a review paper [38] centered its attention on a new solvent-free MAE technique (SFME) that has additional advantages since it reduces energy consumption, solvent use, and CO_2 emissions, which is advantageous for the extraction of bioactive compounds from natural sources. For example, SFME could reduce an extraction time of 3 hr to only 15 min, achieving similar yields to those observed by conventional methods [38]. Compared to traditional extraction techniques, the use of microwaves decreases extraction times, consequently also decreasing the amount of solvent needed [32].

Sulphated polysaccharides, such as fucoidans, were extracted from brown seaweed *Fucus vesiculosus* using MAE, by Rodriguez-Jasso *et al.* [23]. Some parameters such as pressure (30–120 psi), extraction time (1–31 min), and alga/water ratio (1/25–5/25 g/mL) were evaluated (Table 4.1) and optimized by the application of an experimental design (2^3 full factorial design with 4 central points). All studied variables affect the extraction process presenting significant main effects, and the optimal extraction conditions were found as 120 psi for 1 min and using 1g of alga per 25 mL of water, which allowed an extraction yield of 18.22%. Balasubramanian *et al.* [24] recovered palmitic acid from green algae *Scenedesmus obliquus*, by MAE, which was performed at different temperatures (80 and 95 °C) for a period of 30 min and using a ratio of green algae/water of 1:1. Balsubramanian *et al.* [24] have found that the extraction yield of oil was higher at 95 °C with a total recovery value of 76–77% (w/w). When compared to a conventional water bath heater, the extraction of palmitic acid by MAE is approximately 10-fold lower in terms of processing time, increasing the heating rates 4× to 5×.

4.1.5 Enzyme Assisted Extraction (EAE)

Taking into account the principles of Green Chemistry (Figure 4.1), the extraction of bioactive compounds through an enzymatic route could be considered as a green extraction methodology. Effectively, the enzymes can accelerate a reaction, lower the use of energy, use alternative starting materials, and reduce both the use of solvents and the production of waste [6]. On the other hand, enzymes can biodegrade under environmental conditions, being considered alternative and renewable chemicals, and improving the extraction yield. The limitation of using enzymes for the extraction of bioactive compounds is related to their dependence on several physicochemical factors. The selection of the appropriate enzyme is crucial to the success of the process of extraction together with their optimal reaction conditions, mainly the combination of factors incubation time and temperature.

The main advantage of EAE is the discarding of the residual organic solvents or toxic chemicals in the products associated with the target compounds, as observed when working with solvent extraction [39]. Thus, a variety of peptides such as arenastatin A, aplidine, and ziconotide have been isolated from marine animals such as tunicates, molluscs, and sponges [39]. According to Samarakoon and Jeon [40], the isolation of bioactive compounds from enzymatic extracts results in high yield and purity compared to the extractions with organic solvents. Recently, Wijesinghe and Jon [41] have reviewed the potential of the enzyme-assisted extraction of bioactive compounds as a useful approach for recovery of industrially important metabolites from seaweeds. The extraction efficiency of bioactive compounds from seaweeds is reduced due to the presence of the cell wall polysaccharides such as alginates and carrageenans [41]. Thus, the enzymatic degradation of such wall polymers has become a more attractive alternative to chemical and mechanical processes, has improved the extraction efficiency of algal active compounds, and has become an alternative to classical extraction methodologies [42].

The enzymatic hydrolysis is performed at controlled pH, temperature, and substrate concentration in order to reduce the formation of undesirable products and to optimize the activity of the enzymes used such as alcalase, trypsin, pepsin, chymotrypsin, and papain [41,43]. According to Alemán *et al.* [44], the enzymatic hydrolysis could increase the antioxidative activity of the resulting hydrolysate through the enhancement of radical scavenging activity. The molecular size, hydrophobicity, and polar groups of hydrolysates are also influenced by the hydrolysis, affecting the characteristics of a hydrolysate and consequently its functional properties [45,46]. Various bioactive peptides have been isolated from marine organisms through the enzymatic hydrolysis. As recently reviewed by Freitas *et al.* [43], the enzyme hydrolysis is mostly used to obtain antioxidant peptides from marine organisms, such as macroalgae, microalgae, and marine animals. Actually, bioactive peptides were obtained from macroalgae [25] and microalgae [26,27], as well as from the viscera and skins of mackerel [28,29] and the backbones and muscles of breams [30,31]. Cian *et al.* [25] have employed the enzymatic hydrolysis to obtain bioactive peptides containing aspartic acid, alanine, and glutamic acid, from the marine macroalgae *Porphyra columbina*, which is known to have a high protein content (approximately 30% of dry weight), using trypsin and alcalase as enzymes. Cian *et al.* [25] take advantage of the water soluble proteins (phycobiliproteins) of such macroalgae to obtain peptides with immunomodulatory properties, antihypertensive activity, and antioxidant capacity. In turn, Kang *et al.* [26] have used the enzymatic hydrolysis to obtain peptides (1108 Da) from the microalgae *Navicula incerta*, with the enzymes alcalase, α-chymotrypsin, neutrase, papain, pepsin, pronase-E, and trypsin. Kang *et al.* [26] found antioxidant activities from the protein hydrolysates obtained with papain with higher efficiency than that obtained with the other enzymes.

Ko *et al.* [27] have also used the enzymatic hydrolysis to obtain peptides (702 Da) from the microalgae *Chlorella ellipsoidea*, but with papain, trypsin, pepsin, and α-chymotrypsin. Ko *et al.* [27] have found a higher antioxidant activity with the pepsin hydrolysate. According to Ngo *et al.* [47], the efficiency to obtain antioxidant activities from specific protein hydrolysates depends on the hydrolytic reaction, which in turn, depends on the availability of susceptible peptide bonds for the protease attack, as well as the physical structure of the protein. The bioactive peptides as the products of the hydrolysis of marine invertebrates such as shellfish, crustacean, and rotifer have also been isolated as reviewed by Lee *et al.* [48]. In such marine organisms, the muscles are the target material as the source of bioactive compounds such as actin, myosin, and collagen providing antimicrobial, antioxidant, and antihypertensive activities [48]. As stated by Samarakoon and Jeon [40], the bioactive compounds from the algae, such as the inner cellular materials like proteins, obtained through the enzyme-assisted extractions, are accessed due to the degradation of the cell wall polymers (e.g., cellulose, hemicelluloses, lignin, and pectin) by enzymes such as cellulase. However, the efficiency of the extraction is reduced since macroalgae have complex cell wall polysaccharides such as alginates and carrageenans [41]. Thus, specific extraction methodologies such as ultrasonic-assisted extraction and high pressure-assisted extraction have been used to maximize the extraction of bioactive compounds, as reviewed by Yang *et al.* [49].

Wang *et al.* [50] have studied the effect of various protease and carbohydrase treatments on the extraction of bioactive compounds such as polyphenols and other hydrophilic antioxidant compounds from the macroalga *Palmaria palmata*, by EAE methodology. Wang *et al.* [50] concluded that the EAE was effective in the enhancement of the recovery of polyphenols, the proteases being the most effective in comparison with the carbohydrases, to obtain natural antioxidants and functional ingredients.

In the enzyme-assisted extraction, the selection of the extraction solvent is also important since it can be considered the most influencing variable in the extraction process. Green solvents usually are used in the enzyme-assisted extraction methodology, which are the solvents with lowest toxicity such as acetone, ethanol, methanol, 2-propanol, ethyl acetate, isopropyl acetate, methyl ethyl ketone, 1-butanol, and tert-butanol [6].

Figure 4.3 summarizes the main advantages and limitations of the five green extraction methodologies already discussed, according to de la Guardia and Armenta [5], concerning, for example, the time of extraction, the amount of solvent needed, and the amount of sample. The extraction methodologies compared in the figure are commonly used for solid samples, and subsections 4.1.6 and 4.1.7 present two other extraction methodologies commonly used for liquid samples.

As shown in Figure 4.3, only the EAE does not use solvents for the extraction of bioactive compounds, although some of the other extraction methodologies use only small volumes of solvents. However, the main limitation of EAE is the mandatory need of optimization of the extraction parameters since the enzymes are very sensitive, mainly to the combination of extraction temperature and extraction time.

4.1.6 Solid Phase Microextraction (SPME)

By definition, the microextraction is a nonexhaustive sample preparation where a very small volume of extraction phase in the range of microliters or smaller, relative to the sample volume, is used [51]. Among the existing microextraction techniques, the solid phase microextraction (SPME) combines sampling, analyte isolation, and enrichment into one single step and it is based on the sorption of analytes directly from the aqueous or gaseous samples, or in headspace vials by a thin film of an extracting phase immobilized over the surface of a sorbent-coated

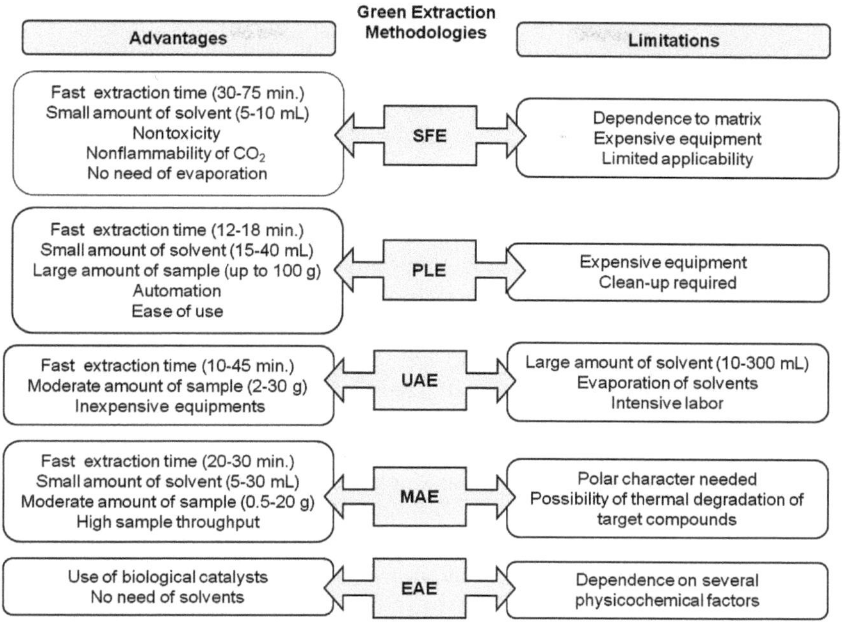

FIGURE 4.3 Schematics of the advantages and limitations of the green extraction methodologies SFE, PLE, UAE, MAE, and EAE.

silica fiber [1]. The extracts can be then thermally desorbed by inserting into a gas chromatographic system, or using small amounts of organic solvents for injection into a liquid chromatographic system coupled to a mass spectrometer. The biggest advantage of SPME is the reduction of time and solvent, and the combination of the whole extraction process in one step. Few works have been published dealing with the use of SPME for the extraction of marine bioactive compounds. For example, Zhang *et al.* [52] have used the space-resolved SPME technique to study the tissue-specific bioconcentration of pharmaceuticals and bioactive analytes in live fish (rainbow trout *Oncorhynchus mykiss*), also estimating their accumulation in fish muscle and adipose tissue. Zhang *et al.* [52] found that SPME is a promising technique for sampling heterogeneous living systems due to its simplicity, good spatial and temporal resolution, and the capacity to simultaneously monitor multiple tissues *in vivo* within the same fish over time. Effectively, the *in vivo* analysis is a special application where SPME can play an important role, as reviewed by Ouyang *et al.* [51].

4.1.7 Solid Phase Extraction (SPE)

Solid phase extraction (SPE) is based on the sorption of analytes on a solid phase from the original sample [5]. SPE avoids the use of large amounts of organic solvents in preconcentration and extraction steps, and it has been used mainly for trace enrichment of water samples [1]. The SPE procedure involves three critical steps—sample loading, matrix washing, and analytes recovering—which should be optimized in every particular study in order to find the best extraction conditions to obtain the highest purity of extracts. On the other hand, SPE can easily be automated with relatively simple and inexpensive equipment, which can lead to improved accuracy, precision, and laboratory throughput [5]. Thus, the main advantage of this technique is its automation, providing greater capacity and speed of analysis. The drawback of SPE is that the packing must be uniform to avoid poor efficiency.

Recently, Shen *et al.* [53] have employed SPE with graphene as sorbent, due to its ultrahigh surface area, to extract marine lipophilic toxins in shellfish muscle, such as yessotoxins, okadaic acid, and dinophysistoxin-1. The sample of shellfish tissues was previously extracted by UAE during 15 minutes at 4 °C. Shen *et al.* [53] concluded that the graphene has a greater potential as SPE sorbent since it is more effective than other commonly used sorbents such as C18, Strata-X, and HLB, showing high extraction efficiency.

It is important to highlight that SPE and SPME are extraction techniques commonly used for liquid samples with a preconcentration capacity. However they can also be used as a fractionation technique, mainly the SPE, since the SPE apparatus separates the analytes by affinity to the sorbent used.

Although green techniques have been used to extract bioactive compounds from marine sources, the fractionation/purification step is used to obtain pure bioactive compounds, and this step does not follow the principles of Green Chemistry. In other words, green techniques are used to extract the bioactive compounds, but sometimes classical methodologies are used in the purification of extracts such as TLC, which can also be used as a direct method for untreated samples, as reported in Section 4.2. For example, Uddin *et al.* [10] used SFE with supercritical CO_2 for the extraction of lecithin from squid viscera of *Todarodes pacificus*, but TLC was used to purify lecithin, employing a mixture of chloroform and methanol (2:1, v/v) for the separation of such bioactive compounds. However chloroform is toxic, persistent, and since it is a nonbiodegradable solvent, difficult to recycle, and its final disposal is very expensive [6]. Thus, special attention should be paid to ensure that green techniques are used in the entire process of preparation of extracts and fractions in order to obtain pure bioactive compounds from natural matrices such as marine samples.

Recently, a new type of solvents, room-temperature ionic liquids (RTIL), has been investigated in order to replace the toxic organic solvents. These particular solvents are a combination of organic cations and various anions that are liquid at room temperature, being salts with melting points of approximately 100 °C [5]. Due to their negligible vapor pressure and good thermal stability, the RTIL can be used in various analytical fields such as sample preparation, chromatographic and electrophoretic separation, and detection. The use of RTIL in the extraction of target analytes is in the early stages, but it constitutes a promising application. RTIL have unique properties such as low volatility, adequate viscosity, and immiscibility with water, being a potential extraction solvent. The use of RTIL as extraction solvent has been used in UAE for the extraction of analytes in food samples, and in medicinal plants as recently reviewed, respectively, by Pico *et al.* [37] and Routray and Orsat [54].

Melo *et al.* [55] have studied the combination of supercritical CO_2 and ionic liquids for the extraction of carvacrol, which is an antimicrobial reagent, from plants. The combination of RTIL with supercritical fluids have also been used in extraction processes in order to link the possibility of carrying out a reaction in the most favorable phase (the ionic liquid) while the reaction products are extracted into the supercritical phase, as reviewed by Mendiola *et al.* [56]. An application of RTIL concerning the extraction of marine bioactive compounds was found only in the work of Han *et al.* [57]. They used the 1-butyl-3-methylimidazolium tetrafluoroborate ([Bmim][BF4], 1:15) as an extraction solvent in the ultrasonic extraction of phenolic compounds from macroalgae *Laminaria japonica*, and found that an ionic liquid-based ultrasonic-assisted extraction procedure leads to a higher extraction efficiency with shorter extraction times, when compared to the traditional extraction technique using water and methanol.

4.2 DIRECT ANALYSIS OF UNTREATED SAMPLES TO OBTAIN BIOACTIVE COMPOUNDS

Usually marine samples, which are solid matrices of complex mixtures of sugars, proteins, lipids, vitamins, and pigments, are generally unsuitable for direct analysis due to their complexity [38], and they need previous preparation before the analysis step. The problem with sample preparation is the loss or contamination of the target compounds, which are vulnerable to chemical changes [38]. Thus, the direct analysis of untreated samples removes the sampling and sample treatment steps, avoiding the use of large amounts of acids and organic solvents and strongly reducing waste generation, reagent and energy consumption, and safety risks to the operators [5]. For example, remote sensing or fiber-optic-based methodologies followed by direct analysis of samples inside vials or blisters are preferable methodologies when compared to methods that involve sample damage and others requiring chemical treatments, since such procedures avoid potential risks to operators and the environment through the elimination of dangerous sample treatment steps [5]. On the other hand, the development of energetically self-sufficient portable instruments for the direct analysis of untreated samples has a positive effect on greening the analytical methods and facilitating the direct measurement of the parameters of interest. However, the power requirements of the miniaturized and portable equipment are the key limitation to the use of instrumentation for direct analysis in the field [5].

Concerning the direct analysis of untreated marine samples for obtaining bioactive compounds, very few papers have been published. Beal *et al.* [58] analyzed the lipid content of *Neochloris oleobundans* by liquid state nuclear magnetic resonance (^{13}C NMR), in order to use such algal cultures for the biodiesel production. The advantages of liquid state NMR include the ability to analyze intact algae (thus avoiding lengthy extraction procedures with associated coextraction of unwanted compounds or alteration of the sample), bulk algal samples (i.e., many cells at the same time), and also the ability to obtain detailed chemical structure information (e.g., distinguishing between types of lipids and types of fatty acids in lipids) [55]. Beal *et al.* [58] have concluded that the NMR is a useful diagnostic tool for selectively identifying lipids in algae, with particular relevance to biodiesel production, reducing the multiple processing steps, commonly used in other production pathways such as solvent extraction for gravimetric lipid quantification. Algal lipids in individual cells of three algal species (*Botryococcus sudeticus*, *Chlamydomonas* sp., and *Trachydiscus minutus*) were also determined by Samek *et al.* [59] by Raman microspectroscopy. According to Samek *et al.* [59], the intensities of the Raman spectral peaks that correspond to the saturated and unsaturated carbon–carbon bonds in lipid molecules were used to estimate the degree of nonsaturation in the lipid bodies. Raman microspectroscopy only needed a simple and rapid sample preparation, which consisted of mixing the algal suspension in an aqueous solution of low temperature melting agarose, and analyzing the cells 10 min after the immobilization. Thus, Samek *et al.* [59] have found that Raman microspectroscopy was a fast, versatile, and virtually noninvasive tool for applications in algal lipid engineering and industry.

Noninvasive measurements provide green analytical alternatives that are free of risks to operators and do not require solvents, reagents, and sample handling, thus providing fast, sustainable, and waste-free methodologies [5]. Furthermore, miniaturization reduces the amount of samples and reagents, minimizing the risks. Both NMR and Raman microspectroscopy provide the direct analysis of untreated samples without sample damage; that is, the use of nondestructive methodologies provide several advantages in terms of Green Analytical Chemistry, allowing the storage of the samples for another analysis and reducing the sample consumption and waste generation [5].

As shown in Figure 4.2, the direct analysis of untreated samples by real-time sensors is one of the strategies proposed in order to follow Green Chemistry principles. However, the association of such direct analysis by sensors to the detection of bioactive compounds in untreated marine samples has not been documented yet. Special attention should be paid in future works to the design of sensors for the direct analysis of untreated marine samples used as sources of bioactive compounds, taking into account the advantages of real-time sensors as a potential way to reduce analysis time and consumption of toxic reagents, thus following the 12 principles of Green Chemistry.

REFERENCES

[1] S. Armenta, S. Garrigues, M. de la Guardia, Trends Anal. Chem. 27 (2008) 497–511.

[2] P.T. Anastas, Crit. Rev. Anal. Chem. 29 (1999) 167–175.

[3] P.T. Anastas, J. Warner, Green Chemistry: theory and practice, Oxford University Press, Oxford, 1998.

[4] J. Namiesnik, Green analytical chemistry - some remarks, J. Sep. Sci. 24 (2001) 151–153.

[5] M. de la Guardia, S. Armenta, Green Analytical Chemistry, volume 57: theory and practice, Comprehensive Analytical Chemistry, Elsevier, 2011.

[6] A. Valavanidis, T. Vlachogianni, Green chemistry and Green engineering: from theory to practice for the protection of the environment and sustainable development, Synchrona Themata, 2012.

[7] C.O. Kappe, D. Dallinger, Nat. Rev. Drug Discovery 5 (2006) 51–63.

[8] G. Centi, S. Perathoner, Catal. Today 77 (2003) 287–297.

[9] B.V. Badami, Resonance 13 (2008) 1041–1048.

[10] S. Uddin, H. Kishimura, B.-S. Chun, J. Food Sci. 76 (2012) 350–354.

[11] M.D. Macías-Sánchez, J.M. Fernandez-Sevilla, F.G.A. Fernández, M.C.C. García, E.M. Grima, Food Chem. 123 (2010) 928–935.

[12] R.M. Couto, P.C. Simões, A. Reis, T.L. Da Silva, V.H. Martins, Y. Sánchez-Vicente, Eng. Life Sci. 10 (2010) 158–164.

[13] A. Tanniou, S.L. Esteban, L. Vandanjon, E. Ibanez, J.A. Mendiola, S. Cerantola, N. Kervarec, S. La Barre, L. Marchal, V. Stiger-Pouvreau, Talanta 104 (2013) 44–52.

[14] B. Klejdus, L. Lojkova, M. Plaza, M. Snoblova, D. Sterbova, J. Chromatog. A 1217 (2010) 7956–7965.

[15] S. Santoyo, M. Plaza, L. Jaime, E. Ibañez, G. Reglero, F.J. Señorans, J. Agric. Food Chem. 58 (2010) 8522–8527.

[16] M. Plaza, S. Santoyo, L. Jaime, B. Avalo, A. Cifuentes, G. Reglero, G.G.-B. Reina, F.J. Señoráns, E. Ibáñez, Lebensm. Wiss. Technol. 46 (2012) 245–253.

[17] M.-T. Golmakani, J.A. Mendiola, K. Rezaei, E. Ibáñez, J. Supercrit. Fluids 62 (2012) 109–115.

[18] S.-H. Oh, J. Ahn, D.-H. Kang, H.-Y. Lee, Mar. Biotechnol. 13 (2011) 205–214.

[19] T.-B. Zou, Q. Jia, H.-W. Li, C.-X. Wang, H.-F. Wu, Mar. Drugs 11 (2013) 1644–1655.

[20] G.S. Araujo, L.J.B.L. Matos, L.R.B. Gonçalves, F.A.N. Fernandes, W.R.L. Farias, Bioresour. Technol. 102 (2011) 5248–5250.

[21] V. Romarís-Hortas, A. Moreda-Piñeiro, P. Bermejo-Barrera, Talanta 79 (2009) 947–952.

[22] Z. Li, B. Wang, Q. Zhang, Y. Qu, H. Xu, G. Li, J. Appl. Psychol. 24 (2012) 1527–1536.

[23] R.M. Rodriguez-Jasso, S.I. Mussattoa, L. Pastrana, C.N. Aguilar, J.A. Teixeira, Carbohyd. Polym. 86 (2011) 1137–1144.

[24] S. Balasubramanian, J.D. Allen, A. Kanitkar, D. Boldor, Biores. Technol. 102 (2011) 3396–3403.

[25] R.E. Cian, O. Martínez-Augustin, S. Drago, Food Res. Int. 49 (2012) 364–372.

[26] K. Kang, Z. Qian, B. Ryu, F. Karadeniz, D. Kim, S. Kim, Phytother. Res. 26 (2012) 1555–1563.

[27] S. Ko, D. Kim, Y. Jeon, Food Chem. Toxicol. 50 (2012) 2294–2302.

[28] N.S.S. Kumar, R.A. Nazeer, R. Jaiganesh, Peptides 32 (2011) 1496–1501.

[29] N.S.S. Kumar, R.A. Nazeer, R. Jaiganesh, Amino Acids 42 (2012) 1641–1649.

[30] Y.N. Shabeena, R.A. Nazeer, Int. J. Food Sci. Technol. 46 (2011) 1993–1996.

[31] S.Y. Naqash, R.A. Nazeer, Amino Acids 43 (2012) 337–345.

[32] M. Herrero, J.A. Mendiola, M. Plaza, E. Ibañez, in: J.W. Lee (Ed.), Screening for bioactive compounds from Algae, Springer, 2012.

[33] E. Ibañez, M. Herrero, J.A. Mendiola, M. Castro-Puyana, M. Hayes (Eds.), Marine Bioactive Compounds: Sources, Characterization and Applications, Springer, 2012.
[34] C.G. Pereira, M.A.A. Meireles, Food Bioprocess. Technol. 3 (2010) 340–372.
[35] H.-M. Wang, J.-L. Pan, C.-Y. Chen, C.-C. Chiu, M.-H. Yang, H.-W. Chang, J.-S. Chang, Process Biochem 45 (2010) 1865–1872.
[36] M. Hayes, in: M. Hayes (Ed.), Chapter 4 - Chitin, chitosan and their derivatives from marine rest raw materials: potential food and pharmaceutical applications, Springer, 2012.
[37] Y. Picó, Trends Anal. Chem. 43 (2013) 84–99.
[38] Y. Li, A.-S. Fabiano-Tixier, M.A. Vian, F. Chemat, Trends Anal. Chem. 47 (2013) 1–11.
[39] G.-M. Suarez-Jimenez, A. Burgos-Hernandez, J.-M. Ezquerra-Brauer, Mar. Drugs 10 (2012) 963–986.
[40] K. Samarakoon, Y. Jeon, Food Res. Int. 48 (2012) 948–960.
[41] W.A.J.P. Wijesinghe, Y.-J. Jeon, Fitoterapia 83 (2012) 6–12.
[42] S.U. Kadam, B.K. Tiwari, C.P. O'Donnell, J. Agric. Food Chem. 61 (2013) 4667–4675.
[43] A.C. Freitas, J.C. Andrade, F.M. Silva, T.A.P. Rocha-Santos, A.C. Duarte, A.M. Gomes, Curr. Med. Chem. 20 (2013) 4575–4594.
[44] A. Alemán, B. Giménez, P. Montero, M.C. Gómez-Guillén, LWT Food Sci. Technol. 44 (2011) 407–413.
[45] H.G. Kristinsson, B.A. Rasco, J. Agr. Food Chem. 48 (2000) 657–666.
[46] J.M.S. Kim, Food Sci. Biotechnol. 20 (2011) 1075–1085.
[47] D.H. Ngo, Z.J. Qian, B.M. Ryu, J.W. Prak, S.K. Kim, J. Funct. Food 2 (2010) 107–117.
[48] J.K. Lee, J.-K. Jeon, S.-K. Kim, H.G. Byun, Adv. Food Nutr. Res. 65 (2012) 47–72.
[49] B. Yang, Y. Jiang, J. Shi, F. Chen, M. Ashraf, Food Res. Int. 44 (2011) 1837–1842.
[50] T. Wang, R. Jónsdóttir, H.G. Kristinsson, G.O. Hreggvidsson, J.Ó. Jónsson, G. Thorkelsson, G. Ólafsdóttir, LWT - Food Sci. Technol. 43 (2010) 1387–1393.
[51] G. Ouyang, D. Vuckovic, J. Pawliszyn, Chem. Rev. 111 (2011) 2784–2814.
[52] X. Zhang, K. Oakes, S. Cui, L. Bragg, M.R. Servos, J. Pawliszyn, Environ. Sci. Technol. 44 (2010) 3417–3422.
[53] Q. Shen, L. Gong, J.T. Baibado, W. Dong, Y. Wang, Z. Dai, H.-Y. Cheung, Talanta 116 (2013) 770–775.
[54] W. Routray, V. Orsat, Food Bioprocess Technol. 5 (2012) 409–424.
[55] C.I. Melo, R. Bogel-Tukasik, E. Bogel-Tukasik, J. Supercritical Fluids 61 (2012) 191–198.
[56] J.A. Mendiola, M. Herrero, M. Castro-Puyana, E. Ibanez, in: M.A. Mauricio, J.M. Prado (Eds.), Chapter 6 – Supercritical fluid extraction, Royal Society of Chemistry, 2013.
[57] D. Han, T. Zhu, K.H. Row, Bull. Korean Chem. Soc. 32 (2011) 2212–2216.
[58] C.M. Beal, M.E. Webber, R.S. Ruoff, R.E. Hebner, Biotechnol. Bioeng. 106 (2010) 573–583.
[59] O. Samek, A. Jonáš, Z. Pilát, P. Zemánek, L. Nedbal, J. Tíska, P. Kotas, M. Trtílek, Sensors 10 (2010) 8635–8651.

Chapter 5

Bioassays for Bioactivity Screening

Sofia Isabel G.H.M. Montalvão[1], Vineeta Singh[2] and Shafiul Haque[3]

[1]*Centre for Drug Research, Faculty of Pharmacy, Viikki Biocenter 2, P.O. Box 56 (Viikinkaari 5 E), FIN, University of Helsinki, Finland*

[2]*Microbiology Division, CSIR-Central Drug Research Institute, Sector 10, Jankipuram Extension, Sitapur Road, Lucknow, Uttar Pradesh, India*

[3]*Gene Expression Laboratory, Department of Biosciences, Jamia Millia Islamia (A Central University), New Delhi, India*

Contents

5.1 INTRODUCTION

Over the past few decades tremendous advancements have occurred in marine biology and pharmacological sciences. Currently, nearly 16,000 marine natural products have been isolated from marine sources and communicated in about 6,800 research articles [1]. Despite many developments in analytical and cell

Analysis of Marine Samples in Search of Bioactive Compounds, Vol. 65. DOI: 10.1016/B978-0-444-63359-0.00005-7

culture techniques and the rich tradition of the use of compounds produced from marine sources, negligible efforts have been made to isolate bioactive compounds from marine sources. Possibly, this is due to dearth of proper expertise and sustainable infrastructure for bioactivity screening and requires lengthy waiting time for such pharmacological screenings even if services are outsourced. Thus, it is important to develop in-house bioassays in marine biolaboratories that are economic, produce results quickly, and do not need specialized knowledge. Nowadays there are a large number of marine biological laboratories worldwide have established easy "bench-top" bioassays that can be accomplished even by nonscientific specialists.

Bioassays can be classified into broad groups based on the target life forms; that is, lower organisms (e.g., bacteria, fungi, insects, lower plants, etc.), live cells in tissue culture conditions (such as cancer cells) and tissues of animal or human origin, animals (as a whole), isolated organs of vertebrates, and isolated subcellular systems (e.g., enzymes, receptors, etc.).

Various extracts or samples from marine sources have served as an important source of molecular diversity in variety of drug discovery programs, and some valuable bioactive compounds have already been isolated from marine sources and utilized successfully for various therapeutic purposes. For any marine compound or extract isolation program, where the end-product is a lead compound or an active or a drug, various types of pharmacological evaluation or bioassay screening must essentially be involved to direct the isolation and extraction of a pure bioactive compound. Therefore the bioactivity assessment of extracts of marine sources is an integral and indispensable aspect of any drug discovery program. Innovative and extraordinary developments in the area of *in vitro* cell culture techniques have significantly changed the scenario of natural product biochemistry, which ultimately influenced the bioactivity evaluation techniques. Earlier it took weeks or even months to check a test sample for some bioactivity, but with advancements in *in vitro* cell culture techniques, it's a matter of few hours.

In general, the bioactivity assay screening can be classified into two groups, primary bioassay screens and secondary screens. The primary bioassays are those assays that can be quickly applied to a huge number of samples to evaluate whether any desired bioactivity is present. The basic qualities of these bioassays include having high capacity, being economic and capable of providing quick results, and not being quantitative. The secondary screening procedures involve more exhaustive and comprehensive testing of lead compounds on a greater number of system models to select active compounds for clinical trials. Generally, the secondary screening assays have low capacity and are slow, time consuming, and costly.

The major requirements for primary bioassay screens include the following:

- The outcomes of primary bioactivity assay must indicate some type of therapeutic prospective, either directly or by analogy with clinically efficient drugs that also have been screened by the same methodology.

- Potentially useful pharmacological property must be detected even though the action may be either unique or unexpected.
- The nature of the biological activity of the compound (from any source, including marine) must be predicted for subsequent research studies.
- The screens must be potentially tolerant to several impurities available in a crude extract.
- They should be sensitive and capable of detecting the presence of potentially interesting substances or actives present in low concentrations (the detection limit must be in the range of 0.0001% of an active compound in any marine extract, based on the dry weight of the extracted organism).
- They must be unbiased and it must involve the coding of all marine samples, including both the standard reference material and the unknown test sample.
- The results of the bioassay must be reproducible and reliable.
- The screens must allow the use of crude materials as well as pure isolates so that the methodology can be applied to guide the extraction, isolation, and purification work of the marine natural product chemist/biochemist/biologist.
- They should not need more than 1.0 to 2.0 g of the crude dry marine natural material for the completion of a single test.
- They must have a high throughput (irrespective of low information content; it must be available rapidly).
- They should not require costly equipment or a sophisticated laboratory so that primary bioactivity screening can be performed in sync with the fractionation (purification step) process.
- The bioassay procedure must be well suited to the applications of dimethyl sulfoxide (DMSO) as it is commonly applied for solubilization of extracts and pure polar compounds for screening purposes.
- The assay methodology should be simple so that it can be easily conveyed to lab technicians or other assistants for routine operations.
- If needed in bioassay, the test animals should be easily obtainable, easily handled, easily bred, and infection resistant.
- Most importantly, the bioassay should be inexpensive and economical so that it can be carried out for an extended period of time.

According to the subject experts, a hit rate of 1% or even less is considered a reasonable one, and moves from primary screening to secondary screening for selectivity profiling in order to establish biological activity.

The bioactivity assay directed marine product discovery research has four major approaches to attain its final objectives:

- Application of a single bioassay methodology to search for a specific type of pharmacological activity (e.g., antidiabetic, anti-inflammatory activity, etc.).
- Application of a battery of specific bioassays with each procedure directed to decipher a different type of useful activity.
- Utilization of a single bioassay technique intended to discover multiple activities (nonspecific bioactivity assay); for example, cell-based toxicity

assays can be employed to test a variety of bioactivities such as antimicrobial, antitumor, and insecticidal activities.
- Use of a combination of various bioassays to identify specific activities as well as to discover multiple activities.

The primary bioassay screens can be used for a large number of samples in order to test their bioactivity, and these assays do not need to be quantitative but have high capacity and low running cost. As said earlier, general requirements (e.g., validity, lack of ambiguity, data accuracy, reproducibility, ease, running cost, sensitivity, and selectivity to limit the number of compounds for secondary screening and to exclude false positives) should be taken into consideration while selecting a bioactivity assay. In case a positive sign is detected in the primary pharmacological screening, a secondary bioassay screening, which is more precise and accurate, is carried out. In secondary screens, the pure compounds are examined in multiple models and test conditions in order to select the candidates for clinical trials [2,3]. Figure 5.1 provides a schematic diagram

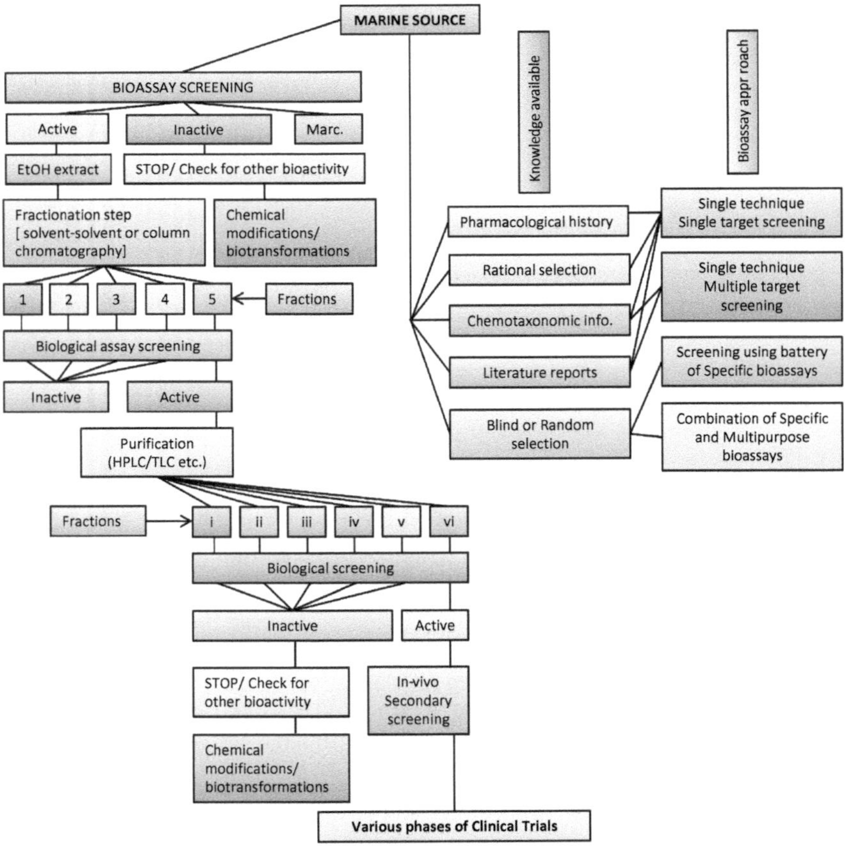

FIGURE 5.1 Bioactivity-directed isolation of marine natural products.

of a bioactivity-directed isolation of marine natural products for therapeutic purpose.

Basically, the selection of the bioactivity screening assay depends mostly upon the target disease as well as the available information about the target marine source/organism. For example, if a marine source has some pharmacological history of use for a specific disease, we would rationally use a particular bioassay, which can show the known therapeutic activity in order to isolate the lead or active, which is accountable for that bioactivity. Likewise, based on the chemotaxonomic information of related species, we can easily choose bioassay(s) based on the known or reported bioactivity of related species. The selection of bioactivity assay depends on knowledge about marine organisms/sources and their behavioral characteristics in particular conditions. For example, many plants exhibit insect resistance, therefore they can be tested for insecticidal compounds by employing any pesticidal bioassay. In the case of random collection of common and/or uninvestigated organisms, however, it is advisable to apply a series of bioassay screens at extraction and trail the most significant activity through a specific bioassay.

5.2 SCREENING MODELS

5.2.1 Antimicrobial Assays

Marine organisms are a rich source of structurally novel and biologically active compounds and have been largely studied for the identification of secondary metabolites. Screening of extracts from marine resources is a first approach to identify compounds of biomedical importance. The fast development of resistance in antimicrobial therapy has motivated researchers for the search of new potential novel clinically useful drugs against human pathogens. The antimicrobial activity can depend of two factors: type of organism (macro- and microalgae, sponge, cyanobacteria, etc.) and the yield of efficiency in the extraction of bioactive compounds. Organic solvents such as methanol, ethanol, and dichloromethane have been used for the extraction of possible active metabolites from marine organisms. The assays can be divided in two categories: primary and secondary. Generally, in a primary assay, the test compound is usually screened at a single concentration. In the secondary assay, there is a range of concentrations established to find the minimum inhibitory concentration (MIC) and other parameters.

5.2.1.1 Antibacterial Assays

Macroalgae are an important marine source since they have an efficient activity against a range of gram positive and gram negative bacteria [4–6]. Their potential is due to the ability of synthesizing, for example, halogenated terpenoids [7,8]. Other compounds such as steroids, phenols, and heterocyclic have antibacterial activity as well and can be found in algae [9].

5.2.1.2 Antifungal Assay

Sponges, corals, and marine bacteria have been shown to be an important source of marine organisms in the study of antifungal activities. Swinhoeiamide A, a compound isolated from *Theonella swinhoei* sponge, showed strong antifungal activity against *Candida albicans* and *Aspergillus fumigatus* [8]. Fractions of red-algae *Bostrychia tenella* also have shown potential antifungal activity against *Cladosporium* sp.

Antimicrobial susceptibility assay methodology can be divided into three major types: (1) Diffusion: Kirby-Bauer method and Stokes method; (2) Dilution: Broth dilution method, broth microdilution method, and Agar dilution method; and (3) Diffusion and Dilution: E-Test method. All the methods should be written and followed according to CLSI (Clinical and Laboratory Standards Institute) guidelines [10,11] and/or EUCAST (European Committee on Antimicrobial Susceptibility Testing) guidelines [12]. CLSI is an international, nonprofit, nongovernmental, and interdisciplinary organization composed of researchers, medical professionals, industry, and so on, and promotes accurate antimicrobial susceptibility testing and suitable reporting, developing standard reference methods, creating quality control parameters for standard test methods and reporting clinically relevant strategies. EUCAST is a network supported by the European Society of Clinical Microbiology and Infectious Diseases(ESCMID) and the European Centre for Disease Prevention and Control (ECDC) to standardize antimicrobial susceptibility testing under European standards.

The three methodology types are described as follows:

Diffusion

Diffusion methods are generally used for microorganisms with fast growth. The Kirby-Bauer method [10] is a viable alternative to broth dilution methods, and is considered simple, economic, and reliable (Figure 5.2(a)). In the first step, a standardized inoculum (compared to McFarland 0.5 OD) is inoculated in the surface of the agar plate with the help of a sterile cotton swab. The disc(s) of antimicrobial agents are placed on the surface of the agar plate and this one(s) is incubated. Afterward, the susceptibility of the agents is determined on the basis of zone of inhibition and is affected by the ability of diffusing through agar and type of media used. The results are interpreted depending on the diameter of the growth inhibition zone around each disc and comparing with the minimum inhibitory concentration (MIC) interpretive standard of well-known antibiotics. When the zone diameter of the antimicrobial agents are larger than or equal to the control, the bacteria is sensitive; if the zone diameter is lower than the control or zero, the bacteria is resistant.

The Stokes method is similar to the Kirby-Bauer method, except that the control culture and test culture are grown in the sample plate (Figure 5.2(b))

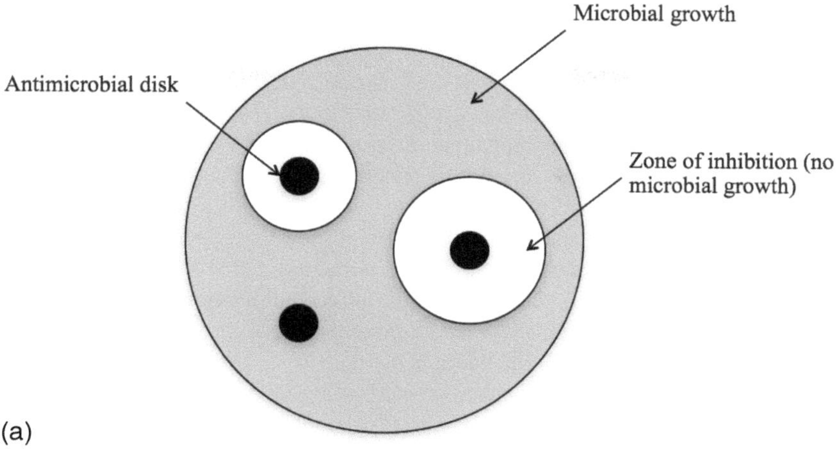

(a)

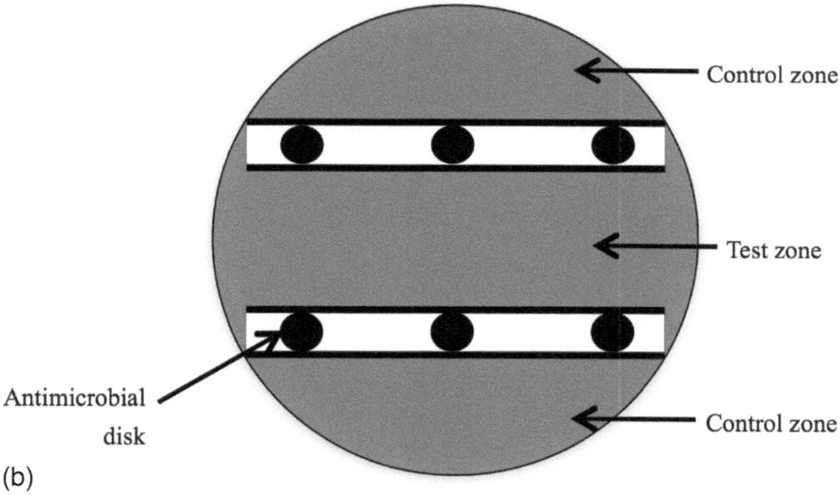

(b)

FIGURE 5.2 (a) Antimicrobial disk diffusion test (Kirby-Bauer method); (b) Stokes method of antimicrobial susceptibility.

and therefore, there is a comparison of the zones of inhibition produced in two cultures by the same disc. In this method, the inoculum of the control strain (the standard drug) is spread on the agar plate and the tested bacterium is inoculated. An uninoculated strip is placed between the tested bacteria zone and control areas, for placing the discs (in this case, test samples that can be natural, semisynthetic, or synthetic). These are put on each side (bacteria zone and control zone) and the plate is incubated. The interpretation of results is made by measuring and comparing the zones of inhibition of control and bacterium used [13].

Dilution

The Dilution method is used to determine the minimal inhibitory concentration of an antimicrobial to inhibit or kill the bacteria/fungi and is the reference for antimicrobial susceptibility testing. For broth dilution methods [14] initially, serial dilutions of the antimicrobial agent are prepared and, once ready, are inoculated in tubes with a known number of tested bacterium/fungi. A control strain set with known susceptibility is made separately. The tubes are incubated at a specific temperature, depending on the optimal growth temperature of bacteria or fungi, and for a period of time. The end points are taken as MIC (minimum inhibitory concentration, i.e., amount of antimicrobial agent required to inhibit growth) or MBC (minimum bactericidal concentration, i.e., amount of antimicrobial agent required to kill the bacteria).

Another dilution method is broth microdilution [14], which is performed in 96-well microdilution plates (capacity: 200 μl). Generally, there exists a positive control (PC; i.e., antibiotic), a MAX (maximum growth of bacteria/fungi), and MIN (minimum growth of bacteria/fungi with the presence of antibiotic) as controls. Similarly, the agar dilution assay [15] is performed in Petri dishes, having the possibility of testing several organisms on each plate. An antibiotic control is necessary and the results are accepted only if the control plate shows good growth. No growth of the test organism indicates that there is susceptibility at the antimicrobial concentration incorporated in the medium.

Diffusion and Dilution

The E-test [16] is known as the epsilometer test or "exponential gradient" test. This is a quantitative method that applies both the dilution and diffusion of a drug into the medium. In the first step of this test, a predefined thin, inert, and nonporous plastic strip with an antimicrobial gradient of a test sample (natural, semisynthetic, or synthetic) is applied in an agar plate for the release of drug. Second, the plates are incubated at a specific temperature, depending on the optimal growth temperature of bacteria or fungi, and for a specific period of time. After the incubation, a symmetrical inhibition ellipse centered along the strip is seen. The MIC value is indicated when the inhibitory zone edge intersects the strip. When growth happens along the entire strip (no inhibition ellipse seen), MIC is reported as greater than the highest value on the MIC scale. But if the inhibition ellipse is below the strip, where the zone edge doesn't intersect it, MIC is reported as less than the lowest value on the MIC scale. This method is extremely useful in determining MIC for a wide spectrum of pathogens as glycopeptide resistant *Enterococci* sp., glycopeptide intermediate *S. aureus*, and fastidious growing *M. tuberculosis*.

5.2.2 Anticancer Assays

After cardiovascular disease, cancer is the second cause of death worldwide. Cancer is a group of diseases characterized by uncontrolled cellular growth, local tissue invasion, and so on [17]. Over 60% of anticancer drugs available in the market are of natural origin. Natural products are also the lead molecules for many of the drugs that are in use [18]. The chemical and biological diversity of the marine environment is immeasurable and therefore is an extraordinary resource for the discovery of new anticancer drugs. An exciting "marine pipeline" of new anticancer clinical and preclinical agents has emerged from intense efforts over the past decade to more effectively explore the rich chemical diversity offered by marine life [19].

5.2.2.1 In Vitro *Anticancer Activity*

In vitro anticancer assays are primarily of two types: molecular assays or cellular assays. Molecular assays are highly particular, defined for specific subcellular targets. Due to specificity, these assays result in a low hit rate from a large number of compounds screened. The major disadvantage of this assay is that bioactive compounds with different mechanisms will be missed.

Cellular assays

Cellular assays are based on cytotoxicity assays and morphological assays. Cytotoxicity assays are based on IC_{50}; that is, 50% growth inhibitory concentration against a single cell line, which leads to a large number of active materials.

Using both types of assays in conjunction with one another provides the best bioactive compounds. For example, cytotoxicity assays in the first step give a large number of bioactive compounds, which are then further screened by molecular assays to select compounds acting through mechanisms of interest.

Sulforhodamine B (SRB) assay

The sulforhodamine B (SRB) assay is used for *in vitro* cytotoxicity screening through cell density determination, which is based on the measurement of cellular protein content [20]. The assay relies on the ability of the SRB to bind with basic amino acid residues of trichloro acetic acid (TCA) fixed protein contents of the cells. SRB is a bright-pink aminoxanthene dye with two sulfonic groups that bind to basic amino acid residues under mild acidic conditions, and dissociate under basic conditions. In this assay, human tumor cell lines (like vero, liver HEP-2, neuroblastoma IMR-32, colon HCT-15, HT-144 malignant melanoma, etc.) used for assay first are grown in a defined medium and incubated in a CO_2 incubator. After harvesting, cells are washed with a phosphate buffer solution. The buffer is discarded and cells are seeded at a defined concentration in a 96-well plate. Serial dilutions of the compounds, whose activity has to be examined, and standard drugs (adriamycin, 5-fluorouracil, mitomycin C, doxorubicin, etc.) are prepared.

A specific volume of serially concentrated biologically active metabolites is added to the cell lines and incubated in a humidified atmosphere with 5% CO_2 in a CO_2 incubator. Each concentration should be repeated three times. Positive and negative control has standard drugs and no drug, respectively. After incubation, the medium is removed for the wells and 10% TCA is added. Plates are kept at 4 °C for a specific time. TCA is removed, plates are washed gently under tap water, and then air-dried at room temperature. SRB reagent is added into each well and left for 15 min. SRB is removed and wells are washed four times with 1.0% acetic acid and air-dried. Stain is solubilized with 10 mM unbuffered Tris base and absorbance is measured at a specific wavelength. IC_{50} value of compounds possessing cytotoxic activity is calculated.

$$\%\,cell\,growth = \frac{mean\,OD_{sample} - mean\,OD_{day0}}{mean\,OD_{neg\,control} - mean\,OD_{day0}}$$

$$\%\,Growth\,inhibition = 100 - \%\,of\,control\,cell\,growth$$

Propidium iodide (PI) assay

This is an example of a cellular biomass assay where Propidium iodide acts as a general biomass stain that binds to RNA, DNA, proteins, and glycosaminoglycans. Other biomass stains are Thionin, Azure A, and Toluidine Blue O, which approach the protein stains in sensitivity. In this assay [21], human tumor cell lines are grown in a defined medium such as RPMI 1640 at a specific time incubation in a CO_2 incubator. After harvesting, cells are seeded at a concentration of $1.5 - 3.0 \times 10^4$ cells/ml in a 96-well plate. Serial dilutions of the compounds whose activity has to be examined and standardized are prepared. A specific volume of serially concentrated biologically active metabolites is added to the cell lines and incubated in a humidified atmosphere with 5% CO_2 in a CO_2 incubator. Each concentration should be repeated three times. Positive and negative control have standard drug(s) and no drug, respectively. Plates are kept at a determined negative temperature for a specific amount of time and posteriorly thawed. A specific volume of PI stock solution is added to each well so that final PI concentrations are 400 μg/ml and the plates are incubated in the dark at room temperature. Fluorescence is read at a fixed wavelength in a fluorescent plate reader and the effective dose is calculated.

MTT assay

MTT (3-(4,5-dimethylthiazole-2-yl)-2,5-diphenyl tetrazolium bromide) assay is used to measure the metabolic activity of living cells by their ability to reduce the tetrazolium salt, MTT, to form formazan using the dehydrogenase enzymes in the mitochondria. For this assay [22], human tumor cell lines are grown in defined medium such as RPMI 1640 in a CO_2 incubator. After harvesting, cells are seeded at a specific concentration in a 96-well plate. Serial dilutions of the compounds assuming to have positive activity along with the standard drugs are prepared. A specific volume of serially concentrated biologically active metabolites are added to the cell lines and incubated in a humidified atmosphere with 5% CO_2. Each concentration should be repeated three times. MTT solution (5 mg/ml in phosphate buffer) is added to each well and incubated. After the incubation, the medium is removed and the remaining is dissolved in DMSO and the absorbance is taken at a specific wavelength. The growth inhibition and cell viability is determined using

$$Growth\,inhibition = \left(Control\,O.D. - \frac{Sample\,O.D.}{Control\,O.D.} \right)$$

$$Cell\,viability\,(\%) = \frac{mean\,O.D.}{Control\,O.D.} \times 100$$

5.2.2.2 In Vivo *Anticancer Activity*

Compounds showing positive results in *in vitro* studies are further examined in *in vivo* studies. *In vivo* target modulation studies attempt to determine the effect of compound (inhibitor) treatment on the phosphorylation of specific target of tumor cells like receptor tyrosine kinase signaling (RTKs), A375 melanoma cells for KDR, SF767T glioma cells for PDGFR, and A431 epidermoid cells for EGFR [23]. *In vivo* anticancer activities of the compounds are examined using animal models. Animals used in the study should be acclimatized to the laboratory conditions and are inoculated with 1×10^6 cells per mouse on day 0. Drugs doses of test samples and the standard drugs are prepared in suitable solvent, like NaCl. Treatment is started 24 h after inoculation at a specific dose. The control group is treated with the sodium chloride solution. All the treatments are continued for more than a week and observations are carried out for more than one month. The antitumor efficacy is compared with 5-fluorouracil [intraperitoneally (i.p.)]. The Mean Survival Time (MST) of the treated group is compared with that of the control group using the following formula:

$$MST = \frac{T - C}{C} \times 100,$$

where T = Treated group and C = Control group.

5.2.3 Antiviral Assays

In the twenty-first century, there is still a lot to be done concerning the research of new drugs for an efficient battle of an enormous variety of extremely important and dangerous viruses that still affect our society such as HIV, hepatitis, and dengue fever, among others. In recent studies, several compounds with antiviral activity isolated from marine algae have been described by Mayer *et al.* [24]. A green algae, *Ulva lactuca*, had shown antiviral activity with human and avian influenza. The polysaccharide isolated contains arabinose, xylose, rhamnose, galactose, glucose, mannose, and a nonidentified sugar [25]. Compounds isolated from a marine sponge, *Tethya crypta*, also had shown antiviral activity [26,27]. Particularly, one of the compounds, Ara-A (vidarabine), is considered the first antiviral drug approved by the US FDA (Food and Drug Administration). The mechanism of action of Ara-A is based in the inhibition of viral DNA synthesis by conversion into adenine arabinoside triphosphate, which inhibits viral DNA polymerase and synthesis of viruses such as *Varicella zoster* and herpes.

In antiviral assays, the activity is determined by the combination of the virus (hepatitis, HIV virus, influenza virus, among others) and a living cell or animal host, with the test material to screen. The primary screening only involves one step and is evaluated by the comparison of response to viral infection of the treated and untreated host systems. However, in the secondary screening, there are usually multiple doses of the test substance against several doses of the virus. For this *in vitro* antiviral assay, the properly maintained cell line is infected with a virus. Afterward, that same cell line is exposed to the compound (test sample—natural, semisynthetic, or synthetic) and incubated for suitable time period. After the incubation time and washing of cells with a proper buffer, the cell monolayer morphology is observed microscopically to determine if any antiviral activity can be seen. However it should also measure the toxicity of the test material on cells (e.g., with a control using only uninfected cells and compound). The therapeutic index is calculated with the observed ED_{50} (dose required to reduce the yield of virus by 50%, i.e., effective antiviral dose), LD_{50} (amount of toxic agent sufficient to kill 50% of cells), and selectivity ratio (LD_{50}/ED_{50}). The ideal model to be used should have a virus with minimal alteration by adaptation, a natural route of infection, and a similarity of this with pathogenesis, host response, drug toxicity, and drug metabolism. Bioassays models are described in literature [28].

5.2.4 Anti-Inflammatory Assays

Inflammation is an important body response to various harmful stimuli such as microbial invasion, immune reactions, physical damage, ultraviolet (UV) radiation, among others. However, if the inflammation is prolonged for a long time it can contribute to the pathogenesis of other harmful diseases such as rheumatoid, asthma, multiple sclerosis, among others. Inflammation can be classified into two groups: acute or chronic. Acute inflammation happens almost at the same moment of the injury and results in the healing of the tissue when the injury/infection is removed, having symptoms such as pain, redness, and swelling. The duration is short (hours or days). Chronic inflammation can exist after the acute. It is a process where there is tissue breakdown and healing of the tissue simultaneously, and it can last for months or even years.

In marine research, it is found that some natural products contain antioxidants with anti-inflammatory activities [29]. A compound isolated from green algae, *Dunaliella bardawil* (rich in β-carotene), had shown protective effects against acetic-acid-induced small bowel inflammation in rats [30]. Manoalide, a sesterterpenoid isolated from *Luffarella variabilis* sponge and that contains a γ-lactone function, had shown anti-inflammatory activity [31]. Anti-inflammatory screening tests are being constantly optimized to studies *in vitro*. Two examples of the methods used, which depend of the type of inflammation that is being studied (acute or chronic), are described next.

5.2.4.1 Acute Assay

In carrageenin-induced paw edema assay [32], which was previously developed by Srimal, the model animals (mice) are divided into groups according to the testing material administered; one group is fed with the testing material. The animals used should be previously acclimatized to the laboratory conditions. After one hr of feeding, a solution of the sulfated-polysaccharide carrageenin in normal saline is injected (subcutaneously) into the hind paw of mice. Posteriorly, after some hours, the mice are killed with ether and the paws are cut at the ankle joint and weighed. The amount of edema (or dropsy) developed is determined by the difference between the weight of the two paws, and the group mean is calculated and compared with the control group. The percent inhibition of the dropsy by the test substance is calculated and compared with the group that received a standardized compound (for example, hydrocortisone). If the percent inhibition is less than 20, the activity is rejected.

5.2.4.2 Chronic assay

Adjuvant-induced arthritis [33] (like in the acute assay) animals (i.e., mice, properly acclimatized to the laboratory conditions) are divided into groups—the arthritis-induced group (+) and the control group (–)—with testing material administrated. In one group, arthritis is induced by intradermal injection of Freund's adjuvant (solution of antigen emulsified in mineral oil, generally containing killed *M. tuberculosis*) into one of the hind paws. The control group is injected with incomplete Freund's adjuvant; that is, solution without a mycobacterial component. During the experiment, the volume of the paws is measured with a plethysmograph, an instrument that measures the variation of the paws on the basis of the amount of blood passing or present in them.

5.2.5 Analgesic Assays

For humans, pain has been one of the most common symptoms of disease. Many therapies have been installed (e.g., anti-inflammatory drugs (NSAIDs)), but even if the results are positive, the side effects tend to be negative with severe gastrointestinal and cardiovascular problems. Therefore, one of the new challenges is the identification of new analgesic natural products. Recent studies have shown that neorogioltriol, a tricyclic brominated diterpenoid isolated from the red algae *L. glandulifera*, had an efficient analgesic activity [34] when injected intraperitoneally in doses of 1 mg/kg. A possible mechanism of action for this compound could be the inhibition of inflammatory mediators, like prostaglandins. The main tests for analgesic activity can be divided in the writhing syndrome method and the hot plate method.

5.2.5.1 Writhing Syndrome Method (or Acetic acid-induced Writhing Syndrome Method)

This assay is generally based on the method of Koster *et al.* [35]. For this assay, the test sample (crude extract, pure natural product, or synthetic compounds) is administered to groups of mice (properly acclimatized to the laboratory conditions) intraperitoneally (i.p.). Thirty minutes after the treatment, a specific volume of acetic acid solution is injected in the mice, inducing contortions in the body (writhing). The numbers of contortions are recorded between intervals of time. The results are compared with a control group of mice (20% Tween 80).

5.2.5.2 Hot Plate Method

The original method was described by Woolfe and MacDonald [36]. However, this assay has been modified and proven to be suitable for evaluation of centrally acting analgesics. In this, groups of mice (properly acclimatized to laboratory conditions) are used for each dose, with a determined range of initial weight. The animals are left in commercialized hot plates (having a controlled temperature) for a period of time. The end point is accepted when the mice start licking or jumping. The latency is noted before and after an established period of time and posteriorly the mice can be injected with a standard analgesic (e.g., morphine) to compare the activity.

5.2.6 Antiallergic Activity

Allergy is an immunological reaction to a foreign particle (antigen/allergen) such as dust mites, pollen, cosmetics, food, and mold spores that causes tissue inflammation and organ dysfunction. The allergic mediators induce various physiological effects that cause allergic diseases. Allergic disorders such as rhinitis, sinusitis, atopic dermatitis, asthma, pollenosis, and food allergy are the most common causes of human diseases.

Hypersensitivity of the immune system to specific antigens plays a central role in the specific initiation of asthma and allergic rhinitis [37]. For example Th2 lymphocytes are a major cellular infiltrate in the asthmatic lung [38]. Antibody immunoglobulin E (IgE) is overproduced in most of the patients suffering from an allergic condition [37]. The mediators such as histamine and serotonin are released from mast cells and basophils within minutes—the early phase reaction [39]. When granules in mast cells or basophiles are activated an enzyme beta hexosaminidase stored in secretary granules is released along with histamine. This enzyme activity is used as a marker of mast cells and basophiles degradation [40].

Effective procurement of allergic diseases relies on accurate diagnosis, which can help to confirm the cause of allergy. Accurate diagnosis is based on valid allergy test results. For assessing the presence of allergen-specific IgE antibodies, two different methods can be used: a skin prick test or an allergy blood test. With the increase in allergy cases and their type, new antiallergic drugs are required. Antiallergic activity is an essential step in the development of antiallergic agents.

5.2.6.1 Antiasthmatic Assay

Asthma is one of the most common allergic disorders and is characterized by airway inflammation. To evaluate antiasthmatic activity of the new compounds, milk-induced leucocytosis and eosinophilia, egg albumin-induced mast cell degranulations, and passive cutaneous anaphylaxis in animal models is generally observed.

Milk-induced leucocytosis and eosinophilia

Antiasthmatic activities [41] of the new compounds are determined using animal models. Animals used in the study should be acclimatized to the laboratory conditions. Swiss albino mice of either sex can be used for experiments, and are housed in standard temperature conditions, relative humidity, and light (12 h light/dark cycles). Animal models are divided into groups ($n = 6$ in each group). Blood samples are collected from the retro-orbital plexus. In the next step, control group animals are intraperitoneally treated with 1% Tween-80 solution; test group animals are treated with compounds of interest at different doses. The standard group receives dexamethasone. Boiled and cooled milk is injected to all the groups 30 min after treatments. Total leukocyte and eosinophile count is done in each group before treatment and 1 day after milk injection. Difference in total leucocytes and eosinophile count before the treatment and 24 h later is calculated.

Egg albumin induced activity

For antiasthmatic activity [42] animal models as used in the previous experiment are sensitized with three doses of egg albumin intraperitoneally as well as subcutaneously. Doses of egg albumin are given on alternate days. Three weeks after the last dose, animals are randomly allocated into groups. Group I (control), Group II (standard), and Group III receive the compounds in different doses orally. Thirty minutes later, an intravenous injection of Evans blue saline solution and an intraperitoneal injection of acetic acid are given to mice. After a specific time, the dye that leaked into the peritoneal cavity is collected with distilled water and is transferred to a volumetric flask through glass wool. To each flask, a specific volume of sodium hydroxide solution is added and then distilled water is added to fill to the mark, followed by measurement of absorbance at a specific wavelength.

5.2.6.2 Allergic Pleurisy

Raj *et al.* [37] described that pleurisy, also known as pleuritis, is an inflammation of the pleura, the lining surrounding the lungs, mainly caused by viral infections. The inflamed pleural layers rub against each other every time the lungs expand to breathe in air. This can cause sharp pain when breathing, also called pleuritic chest pain. In this assay, Swiss albino mice of either sex used for experiments are randomly allocated into five groups (I, II, III, IV, and V), generally each containing six animals. Active sensitization of animals is achieved with a subcutaneous injection of Freund's complete adjuvant emulsion containing egg albumin. In the next step, after 14 days, animals are treated with an intrathoracic injection of egg albumin. Briefly, an adapted needle inserted into the right side of the thoracic cavity of egg albumin sensitized animals to permit the intrapleural administration of egg albumin diluted in sterile pyrogen free saline. Twenty-four hours after the stimulus, animals are anesthetized and their thoracic cavities are rinsed with phosphate buffer saline containing 10 mM EDTA, pH 7.4. Total leukocyte counts are made.

Finally differential cell counts by light microscopy are stained with Leishman's stain. Groups IV and V animals fasted overnight and received EtMe and EtMe orally one hour before stimulation. In Group III, dexamethasone is given intraperitoneally 24 hr and 1 hr before stimulation. Group II animals are similarly treated with vehicle alone. Group I is considered the control.

5.2.7 Diuretic Activity

Diuretics are substances that remove fluid from the body. In another words they increase the rate of urine flow and sodium excretion (Figure 5.3). Therapeutically

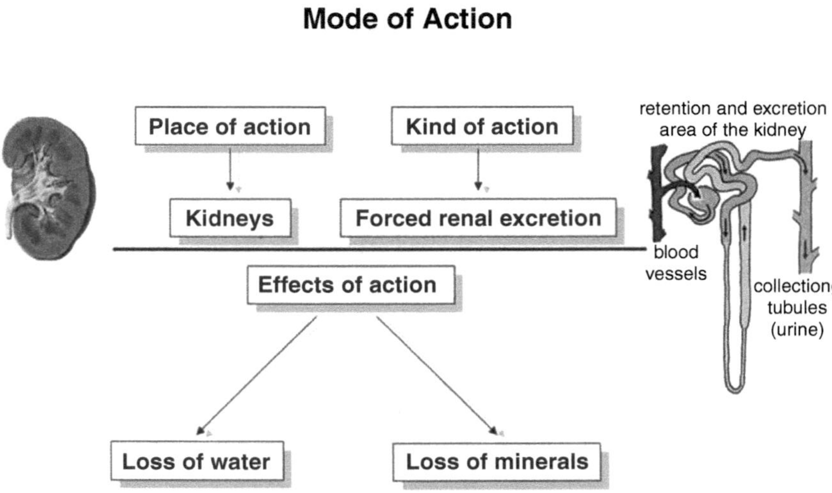

FIGURE 5.3 Mode of action of diuretics (adopted from [91]).

they should increase the output of sodium as well as water, which is required to remove edema fluid composed of water and solute of which sodium is an important part. Diuretics not only alter the excretion of sodium but also modify renal handling of other cations K^+, H^+, Ca^{2+}, Mg^{2+}, and anions Cl^-, HCO^{3-}, H_2PO_4, and uric acid [36]. Diuretics increase urine excretion of both water and electrolytes, and are commonly called water tablets. In general, they inhibit electrolyte reabsorption from the lumen of the nephron, increasing osmolarity and enhancing water excretion. The method described by Lipschitz *et al.* [43] estimates diuretic activity.

Diuretic activity can be performed in three simple steps using animal models. Animals used for activity measurement are generally Male Wistar rats (196 ±1 g) and should be housed and kept in a 12 hr light–dark cycle and allowed to access water and ad libitum except for the brief period during the experiment. In the first step rats are divided in groups like negative control where they received only normal saline in a specific amount, positive control of any standard drug like furosemide, and groups treated with new metabolites supposed to act as antidiuretic drugs. All the doses of drugs should be given with normal saline of the same concentration used earlier. In the second step after dosing, rats should be placed in metabolic cages individually at the room temperature. In the final step urine samples are collected up to 5 hr after dosing and the volume is measured. Sodium potassium concentration is calculated through a flame photometer.

5.2.8 Adaptogenic Assays

Adaptogens cause an adaptive reaction to a disease and are useful in many unrelated illness. They appear to produce a state of nonspecific increased resistance during stress resulting in stress protection [44]. The adaptogens not only help in coping with stress but help in enhancing a general state of well-being. Studies on the mechanism of action of adaptogenic drugs revealed that they produce immunostimulation [45]. These drugs strengthen the defense mechanism against the free radical induced damage inducing stress [46].

5.2.8.1 Antistress Activity

Antistress activity can be performed according to the procedure described by Mehta *et al.* [47]. For the assay, rats of either sex weighing between 150 and 250 gm are divided into six groups (I, II, III, IV, V, and VI), each containing six animals. The 24 urine samples from each group are collected into two different beakers, one containing 10% oxalic acid for spectrophotometric determination of ascorbic acid and the other containing 6N hydrochloric acid, for the determination of vanillylmandellic acid (VMA) spectrometrically. After collecting one-day urine samples in all the six groups, samples are subjected to analysis and both VMA and ascorbic acid and the normal values are recorded for seven consecutive days. In the next phase of experiment, animals in each group are subjected to fresh water swimming stress individually. In this method, rats are forced to swim until exhausted (3–4 min) in a cylindrical vessel containing water at room temperature. The 24 h urinary levels of VMA and ascorbic acid under stressed conditions are determined again as described earlier daily for seven consecutive days. After recovering the data for normal conditions, animals of the same groups are subjected to drug administration. Groups II, III, IV, V, and VI are administered five daily doses (body weight), orally, of test compounds in the suitable solvent for seven consecutive days while group I serves as control (without any compound). The urine samples are collected after every day and the levels of both VMA and ascorbic acid are determined.

The final phase of the experiment consists of the administration of test compounds in the suitable solvent to the same groups of animals after a recovery period of one week. Groups II, III, IV, V, and VI are administered orally with test compounds at five different doses (body weight), respectively, an hour prior to the daily induction of stress for seven consecutive days while group I serves as control. The urine samples are collected after every 24 h for seven consecutive days and analyzed for VMA and ascorbic acid to study the influence of the test compounds on the stress-induced biochemical changes.

5.2.9 Immunomodulation Activity

Immunomodulation is a therapeutic approach in which we try to intervene in the autoregulating processes of the defense system. In other words we can say that immunomodulation is the condition where we can modulate immunity using various natural or synthetic compounds. Compounds that are capable of interacting with the immune system to regulate up or downward to specific aspects of the host response are known as immunomodulators or biologic response modifiers. This regulation is a normalization process, so that immunomodulators help in optimizing the immune response. These are biological response modifiers; they exert their effects by improving host defense mechanisms against diseases [48] (Figure 5.4 has been taken from [49]). Enhancement or suppression of immune responses through compounds can depend on a number of factors, such as including dose, route of administration, and timing of administration of the compound in question. The type of activity these compounds exhibit can also depend on their mechanism of action or the site of activity. It appears that the normal way by which the immune system works is through its own modulation by factors usually synthesized by the immune cells. The immune system is involved in the etiology as well as pathophysiologically.

Immunomodulators are becoming very popular in the worldwide natural health industry as people start to realize the importance of a healthy immune system in the maintenance of health and the prevention of diseases. Allopathic drugs are available for counteracting the oxidative stress and hence improve immunity, but the side effects and prohibitive cost of these allopathic drugs makes it necessary to search for an alternative. Although extensive work has been carried out in the field of chemotherapy during this century, it is only in the last two

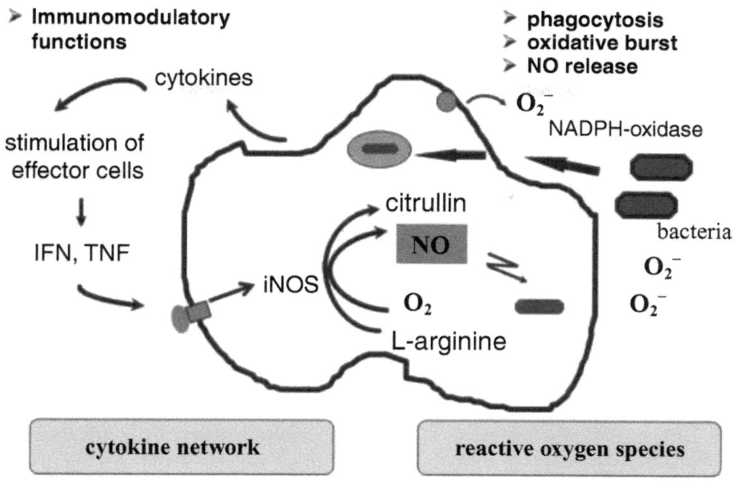

FIGURE 5.4 Immunomodulation illustration of cytotoxic defense mechanisms [49].

decades that a number of compounds with immunomodulatory activity have been identified from the plant materials. Among immunomodulatory activity evaluation methods, the Polymorphonuclear (PMN) function test is one of the frequently used methods.

In order to examine the immunomodulation activity, peripheral venous blood is collected in sterile heparinised tubes. From this sample, neutrophils are isolated by density-based gradient sedimentation. Further red blood cells (RBC) and PMN from the pellet are separated by dextran sedimentation. As a result, supernatant contain PMN cells having a cell density of 1×10^6 cells/mL is used for the experiments. PMN cells are mixed with 1×10^6 cells/mL of test culture like *Candida albicans* and the like. Compound solutions whose immunomodulating activity has to be examined is added to the above mixer and incubated at a specific temperature or accordingly for 1 h in a CO_2 incubator. After incubation, cytosmear of the fraction is prepared and stained with Giemsa stain and observed under 100× oil immersion objective to determine the phagocytic activity of PMN cells.

5.2.9.1 Determination of Cell Mediated Immune Response

Gaur *et al.* [50] explain a simple method for the determination of cell mediated immune response. For this, an animal model, like BALB/c female mice (six weeks old), are used. The animals are immunized by injecting sheep red blood cells (SRBC) suspension containing 1×10^8 cells, intraperitoneally, on day 0. In the next step, at day 7, they are challenged with a volume of 2×10^8 SRBC in the right hind foot pad. The contra lateral paw receives an equal volume of saline. After a 24 h interval, the foot thickness is measured and the difference in the thickness of the right hind paw and left hind paw is used as a measure of the Delayed-type Hypersensivity (DTH) reaction.

5.2.9.2 *Determination of Phagocytic Ability*

The phagocytic ability of macrophage is measured by neutral red uptake [51]. RAW 264.7 cells are cultured in 96-well plate at a density of 5×10^4 cells/mL at a specific temperature in a 5% CO_2 atmosphere for 24 h. The cells are incubated with various concentrations of compounds and LPS at a specific temperature for 48 h. A small amount of neutral red solution is added to this mixture and incubated for one hour. After incubation, supernatant is discarded and the cells are washed twice with PBS. A fixed volume of cell lysate solution (ethanol and 0.01% acetic acid at the ratio of 1:1) is added into a 96-well plate to lyse cells at room temperature for 2 h. The optical density is measured at a determined wavelength.

5.2.9.3 *Measurement of Nitric Oxide Production*

Guan *et al.* [51] demonstrated a simple procedure for the estimation of accumulated nitric oxide production where nitrite accumulation is measured by a Griess reagent and used as an indicator of nitric oxide (NO) production in the medium. In order to measure the nitric oxide production, macrophage cells (1×10^5 cells/mL) are dispensed into a 6-well plate for a day and cells are stimulated with a dose LPS (lipopolysaccharide); various concentrations of compounds are incubated for 24 h. After incubation, a specific volume of culture supernatants are mixed with an equal volume of Griess reagent in a 96-well plate and incubated at room temperature for 10 min. The absorbance is measured on a microplate reader. $NaNO_2$ is used as standard to calculate nitrite concentrations.

5.2.10 Choleretic and Anticholestatic Activities

The liver occupies a central position in body metabolism, therefore it is an important organ for pharmacological and biochemical studies. The liver is also exposed to a variety of xenobiotics and therapeutic agents, thus, the disorders associated with this organ are numerous and varied. One example, cholestasis, is a reduction in bile flow that leads to the intrahepatic accumulation of bile acids and other toxic compounds with progression of liver pathology, including hepatocellular injury and fibrosis [52]. The substance or agents that stimulate bile production by the liver are known as choleretic.

For the study of intact liver functions under controlled conditions, temporary biliary fistula in anesthetized rats [53] is used to measure the biliary flow. Animals pretreated with the respective doses and fasted overnight are anesthetized with urethane prior to the experimental procedure and fixed onto a surgical stage. The abdomen is opened and the common bile duct cannulated by polyethylene tubing. The duct that is formed by the union of the hepatic and cystic ducts and discharges into the duodenum. Also called *gall duct.* cannulated by polyethylene tubing. Abdominal temperature is maintained at a specific temperature throughout the experiment with the help of a lamp. The bile is collected on ice and the total bile volume/100 g animal is measured every half an hour for a total of 3 h. The samples are frozen until analysis. Bile flow rate is expressed in mL/100 g body weight/h. Likewise, for anticholestatic activity, animals are treated separately with a test substance and cholestatic agent. Study on bile flow is carried out a day or two posttreatment with a toxic agent. The details of test methods are described by Dhawan and Srimal [54].

5.2.11 Acute Toxicity and CNS Activities

The central nervous system (CNS) is the part of the nervous system that includes the brain and spinal cord. Substances that speed up physical and mental processes are known as central nervous system (CNS) stimulants and are used to treat conditions characterized by lack of adrenergic stimulation, including narcolepsy and neonatal apnea. For acute toxicity, LD_{50} is determined in mice using the method of Horn *et al.* [55]. Doses of the test substance are given to groups of four to five mice each in a geometrical progression or regression starting with a specific i.p. dose. The animals are observed continuously for three hours after administration of the test substance, then every 30 min for next 3 h and finally after 24 h. Mortality at 24 h is recorded. The LD_{50} with fiducial limits is read out from the table. CNS stimulation is observed by increased spontaneous motor activity (SMA), piloerection, exophthalmos, clonic and/or tonic convulsions, CNS depression by reduced SMA, sedation, ptosis, crouching, catalepsy and autonomic effects by piloerection, urination, defecation, salivation, lachrymation, and so on.

5.2.11.1 Effect on Hexabarbital Sleeping Time

Bhakuni and Rawat described methods of observing the effect on hexabarbital sleeping time[1]. Hypnosedatives and neuroleptic agents are reported to increase the barbiturate sleeping time in mice. Graded doses of the test substance are administrated (i.p.) in groups of five mice each and after an hour, hexobarbitone is administered. Saline treated controls and hexobarbitone-administered mice are run concurrently. The sleeping time for each animal is the period for which the righting reflex is absent. The prolongation of sleeping time with respect to the control is then calculated. The significance of drug-induced prolongation of sleeping time is judged using the student's t-test.

5.2.11.2 Acute Toxicity Study

An acute toxicity study can be carried out according to the Organization for Economic Co-operation and Development (OECD), Guidelines No. 423 [56]. Three animals are used for each step. In the first step the dose level for the starting dose is selected from one of four fixed levels. As per the OECD recommendations, the starting dose level should be that which is most likely to produce mortality in some of the dosed animals; when there is no information available on a substance to be tested in this regard, for animal welfare reasons, the recommended starting dose is 300 mg/kg body weight. In the next step, clinical signs of toxicity and mortality in the test sample treated animals are observed. Generally one fifth of the lethal dose is taken as the effective dose for the study.

5.2.12 Isolated Tissues

The major benefits with this type of isolated system is that it is fairly trouble-free to set up for a single animal, only a minute amount of the testing material is needed, and the drug effect is checked straightaway without considering other factors like absorption, metabolism, excretion, or interference (perhaps due to nerve reflexes). The water insoluble materials are not suggested for testing in the *in vitro* systems, and if the material is dissolved in a solvent other than water or saline, a control must be included with the solvent. Generally, the most common sources of isolated tissues are guinea pigs, rats, rabbits, cats, and dogs. Often, mice are also employed for testing purposes. An earlier report by Dhawan and Srimal [54] have properly described and depicted the technique of taking out particular organs and/or tissues from animals, performing tests on guinea-pig ileum, mouse vas deferens, rat hind-limb, rat uterus, rat phrenic nerve diaphragm, rabbit heart perfusion, and guinea-pig right ventricular papillary muscles.

5.3 TOXICITY EVALUATION

The toxicological evaluation of the marine test material is mostly performed by the oral or intraperitoneal route in two or three adult albino mice of either sex (weight 15–20 g). For this evaluation, the marine test material is suspended in distilled water in 10% gum acacia or 0.1% agar. The concentrations are maintained in such a way that a 20 g mouse gets 0.2 mL of liquid. Doses greater than 1000 mg/kg are not recommended. The control animals are maintained only on the vehicle dose. The involved mice are observed for about 5 to 6 h after the dosage for any type of toxicity symptoms, and if death occurs within this duration, the possible cause of death is noted. The roughly estimated LD_{50} is calculated and the highest tolerable dose is also noted for subsequent toxicity evaluations of marine samples.

The regulatory toxicity studies of marine test material are performed once its biological or pharmacological activity is established; its baseline data is generated pertaining to the effective dose, highest tolerable dose, and lethal dose. These toxicological studies are performed with the potential drug compound and then passed to clinicians for Phase I and II clinical trials. At this point the desired toxic studies are (1) acute toxicity, (2) repeated dose, (3) subacute toxicity, (4) reproductive, (5) teratological, (6) pre- and postnatal, (7) carcinogenic, and (8) mutagenic.

5.3.1 Reproductive Toxic Evaluation Studies

The reproductive studies are very important for Phase II clinical trials and include (1) toxic evaluation of fertility and general reproductive performance, (2) teratological toxic studies, and (3) prenatal and postnatal toxicity evaluation studies, which are useful to evaluate the effect of a given marine drug on gonadal actions, estrus cycles, mating behavior, conception rates, and early stages of gestation. The important observations during reproductive toxicity evaluation include number and distribution of embryos in each uterine horn, presence of empty implantation sites, abnormal conditions in the uterus, embryo undergoing resorptions, gestation period, litter size, stillborn, and litter weight.

5.3.1.1 Teratological Toxic Evaluation Studies

The teratological evaluation of the marine test material is required to examine the potential of a test sample for embryo-toxicity and its teratogenic actions. The significant observations include number of corpora lutea in the ovary, number of implantations in each horn, relationship of the fetal placement in each horn with the number of corpora lutea in each ovary, number of fetuses, number of live and dead fetuses, number of resorptions, weight of fetus, and visceral/skeletal or any other external anomaly of the fetus.

5.3.1.2 Pre- and Postnatal Toxic Evaluation Studies

Pre- and postnatal toxicological evaluation assay is needed to detect the effects of marine test material on the newborn when given to the mother in the last trimester of her pregnancy and throughout the period of lactation. Important observations include gestation duration; labor and delivery type; litter size and weight at the time of birth, the fourth day, the twenty-first day, and at weaning; litter morbidity and mortality; and the effect on lactation.

5.3.2 Carcinogenic Evaluation Studies

The carcinogenic assay is mostly required to evaluate the carcinogenicity of the marine test sample in both sexes of two species and is designed to cover the greater part of the animal's life span.

5.3.3 Mutagenic Toxicity Evaluation Studies

These days various mutagenic assays are performed on marine test samples to detect the mutagenic as well as carcinogenic activity. The assays include DNA damage and mutation tests in bacteria, *in vitro* chromosomal aberration tests, dominant lethal tests in rodents, micronucleus test in mice, among others.

5.4 USE OF ANIMALS

Applications of various animals in laboratory experiments have a long history but in recent years producing equivalent alternatives has become a matter of significant consideration across the globe. Although laboratory animal models are not the end point in forecasting the toxicity hazards for human beings, in the coming years it may be possible to dispense completely with animal testing, although it might be more feasible to decrease their numbers through optimal experimental design and exhaustive analysis, and optimal employment of available secondary data related to the toxicity of earlier investigated compounds. However, sincere efforts are being made to substitute laboratory-animal-based experiments with suitable alternative techniques. Nowadays, alternatives to animal-based research can be classified into three major groups: (1) the use of modified animals, (2) the use of living systems, and (3) the use of nonliving systems. Recent developments in life sciences research has changed the scenario and now *in vivo* research studies can be easily modified by merging them with some *in vitro* techniques for better, faster, and more precise results. However, certain limitations of *in vitro* methodologies—for example, physiological dissimilarity, data interpretation, and approval by standard regulatory agencies—are still big challenges.

5.5 CLINICAL TRIALS

Marine organisms developed complex biological mechanisms due to their sedentary life style, necessitating chemical means of defense. Because of that, they developed the ability to synthesize compounds (i.e., secondary metabolites) [57–59]. In recent years, the number of potential compounds isolated from marine microorganisms, such as algae, sponges, mollusks, and others, has risen to over 10,000, with hundreds of new compounds still being discovered every year [60]. A number of promising identified molecules have been in clinical trials and some of them exist already in market (e.g., Yondelis™). But before describing the current status of these bioactive molecules, the meaning of "clinical trials" is described.

5.5.1 Definition and Phases of Clinical Trials

Trial is originated from the Anglo-French *trier* (*to try*). *Clinical* is from *clinic*, from the Greek *klinike*, which refers to the practice of caring for the sick at the bedside.

The clinical trial is, *per se*, the testing done on human beings for determining the value of a treatment for the sick or for preventing disease/sickness, and it involves the application of an experimental variable (i.e., treatment to a person or group of persons, and observation during or following application of the treatment to measure its effect).

The compounds first go through a series of tests in animals before they can be tested in humans, and the investigation is often evaluated in two or more phases simultaneously in different clinical trials. The results from the laboratory and animal research become the basis for an investigational new drug application (INDA) to the Food and Drug Administration. An INDA pursues permission to use the drug in a Phase I trial.

The four phases of clinical trials and its medicine development are:

Phase I: First stage of safety trials on a new medicine. Generates preliminary information on the chemical action and safety of the indicated drug. An attempt is made to find a safe dose range tolerated by volunteers (20–100). The trials are generally conducted in healthy humans but sometimes can be conducted in severely ill patients (cancer field) or less ill patients with pharmacokinetic issues.
Phase IIa: Second stage of testing; clinical trials to evaluate efficacy and safety in selected populations of patients with the disease or condition to be treated (20–300). The objectives are focused on drug dose response, type of patient, and other characteristics of efficacy and safety.
Phase IIb: Well-controlled trials to evaluate efficacy and safety in patients with the disease or condition to be treated.
Phase IIIa: Third (and sometimes the final) stage in a clinical trial, conducted after efficacy of medicine is shown, but prior to regulatory submission of an Investigational New Drug Application (INDA). The trials are conducted in patients for whom the medicine is proposed and usually is designed to include a control treatment and a random assignment to the treatment.
Phase IIIb: Conducted after regulatory submission of an INDA but preceding the medicine's approval and launch. This is the period between submission and approval of a regulatory dossier for marketing authorization.
Phase IV: This fourth stage is controlled and performed after approval of the INDA and it is sometimes performed under circumstances approximating real-world conditions.

There can be different types of evaluations (formulations, dosages, duration of treatment, interactions) and studies about patients (age groups, races). After marketing approval, drugs remain under surveillance for adverse effects.
Post-marketing surveillance: Involves the collection of reports of adverse events via reporting, sample surveys, and observational studies.

5.5.2 Status of Marine Bioactive Molecules in Clinical Trials or Market

Many of the bioactive compounds found in marine microorganisms are being taken for clinical trials, but only some of them are already in the market (for example, cytarabine) or being approved soon. Pharmaceutical companies worldwide have been conducting more and more research about marine resources, since it has been proven that oceans and seas are some of the major frontiers for medical research. For example, the companies PharmaMar (Spain, United States), AquaPharm Biodiscovery Ltd. (United Kingdom), and Nereus Pharmaceutical (United States) have several compounds in clinical trials or preclinical trials, and some already on the market. The selected marine bioactive molecules in trials or already commercialized are shown in Table 5.1 (adopted from [61]) and described in the following with their mechanism of action.

TABLE 5.1 Marine Natural Products in Different Phases of Clinical Trials

Product	Source	Chemical Class	Company	Application Area	Status
Compounds targeting ion channels					
Ziconotide	Cone snail	Peptide	Neurex	Chronic pain	Phase III
AM336	Cone snail	Peptide	AMRAD	Chronic pain	Phase I/II
GTS21	Nemertine worm	Anabaseine-derivate	Taiho	Alzheimer's disease and Schizophrenia	Phase I/II
Compounds targeting enzymes					
LAF389	Sponge	Amino acid derivate	Novartis	Cancer	Phase I
Bryostatin-1	Bryozoan	Polyketide	GPC Biotech	Cancer	Phase II
OAS1000	Soft coral	Diterpene-pentoseglycoside	OsteoArthritis Sciences	Wound healing, Inflammation	Phase I/II
Microtubule-interfering agents					
Dolastatin-10	Sea slug	Peptide	NCI/Knoll	Cancer	Phase II
ILX651	Sea slug	Peptide	Ilex Oncology	Cancer	Phase I
Cemadotin	Sea slug	Peptide	Knoll	Cancer	Phase II
Disodermolide	Sponge	Polyketide	Novartis	Cancer	Phase I
HTI286	Sponge	Tripeptide	Wyeth	Cancer	Phase I

DNA-interactive agents					
Yondelis™	Sea squirt	Isoquinolone	PharmaMar/Johnson & Johnson	Cancer	Phase II/III
Oxidative stress inducers					
Aplidin™	Sea squirt	Cyclic depsipeptide	PharmaMar	Cancer	Phase II
Lysosomotropic compounds					
Kahalalide F	Sea slug/alga	Cyclic depsipeptide	PharmaMar	Cancer	Phase I
Immunostimulatory agents					
KRN7000	Sponge	α-galactosylceramide	Kirin	Cancer	Phase I
Calcium-binding protein antagonists					
Squalamine lactate	Shark	Aminosteroid	Genaera	Cancer	Phase II
Compounds with unknown mechanism of action					
IPL512602	Sponge	Steroid	Inflazyme/Aventis	Inflammation Asthma	Phase II

5.5.2.1 Molecules Targeting ion Channels

The first two compounds, Ziconotide and AM336, are conotoxines (group of neurotoxic peptides) isolated from the venom of the marine cone snail from the genus *Conus*. This group can selectively block ion channels and neuronal receptors and because of that, over the last few decades, they have been the subject of pharmacological interest [62]. Ziconotide, the first FDA-approved analgesic since morphine, is a synthetic form of a ω-conotoxin peptide isolated from the Magician's cone snail, *Conus magnum* [63]. This powerful analgesic has a unique mechanism of action where the selective block of N-type calcium channels control neurotransmission at many synapses. AM336 (or CVID) is another ω-conotoxin that had completed a Phase II clinical trial for the treatment of neuropathic pain [64] but the continuation of studies have stopped due to changes in the company's regulations. GTS-21 (also denominated DMXB-A) acts as a partial agonist at the neural nicotinic acetylcholine receptors, is used to measure effects of memory and learning, and is effective in promoting cell survival in *in vitro* and *in vivo* studies [65]. It is being investigated specially for the treatment of Alzheimer's disease and schizophrenia [66–68].

5.5.2.2 Compounds Targeting Enzymes

LAF389 is an analogue of bengamide B, isolated from Jaspidae sponges. The synthetic bengamide was developed as an anticancer agent due to its capacity of inhibiting cell proliferation [69]. From the sea plant *Buglia neritina*, the compound Bryostatin-1 was isolated, and has been showing anticancer activity with melanomas, sarcomas, and lymphomas [70,71]. OAS1000 is a semisynthetic compound derivated from pseudopterosine A (sea whip *Pseudopterogorgia elizabethae*) and has shown anti-inflammatory activity [72].

5.5.2.3 Microtubule-interfering Agents

Dolastin-10 and Dolastin-15 were extracted and isolated from the marine mollusk *Dolabela auricularia*. These two pentapeptides have shown promising anticancer/antiproliferative activity [73] but Dolatin-10 has been studied further due to its potency [74]. ILX651 or Tasitodin is a third-generation synthetic microtubule-targeted derivate of Dolastadin that has successfully completed Phase I trials. Cematodin (LU-103793), another analogue from Dolastadin, was found to be effective for the treatment of breast cancer in Phase I, but its studies didn't continue due to prior unexpected results [75]. From the Caribbean marine sponge *Discodermia dissoluta*, the compound disodermolide was isolated, which has shown immunosuppresant [76] and antifungal activities. It has also been shown that it can stabilize microtubules faster and that it is more potent than other microtubules agents [77], and that it is a potent inhibitor of tumor cells *in vitro*. Harbor Branch Oceanographic Institution, which has isolated disodermolide, licensed its synthetic route to Novartis that began a Phase I trial. However, the studies were disrupted due to compound toxicity. HTI286 (or Taltobulin) is a synthetic analog that inhibits the polymerization of tubulin, induces mitotic arrest and apoptosis, and interrupts microtubule dynamics [78]. Currently it is in clinical development by Novartis.

5.5.2.4 DNA Interactive Agents

Trabectedin or Yondelis™ is a natural product derived from the marine tunicate *Ecteinascidia turbinata*. It has shown potent antitumor activity [79,80] and antiproliferative activity on drug-resistance [81]. PharmaMar markets Yondelis® in Europe and Japan and new clinical trials are being carried out for breast and prostate cancer.

5.5.2.5 Oxidative Stress Inducers

Aplidin (or plitidepsin) is an antitumoral agent [82,83], isolated from a sea squirt, that is being developed by PharmaMar. It is currently in Phase II trials in a variety of solid tumors (lymphoma and myeloma) and hematologic malignancies.

5.5.2.6 Lysosomotropic Compounds

Kahalalide F is a cyclic depsipeptide, derived from the sea slug *Elysia rufescens*. This metabolite has shown anticancer activity *in vitro* and *in vivo* [84].

5.5.2.7 Immunostimulatory Agents

KNR7000 is a synthetic analog of α-galactosylceramide and marine product agelasphin. It is known as an antitumor agent [85,86] and activity in other diseases like lupus, diabetes, tuberculosis, and malaria [87,88].

5.5.2.8 Calcium-binding Protein Antagonists

Squalamine lactate is a synthetic molecule, derived from dogfish shark, that has been shown as an antimicrobial [89] and antiviral agent [90]. It is considered as an antiangiogenic substance since it directly interrupts and reverses multiple facets of angiogenesis (creation of new blood vessels).

5.6 CONCLUSIONS AND FUTURE PERSPECTIVES

Bioassay or biological assay, or bioactivity assay or biological screening, is any qualitative or quantitative analysis of substances that uses a living system, such as an intact cell, as a component. In the case of marine samples, there are some very unique issues encountered in bioassay screening. As we know that the active components in the crude extract are present in a very low concentration (i.e., 1:1000), the bioassay screening for their detection must be very sensitive to produce effective results; also the biological screening process must be very specific and selective. The assay methodology should be adaptable even to marine substances, which are often dark colored, tarry, or poorly soluble in water. Also, it is mandatory that the biological screening must be unique and insensitive to other ubiquitous marine compounds. The success of screening new bioactive compounds from marine sources is generally based upon the quality of the marine sample, collection methodology, storage procedure, cultivation method, extraction procedure, separation and purification, and most importantly the test sample preparation. Still, there is no specific method that could be adopted for the separation of the active components of the complex mixture present in the extract of the marine source (e.g., fungi), as the chemistry of the bioactive component is not known to be a priority. Also, the availability of many bioactive compounds are those that are chemically similar, difficult to separate, and mostly complicate the isolation and purification process.

With unmatched advancements in the cellular and molecular biology and imaging technologies, there has been a tremendous change in the bioassay technology, but still there is a long way to go to save animal life (and ultimately human life), and to provide a significant reduction in assay cost and time involved. Still, much research is being conducted at the global level, where scientists are trying hard to find suitable alternatives for laboratory use of animals. Additionally, research in imaging and chemical sciences are proactively ongoing, to develop sensitive dyes or chemicals that can easily and effectively bind to the active constituent/cell component and give precise visual results with decreased chances of false positives. Researchers are even trying to merge the separation techniques with biological evaluation of the sample. Although the pace of current research in that direction is modest, we hope that in coming years we will have more efficient imaging and cell culture-based technologies for better evaluation of biological screening with more precision, low sample requirements, visibility, speed, lower cost, and longer lasting results with less false positives or negatives. In recent years, most countries (including some of the developed and developing countries; e.g., Finland, India, United States, Germany, Spain, among others) of the world have developed a targeted drug development program through various schemes and research projects covering all areas of drug development including biological evaluation. And, it is expected that such programs will help in the discovery of novel bioactive compounds of natural origin, especially from marine sources. It is hoped that investigated compounds will ultimately become clinically useful drugs in the near future.

ACKNOWLEDGEMENTS

We thank Prof. Arif Ali (Gene Expression Laboratory, Jamia Millia Islamia, New Delhi, India) for his insightful discussions and scientific suggestions in the preparation of the chapter, and Dr. Päivi Tammela (Bioactivity Screening Group, Centre for Drug Research, Faculty of Pharmacy, University of Helsinki, Finland) for her introduction and guidance in ongoing marine biology research.

REFERENCES

[1] D. Bhakuni, D. Rawat, Bioactive Marine Natural Products, Springer-Verlag, Berlin/Heidelberg, 2005. doi:10.1007/1-4020-3484-9.

[2] E.A. Martis, R. Radhakrishnan, High-Throughput Screening: The Hits and Leads of Drug Discovery- An Overview, J. Appl. Pharm. Sci. 01 (2011) 2–10.

[3] K.B. Glaser, A.M.S. Mayer, A renaissance in marine pharmacology: From preclinical curiosity to clinical reality, Biochem. Pharmacol. 78 (2009) 440–448. doi:10.1016/j.bcp.2009.04.015.

[4] N. Xu, X. Fan, X. Yan, X. Li, R. Niu, C.K. Tseng, Antibacterial bromophenols from the marine red alga Rhodomela confervoides, Phytochemistry 62 (2003) 1221–1224. doi:10.1016/S0031-9422(03)00004-9.

[5] G. Johnsi-Christobel, A.P. Lipton, M.S. Aishwarya, A.R. Sarika, A. Udayakumar, Antibacterial activity of aqueous extract from selected macroalgae of southwest coast of India, SEAWEED Res. Util. 33 (2011) 67–75.

[6] B.H. and K.M. Chiheb Ibtissam, Riadi Hassane, Martinez-Lopez José, Dominguez Seglar Josè Francisco, Gomez Vidal Josè Antonio, Screening of antibacterial activity in marine green and brown macroalgae from the coast of Morocco, African J. Biotechnol. 8 (2009) 1258–1262. doi:10.4314/ajb.v8i7.60101.

[7] A.J. Smit, Medicinal and pharmaceutical uses of seaweed natural products: A review, J. Appl. Phycol. 16 (2004) 245–262. doi:10.1023/B:JAPH.0000047783.36600.ef.

[8] R.A. Edrada, R. Ebel, A. Supriyono, V. Wray, P. Schupp, K. Steube, et al. Swinhoeiamide A, a new highly active calyculin derivative from the marine sponge Theonella swinhoei, J. Nat. Prod. 65 (2002) 1168–1172. doi:10.1021/np020049d.

[9] H. Gross, G.M. König, Terpenoids from Marine Organisms: Unique Structures and their Pharmacological Potential, Phytochem. Rev. 5 (2006) 115–141. doi:10.1007/s11101-005-5464-3.

[10] M.A. Wikler, F.R. Cockerill, W.A. Craig, M.N. Dudley, G.M. Eliopoulos, D.W. Hecht, et al. Performance standards for antimicrobial disk susceptibility tests; approved standard — ninth edition, CLSI 26 (2006) 1–35.

[11] C. and L.S. Institute, Performance Standards for Antimicrobial Susceptibility Testing; Twenty-Second Informational Supplement, 2012.

[12] G. Kahlmeter, D.F.J. Brown, F.W. Goldstein, A.P. MacGowan, J.W. Mouton, I. Odenholt, et al. European Committee on Antimicrobial Susceptibility Testing (EUCAST) Technical Notes on antimicrobial susceptibility testing, Clin. Microbiol. Infect. 12 (2006) 501–503. doi:10.1111/j.1469-0691.2006.01454.x.

[13] S.C. Parija, Textbook of Microbiology and Immunology, 2/e, Elsevier India, 2009.

[14] EUCAST, ESCMID, EUCAST DISCUSSION DOCUMENT E. Dis 5.1 MARCH 2003 Determination of minimum inhibitory concentrations (MICs) of antibacterial agents by broth dilution, Clin. Microbiol. Infect. 9 (2003) 1–7.

[15] EUCAST, ESCMID, EUCAST Definitive D ocument E.DEF 3.1, June 2000: Determination of minimum inhibitory concentrations (MICs) of antibacterial agents by agar dilution, Clin. Microbiol. Infect. 6 (2000) 509–515.

[16] S.J. Cavalieri, Manual of antimicrobial susceptibility testing, American Society for Microbiology, 2009.

[17] S. Kameshwaran, V. Suresh, G. Arunachalam, S.K. Kanthlal, M. Mohanraj, In vitro and in vivo anticancer activity of methanolic flower extract of Tecoma stans flower, Int Res J Pharm. 3 (2012) 246–252.

[18] G.M. Cragg, D.J. Newman, K.M. Snader, Natural products in drug discovery and development, J. Nat. Prod. 60 (1997) 52–60. doi:10.1021/np9604893.
[19] T.L. Simmons, E. Andrianasolo, K. McPhail, P. Flatt, W.H. Gerwick, Marine natural products as anticancer drugs, Mol. Cancer Ther. 4 (2005) 333–342.
[20] V. Vichai, K. Kirtikara, Sulforhodamine B colorimetric assay for cytotoxicity screening, Nat. Protoc. 1 (2006) 1112–1116. doi:10.1038/nprot.2006.179.
[21] W.A. Dengler, J. Schulte, D.P. Berger, R. Mertelsmann, H.H. Fiebig, Development of a propidium iodide fluorescence assay for proliferation and cytotoxicity assays, Anticancer. Drugs 6 (1995) 522–532. doi:10.1097/00001813-199508000-00005.
[22] K.N. Prasad, J. Hao, C. Yi, D. Zhang, S. Qiu, Y. Jiang, et al. Antioxidant and anticancer activities of wampee (Clausena lansium (Lour.) Skeels) peel, Biomed Res. Int. 2009 (2009).
[23] D.B. Mendel, A.D. Laird, X. Xin, S.G. Louie, J.G. Christensen, G. Li, et al. In vivo antitumor activity of SU11248, a novel tyrosine kinase inhibitor targeting vascular endothelial growth factor and platelet-derived growth factor receptors: determination of a pharmacokinetic/pharmacodynamic relationship, Clin. Cancer Res. 9 (2003) 327–337.
[24] A.M.S. Mayer, A.D. Rodríguez, R.G.S. Berlinck, M.T. Hamann, Marine pharmacology in 2003-4: marine compounds with anthelmintic antibacterial, anticoagulant, antifungal, anti-inflammatory, antimalarial, antiplatelet, antiprotozoal, antituberculosis, and antiviral activities; affecting the cardiovascular, immune and nervous systems, and other miscellaneous mechanisms of action, Comp. Biochem. Physiol. C. Toxicol. Pharmacol. 145 (2007) 553–581. doi:10.1016/j.cbpc.2007.01.015.
[25] J.F. Aswell, G.P. Allen, A.T. Jamieson, D.E. Campbell, G.A. Gentry, Antiviral activity of arabinosylthymine in herpesviral replication: mechanism of action in vivo and in vitro, Antimicrob, Agents Chemother. 12 (1977) 243–254. doi:10.1128/AAC.12.2.243.
[26] W. Bergmann, R.J. Feeney, Contributions to the study of marine products. XXXII. The nucleosides of sponges, J. Org. Chem. 16 (1951) 981–987. doi:10.1021/jo01146a023.
[27] W. Bergmann, R.J. Feeney, THE ISOLATION OF A NEW THYMINE PENTOSIDE FROM SPONGES 1, J. Am. Chem. Soc. 72 (1950) 2809–2810. doi:10.1021/ja01162a543.
[28] S.C. Gad, Handbook of Pharmaceutical Biotechnology, Chem. Biol. 2 (2007) 2003–2005. doi:10.1345/aph.1D276.
[29] M.J. Abad, L.M. Bedoya, P. Bermejo, Natural marine anti-inflammatory products, Mini Rev. Med. Chem. 8 (2008) 740–754. doi:10.2174/138955708784912148.
[30] A. Lavy, Y. Naveh, R. Coleman, S. Mokady, M.J. Werman, Dietary Dunaliella bardawil, a beta-carotene-rich alga, protects against acetic acid-induced small bowel inflammation in rats, Inflamm. Bowel Dis. 9 (2003) 372–379.
[31] A. Soriente, M. de -Rosa, A. Scettri, G. Sodano, M.C. Terencio, M. Paya, et al. Manoalide, Curr. Med. Chem. 6 (1999) 415–431.
[32] C.A. Winter, E.A. Risley, G.W. Nuss, Carrageenin-induced edema in hind paw of the rat as an assay for antiiflammatory drugs, Proc. Soc. Exp. Biol. Med. 111 (1962) 544–547.
[33] C.M. Pearson, Development of Arthritis, Periarthritis and Periostitis in Rats Given Adjuvant, Exp. Biol. Med. 91 (1956) 95–101. doi:10.3181/00379727-91-22179.
[34] R. Chatter, M. Kladi, S. Tarhouni, R. Maatoug, R. Kharrat, C. Vagias, et al. Neorogioltriol: A brominated diterpene with analgesic activity from Laurencia glandulifera, Phytochem. Lett. 2 (2009) 25–28. doi:10.1016/j.phytol.2008.10.008.
[35] R. Koster, M. Anderson, E.J. Debeer, Acetic acid for analgesic screening, in: Fed. Proc., 1959: p. 412.
[36] G. Woolfe, A.D. MacDonald, The evaluation of the analgesic action of pethidine hydrochloride (Demerol), J. Pharmacol. Exp. Ther. 80 (1944) 300–307.

[37] N.R. Raj, S.K. Jain, C.N. Raj, A.B. Panda, Various Screening Methods for Anti-allergic Activity: An Overview, Int. J. Pharm. Sci. Nanotechnol. 3 (2010) 906–911.
[38] C. Walker, E. Bode, L. Boer, T.T. Hansel, K. Blaser, J.C. Virchow, Allergic and nonallergic asthmatics have distinct patterns of T-cell activation and cytokine production in peripheral blood and bronchoalveolar lavage, Am. Rev. Respir. Dis. 146 (1992) 109–115. doi:10.1164/ajrccm/146.1.109.
[39] H. Matsuda, S. Tewtrakul, T. Morikawa, M. Yoshikawa, Anti-allergic activity of stilbenes from Korean rhubarb (Rheum undulatum L.): structure requirements for inhibition of antigen-induced degranulation and their effects on the release of TNF-alpha and IL-4 in RBL-2H3 cells, Bioorg. Med. Chem. 12 (2004) 4871–4876. doi:10.1016/j.bmc.2004.07.007.
[40] S. Kraithep, K. Oungbho, S. Tewtrakul, Anti-allergic activity of Thai medicinal plants used in longevity formulation, Songklanakarin J. Sci. Technol. 30 (2008).
[41] D.J. Taur, R.Y. Patil, Antiasthmatic activity of Ricinus communis L. roots, Asian Pac. J. Trop. Biomed. 1 (2011) S13–S16 doi:10.1016/S2221-1691(11)60113-5.
[42] A.N. Paranjape, A.A. Mehta, Investigation into the Mechanism of Action of Abutilon indicum in the Treatment of Bronchial Asthma, Glob. J. Pharmacol. 2 (2008) 23–30.
[43] W.L. Lipschitz, Z. Hadidian, A. Kerpcsar, Bioassay of diuretics, J. Pharmacol. Exp. Ther. 79 (1943) 97–110.
[44] A. Panossian, G. Wikman, Effects of Adaptogens on the Central Nervous System and the Molecular Mechanisms Associated with Their Stress—Protective Activity, Pharmaceuticals 3 (2010) 188–224. doi:10.3390/ph3010188.
[45] S. Azmathulla, A. Hule, S.R. Naik, Evaluation of adaptogenic activity profile of herbal preparation, Indian J. Exp. Biol. 44 (2006) 574–579.
[46] R. Govindarajan, M. Vijayakumar, P. Pushpangadan, Antioxidant approach to disease management and the role of "Rasayana" herbs of Ayurveda, J. Ethnopharmacol. 99 (2005) 165–178. doi:10.1016/j.jep.2005.02.035.
[47] S.K. Mehta, N. Nayeem, N. Bains, Adaptogenic activity of methanolic extract of Buchanania lanzan leaves an experimental study in rat model, Der Pharm. Sin. 2 (2011).
[48] S.S. Agarwal, S.C. Khadase, G.S. Talele, Studies on Immunomodulatory activity of Capparis zeylanica leaf extracts, Int. J. Pharm. Sci Techn. 3 (2010) 887–892.
[49] H. Kolodziej, Antimicrobial, Antiviral and Immunomodulatory Activity Studies of Pelargonium sidoides (EPs® 7630) in the Context of Health Promotion, Pharmaceuticals 4 (2011) 1295–1314. doi:10.3390/ph4101295.
[50] M.L.K. Kalpesh Gaur, Comparative Screening of Immunomodulatory Activity of Hydro-alcoholic Extract of Hibiscus rosa sinensis Linn. and Ethanolic Extract of Cleome gynandra Linn, Glob. J. Pharmacol. 3 (2009) 85–89.
[51] D. Guan, Z. Zhang, Y. Yang, G. Xing, J. Liu, Immunomodulatory Activity of Polysaccharide from the Roots of Actinidia kolomikta on Macrophages, Int. J. Biol. 3 (2011) 3. doi:10.5539/ijb.v3n2p3.
[52] F. Jin, D. Cheng, J.-Y. Tao, S.-L. Zhang, R. Pang, Y.-J. Guo, et al. Anti-inflammatory and anti-oxidative effects of corilagin in a rat model of acute cholestasis, BMC Gastroenterol 13 (2013) 79. doi:10.1186/1471-230X-13-79.
[53] T. Saénz Rodriguez, D. García Giménez, R. de la Puerta Vázquez, Choleretic activity and biliary elimination of lipids and bile acids induced by an artichoke leaf extract in rats, Phytomedicine 9 (2002) 687–693. doi:10.1078/094471102321621278.
[54] B.N. Dhawan, R.C. Srimal, The Use of pharmacological techniques for the evaluation of natural products, UNESCO, Lucknow, 1984.
[55] H.J. Horn, Simplified LD 50 (or ED 50) calculations, Biometrics 12 (1956) 311–322.

[56] U.A. Bhosale, R. Yegnanarayan, P.D. Pophale, M.R. Zambare, R.S. Somani, Study of central nervous system depressant and behavioral activity of an ethanol extract of Achyranthes aspera (Agadha) in different animal models, Int. J. Appl. Basic Med. Res. 1 (2011) 104–108. doi:10.4103/2229-516X.91154.

[57] D.J. Faulkner, Marine natural products, Nat. Prod. Rep. 19 (2002) 1–48. doi:10.1039/B009029H.

[58] J. Jimeno, G. Faircloth, J.F. Sousa-Faro, P. Scheuer, K. Rinehart, New Marine Derived Anti-cancer Therapeutics – A Journey from the Sea to Clinical Trials, Mar. Drugs. 2 (2004) 14–29. doi:10.3390/md201014.

[59] M.H.G. Munro, J.W. Blunt, E.J. Dumdei, S.J.H. Hickford, R.E. Lill, S. Li, et al. The discovery and development of marine compounds with pharmaceutical potential, Prog. Ind. Microbiol. 35 (1999) 15–25. doi:10.1016/S0079-6352(99)80093-9.

[60] P. Proksch, R.A. Edrada, R. Ebel, Drugs from the seas - current status and microbiological implications, Appl. Microbiol. Biotechnol. 59 (2002) 125–134. doi:10.1007/s00253-002-1006-8.

[61] B. Haefner, Drugs from the deep: Marine natural products as drug candidates, Drug Discov. Today 8 (2003) 536–544. doi:10.1016/S1359-6446(03)02713-2.

[62] L. Mander, H.-W. Liu, Comprehensive Natural Products II: Chemistry and Biology, Newnes, 2010.

[63] L. Webster, K. Fakata, Ziconotide for chronic severe pain, Pract. Pain Manag. 2005.

[64] R.J. Lewis, K.J. Nielsen, D.J. Craik, M.L. Loughnan, D.A. Adams, I.A. Sharpe, et al. Novel {omega}-Conotoxins from Conus catus Discriminate among Neuronal Calcium Channel Subtypes, J. Biol. Chem. 275 (2000) 35335–35344.

[65] H. Kitagawa, T. Takenouchi, R. Azuma, K.A. Wesnes, W.G. Kramer, D.E. Clody, et al. Safety, pharmacokinetics, and effects on cognitive function of multiple doses of GTS-21 in healthy, male volunteers, Neuropsychopharmacology 28 (2003) 542–551. doi:10.1038/sj.npp.1300028.

[66] W.R. Kem, The brain alpha7 nicotinic receptor may be an important therapeutic target for the treatment of Alzheimer's disease: studies with DMXBA (GTS-21), Behav. Brain Res. 113 (2000) 169–181. doi:10.1016/S0166-4328(00)00211-4.

[67] L.F. Martin, W.R. Kem, R. Freedman, Alpha-7 nicotinic receptor agonists: potential new candidates for the treatment of schizophrenia, Psychopharmacology (Berl) 174 (2004) 54–64. doi:10.1007/s00213-003-1750-1.

[68] A. Olincy, K.E. Stevens, Treating schizophrenia symptoms with an alpha7 nicotinic agonist, from mice to men, Biochem, Pharmacol. 74 (2007) 1192–1201. doi:10.1016/j.bcp.2007.07.015.

[69] D.D. Xu, L. Waykole, J.V. Calienni, L. Ciszewski, G.T. Lee, W. Liu, et al. An Expedient Synthesis of LAF389, a Bengamide B Analogue, Org, Process Res. Dev. 7 (2003) 856–865. doi:10.1021/op0341162.

[70] R.L. Hornung, J.W. Pearson, M. Beckwith, D.L. Longo, Preclinical evaluation of bryostatin as an anticancer agent against several murine tumor cell lines: in vitro versus in vivo activity, Cancer Res. 52 (1992) 101–107.

[71] L.M. Schuchter, A.H. Esa, S. May, M.K. Laulis, G.R. Pettit, A.D. Hess, Successful treatment of murine melanoma with bryostatin 1, Cancer Res. 51 (1991) 682–687.

[72] D.S. Scherl, J. Afflitto, A. Gaffar, Influence of OAS-1000 on mediators of inflammation, J. Clin. Periodontol. 26 (1999) 246–251.

[73] J. Poncet, The dolastatins, a family of promising antineoplastic agents, Curr. Pharm. Des. 5 (1999) 139–162.

[74] M.N. Islam, M.N. Iskander, Microtubulin binding sites as target for developing anticancer agents, Mini Rev. Med. Chem. 4 (2004) 1077–1104.

[75] R.K. Singh, S.P. Tiwari, A.K. Rai, T.M. Mohapatra, Cyanobacteria: an emerging source for drug discovery, J. Antibiot. (Tokyo) 64 (2011) 401–412. doi:10.1038/ja.2011.21.

[76] R.E. Longley, D. Caddigan, D. Harmody, M. Gunasekera, S.P. Gunasekera, Discodermolide--a new, marine-derived immunosuppressive compound. II. In vivo studies, Transplantation 52 (1991) 656–661. doi:10.1097/00007890-199110000-00015.

[77] B. Zhang, A. Maiti, S. Shively, F. Lakhani, G. McDonald-Jones, J. Bruce, et al. Microtubule-binding drugs offset tau sequestration by stabilizing microtubules and reversing fast axonal transport deficits in a tauopathy model, Proc. Natl. Acad. Sci. U. S. A. 102 (2005) 227–231. doi:10.1073/pnas.0406361102.

[78] F. Loganzo, C.M. Discafani, T. Annable, C. Beyer, S. Musto, M. Hari, et al. HTI-286, a synthetic analogue of the tripeptide hemiasterlin, is a potent antimicrotubule agent that circumvents P-glycoprotein-mediated resistance in vitro and in vivo, Cancer Res. 63 (2003) 1838–1845.

[79] H.R. Hendriks, H.H. Fiebig, R. Giavazzi, S.P. Langdon, J.M. Jimeno, G.T. Faircloth, High antitumour activity of ET743 against human tumour xenografts from melanoma, non-small-cell lung and ovarian cancer, Ann. Oncol. 10 (1999) 1233–1240.

[80] R. Garcia-Carbonero, J.G. Supko, J. Manola, M. V Seiden, D. Harmon, D.P. Ryan et al., Phase II and pharmacokinetic study of ecteinascidin 743 in patients with progressive sarcomas of soft tissues refractory to chemotherapy, 2004. doi:10.1200/JCO.2004.02.098.

[81] Y. Takebayashi, P. Pourquier, D.B. Zimonjic, K. Nakayama, S. Emmert, T. Ueda, et al. Antiproliferative activity of ecteinascidin 743 is dependent upon transcription-coupled nucleotide-excision repair, Nat. Med. 7 (2001) 961–966. doi:10.1038/91008.

[82] H. Depenbrock, R. Peter, G.T. Faircloth, I. Manzanares, J. Jimeno, A.R. Hanauske, In vitro activity of aplidine, a new marine-derived anti-cancer compound, on freshly explanted clonogenic human tumour cells and haematopoietic precursor cells, Br. J. Cancer 78 (1998) 739–744.

[83] G. Faircloth, A. Hanauske, H. Depenbrock, R. Peter, C.M. Crews, I. Manzanares et al., Pre-clinical characterization of aplidine, a new marine anticancer depsipeptide, in: Proc. Am. Assoc. Cancer Res., 1997: p. 692.

[84] G. Faircloth, W. Grant, B. Smith, J. Supko, A. Brown, A. Geldof et al., Preclinical development of kahalalide F, a new marine compound selected for clinical studies, In: Proc. Am. Assoc. Cancer Res., 2000: p. 600.

[85] E. Kobayashi, K. Motoki, T. Uchida, H. Fukushima, Y. Koezuka, KRN7000, a novel immunomodulator, and its antitumor activities, Oncol. Res. 7 (1995) 529–534.

[86] R. Nakagawa, I. Serizawa, K. Motoki, M. Sato, H. Ueno, R. Iijima, et al. Antitumor activity of alpha-galactosylceramide, KRN7000, in mice with the melanoma B16 hepatic metastasis and immunohistological study of tumor infiltrating cells, Oncol. Res. 12 (2000) 51–58.

[87] A. Banchet-Cadeddu, E. Hénon, M. Dauchez, J.-H. Renault, F. Monneaux, A. Haudrechy, The stimulating adventure of KRN 7000, Org. Biomol. Chem. 9 (2011) 3080–3104. doi:10.1039/c0ob00975j.

[88] R. Nakagawa, K. Motoki, H. Ueno, R. Iijima, H. Nakamura, E. Kobayashi, et al. Treatment of hepatic metastasis of the colon26 adenocarcinoma with an alpha-galactosylceramide, KRN7000, Cancer Res. 58 (1998) 1202–1207.

[89] K. Alhanout, J.M. Brunel, S. Ranque, J.M. Rolain, In vitro antifungal activity of aminosterols against moulds isolated from cystic fibrosis patients, J. Antimicrob. Chemother. 65 (2010) 1307–1309. doi:10.1093/jac/dkq089.

[90] M. Zasloff, A.P. Adams, B. Beckerman, A. Campbell, Z. Han, E. Luijten, et al. Squalamine as a broad-spectrum systemic antiviral agent with therapeutic potential, Proc. Natl. Acad. Sci. U. S. A. 108 (2011) 15978–15983. doi:10.1073/pnas.1108558108.
[91] Technische Universität München. Doping Prevention: Diuretics and other masking agents. *Doping-prevention.sp.tum.de/*. Retrieved 29 April 2014, from http://www.dopingprevention.sp.tum.de/substances-and-methods/diuretics-and-other-masking-agents/diuretics-and-other-masking-agents.html.

Chapter 6

Vibrational Spectroscopy for Structural Characterization of Bioactive Compounds

Keisham S. Singh, Mahesh S. Majik and Supriya Tilvi
Bioorganic Chemistry Laboratory, CSIR-National Institute of Oceanography, Dona Paula, Goa, India

Contents

6.1 VIBRATIONAL SPECTROSCOPY

Vibrational spectroscopy is concerned with the transitions due to absorption or emission of electromagnetic radiation. These transitions appear in the range of 10^2 to 10^4 cm^{-1} and originate from the vibration of nuclei constituting the molecules. A chemical bond in a molecule can vibrate in many ways, and each vibration is called vibrational mode. There are two types of molecular vibrations, stretching and bending. A stretching vibration is characterized by movement along the bond axis with increasing or decreasing of the interatomic distances, whereas a bending vibration consists of a change in bond angles with respect to the remainder of the molecule. The two widely used spectroscopic techniques based on vibrational energy are the Raman spectroscopy and the infrared spectroscopy. Both methods give complementary information and are based on the

Analysis of Marine Samples in Search of Bioactive Compounds, Vol. 65. DOI: 10.1016/B978-0-444-63359-0.00006-9

fact that within any molecules the atoms vibrate with a few definite sharply defined frequency characteristics of that molecule. When a sample is eradiated to a beam of incident radiation, it absorbs energy at frequencies characteristic to that of the frequency of the vibration of chemical bonds present in the molecules. This absorption of energy by vibrating chemical bond results in an Infrared spectrum. Raman spectroscopy is a scattering phenomenon, and arises due to the difference between the incident and scattered radiation frequencies. It utilizes scattered light to gain knowledge about molecular vibration, which can provide information regarding the structure, symmetry, electronic environment, and bonding of the molecule.

Vibrational spectroscopy had been widely used in both academia and industry as an important tool in the structural characterization of compounds and analysis of impurities. In particular infrared spectroscopy in this field shows remarkable progress. One of the great advantages of Fourier transform infrared (FTIR) for natural products analysis over other techniques is that spectra can be obtained from almost any environment and from relatively small quantities of samples. Moreover, infrared (IR) techniques, when coupled with intensity measurement, may be used both for qualitative and quantitative analysis [1]. With the introduction of FTIR spectrometers and lasers as the source for recording the Raman spectra, vibrational spectroscopy has become an effective tool for the elucidation of molecular structure. IR spectroscopic techniques such as ATR (attenuated total reflectance) provide convenient measurement of most liquid samples such as edible oil and plant extract in routine use. Thus, there is no excessive sample preparation, so the use of hazardous solvents and reagents can be avoided [2]. For this reason, FTIR spectroscopy and other vibrational spectroscopic techniques can be taken into consideration as a "green analytical technique" for the analysis of edible fats and oils [3]. Discovery of FTIR and Raman spectroscopy techniques with recent innovations in Raman spectroscopy with laser as the source of light and the availability of a fluorosence free Raman spectra has renaissance the application of vibrational spectroscopy in various fields at a remarkable level [4].

Vibrational spectroscopy has advantages over the widely used NMR spectroscopy techniques in many aspects such as short analysis time, less quantity requirement, and the possibility for situ analysis of biological samples and nondestructive analysis mode. Although it gives less information on the exact chemical structure of the molecule, this technique is an indispensible and widely used fast analytical process and can be used as a fingerprint for identification and comparison of compounds [5], which allows this method to be used not only in structure characterization but also in other fields such as quality control in food industries [6] and forensic sciences [7]. In subsequent sections of the chapter, we will review the basic principles of vibrational spectroscopy and their applications in the structure characterization of compounds, particularly with reference to the biologically significant compounds.

6.2 INTRODUCTION TO INFRARED (IR) AND RAMAN SPECTROSCOPY

IR spectroscopy is an absorption phenomenon in which photons matching the energy difference of vibrational levels are absorbed, resulting in transition from ground to an excited state. IR spectra results from the absorption of energy by vibrating chemical bonds, primarily stretching and bending modes. The infrared spectra are readily obtained by passing infrared radiation through the sample under investigation that possesses permanent or induced dipole moment and determining the fraction of the incident radiation absorbed by the sample at a particular energy. The absorbed energy corresponds to the frequency or wave number that is characteristic for the bond present in the sample. Thus, it allows qualitative identification of the nature of bond types present in the sample. From FTIR spectral analysis, we can discriminate between closely related compounds with sensitivities of impurities even in low concentration. Several innovations have been introduced in the IR techniques over the last decades such as Fourier transform spectroscopy, diffused reflected infrared fourier transform spectroscopy (DRIFT), and ATR-FTIR spectroscopy. Pereira *et al.* [8] had used the ATR-FTIR technique to determine the composition of the different phycocolloids from dried ground seaweed without having to prepare tablets of KBr. IR spectroscopy is widely used for the identification of functional groups in organic and inorganic materials, quantitative determination of compounds in mixtures, nondestructive method of analysis of biological samples, and determination of structure isomers and stereochemistry, among other uses.

Unlike IR spectroscopy, which is an absorption phenomenon, Raman spectra arises from the inelastic scattering of monochromatic light, usually from a laser in the visible, near infrared, and near UV range. Raman spectroscopy provides information on chemical structures, electronic environment, and bonding of molecules derived from this scattered light. Use of vibrational spectroscopy for identification of functional groups and microbes is well documented [9]. However, nonavailability of biological compounds of high purity in bulk quantities and fluorescence associated with the conventional Raman spectra are the major difficulties in undertaking such investigation with Raman spectroscopy. To avoid strong fluorescence and to enhance the sensitivity and selectivity, the Raman measurement can be combined with the absorption of the sample on nanometal particles. This technique is called SERS (surface-enhanced Raman scattering), and allows enhancement of the Raman signals from molecules absorbed on the metal surface by six orders of magnitude or more due to electromagnetic and chemical factors [10]. This recent innovation in Raman spectroscopy instrumentation, which provides fluorescence free spectra, has emerged, making Raman spectroscopy a promising tool for the investigation of biological samples.

Another technique, Resonance Raman spectroscopy (RRS), has also been introduced. With this technique, the energy of the excited photon matched approximately with the energy required for electronic transition [11]. This technique provided analysis of biological samples without degradation [12], which is usually encountered when lasers near the visible range are used for excitation. Raman spectroscopy has several advantages; for example, the technique is excellently suitable for analysis of alcoholic or aqueous solutions in biological systems and *in situ* characterization of samples that are not desirable for moisture-sensitive IR spectroscopy. Further, strong bands can be derived from nonpolar bonds, thus making it suitable for studying C-C modes in the ring. In addition, Raman spectra mostly show simpler patterns because of often-missing overtones and combination modes.

6.2.1 Fourier Transform IR Spectrometers

Fourier transform spectrometers have recently replaced dispersive instruments for most applications due to their superiority in speed and sensitivity. Using a Michelson interferometer and Fourier transform (FT) processors for analysis of scattered light, an enhancement of the recorded intensity by about two orders of magnitude compared to classical dispersive spectrometers is observed [13]. They have greatly extended the capabilities of infrared spectroscopy and have been applied to many areas that are very difficult or nearly impossible to analyze by dispersive instruments. Instead of viewing each component frequency sequentially, as in a dispersive IR spectrometer, all frequencies are examined simultaneously in FTIR spectroscopy. The advantages of FTIR quickly gave rise to adaptation of attenuated total reflectance (ATR) [14] and diffuse reflectance (DRIFTS) [15] for the rapid and reproducibility analysis of liquids and solids. A major advance in the kinds of samples that can be analyzed by FTIR occurred with the introduction of FTIR microscopy in the early 1980s [16]. This allows the IR spectra of samples as small as 10 microns in diameter to be obtained.

6.2.2 Analytical Information

Qualitative: The combination of the fundamental vibrations or rotations of various functional groups and the subtle interactions of these functional groups with other atoms of the molecule results in the unique, generally complex IR spectrum for each individual compound. IR spectroscopy is mainly used in two ways: structural elucidation and compound identification.

Quantitative: IR technique, when coupled with intensity measurements, may be used for quantitative analysis.

Structural Elucidation: Because of complex interactions of atoms within the molecule, IR absorption of the functional groups may vary over a wide range. However, it has been found that many functional groups give characteristic IR absorption at a specific, narrow frequency range regardless of their relationship with the rest of the molecule. Generalized tables of the positions and relative intensities of absorption bands have been established and used to determine the functional groups present or absent in a molecule.

6.2.3 Functional Groups

Functional group determination is the most important application of infrared spectroscopy and it is the key interest of the majority of people working on natural products. Different atoms or groups show characteristic vibrational bands in the infrared region, thus structure information of the compound under investigation can be derived. Multiple functional groups may absorb at one particular frequency range, but a functional group often gives rise to multiple-characteristic absorption. Thus, the spectral interpretations should not be confined to one or two bands—the whole spectrum should be examined. A precise determination of structure of an unknown substance will require other complementary techniques like nuclear magnetic resonance (NMR), mass spectrometry (MS), and so on. Many compilations on the characteristics group frequency shown by functional groups are available [17]. Table 6.1 shows characteristic frequency,

TABLE 6.1 Characteristic IR Bands Shown by Various Groups [17]

Frequency Range	Intensity[a]	Group	Types of Vibrations[b]	Remarks
3550–3500	m	O–H	str.	Free OH, carboxylic acid
3460–3623	m	O–H	str.	Free OH, alcohols
3600–3100	m	O–H	str.	Water of crystallization
~3520	s	N–H	str.	Primary amide
~3500	m	N–H	str. (asym)	Primary amine, free NH
3500–3060	m	N–H	str.	Secondary amine, free NH
~3400	s	N–H	str.	Primary amide
3350	m	N–H	str.	Primary amide bonded
3075–3030	w-m	C–H	str.	C–H of aromatic ring
2960	s	C–H	str. (asym)	C-methyl
2925	s	C–H	str. (asym)	>CH_2, methylene, Ar-CH_3
2900–2880	w	C–H	str.	C–H, methine
2900–2705	w	C–H	str.	–C(=O)H, aldehyde
2580	s	S–H	str.	Thiol, free thiol
2400	w	S–H	str.	Thiol H-bonded
1500–470	s	C=S	str.	–N–C=S
1468	s	C–H	scissoring	Alkane, –CH_2–, –CH_3
1460	m	C–H	bend (asym)	–CH_3
1460–1400	s	C–O	sym. str.	–COO– carboxylate
1050–1020	s	S=O	str.	>S=O, sulfoxide

(Continued)

TABLE 6.1 Characteristic IR Bands Shown by Various Groups [17] *(cont.)*

Frequency Range	Intensity[a]	Group	Types of Vibrations[b]	Remarks
1200–1145	s	S=O	str.	RO_5O_2R', sulfonic ester
960–930		N–O	str.	oxime
2270	vs	N=C=O	asym str.	isocyanate
2260–2240	w	–C≡N	str.	nitrile
2260–2240	w	C≡C	str.	RC≡CR′, acetylenes
1725–1705	s	C=O	str.	ketone
2160–2120	s	N≡N	str.	Azide
1770	s	C=O	str.	γ-lactone
1720–1700	s	C=O	str.	–COOH, aliphatic carboxylic acids
1700–1670	s	C=O	str.	–CONHR, secondary amide, Amide I
1690–1670	s	C=O	str.	$CONH_2$, primary amide, amide
1680–1630	s	C=O	str.	Secondary amide (solid)
1675	s	C=S	str.	thio ester
1670	w	C=N	str.	Aliphatic oxime
1670–1620	s	C=O	str.	Primary amide (solid) H-bonded 2 bands: Amide I
3645–3500	var	O–H	str.	Free O–H, oxime
3640–3623	m	O–H	str.	Free –OH, alcohols
1650–620	s	N–H	def.	Primary amide (solid) Amide II
1350–1200	m	N–H	def.	Secondary amide, Amide III
1220–1020	m	C–N	str.	Aliphatic amines
950–810	s	C–O	str.	epoxide
650	s	C–Br	str.	Bromo-derivatives
550–450	vw	S–S	str.	disulfide
1450–1400	w	–N=N	str.	Azo
~1455	s	C–H	scissoring	Alicyclic $-CH_2-$
1625–1585	m	C=C	str. skeletal	Aromatic C=C
1580–1520	s	C=N	asym. def.	Amine salts, pyrimidines
1370–1250	m	C–O	str.	lactone
1150–1270	s	C–O–C	asym. str.	Aliphatic ether
1320–1210	s	C–O	str.	Carboxylic acid

[a]key: s, strong; m, medium; w, weak; var, varied; vs, very strong
[b]str., stretching; asym, asymmetric; sym, symmetric; def, deformation

shown by various functional groups and groups of atoms. For systematic evaluation, the IR spectrum is commonly divided into three regions based on absorption by different atoms or groups.

6.2.3.1 The Functional Group Region, 4000 to 1300 cm^{-1}

The appearance of strong absorption bands in the region of 4000 to 2500 cm^{-1} usually comes from stretching vibrations between hydrogen and some other atoms with a mass of 19 or less. The O–H and N–H stretching frequencies fall in the 3700 to 2500 cm^{-1} region, with various intensities. Hydrogen bonding has a significant influence on the peak shape and intensity, generally causing peak broadening and shifts in absorption to lower frequencies. The CH stretching bands occur in the region of 3300 to 2800 cm^{-1}.

6.2.3.2 The Fingerprint Region, 1300 to 910 cm^{-1}

Absorptions in this region include the contributions from complex interacting vibrations, giving rise to the generally unique fingerprint for each compound. A good match between the IR spectra of two compounds in all frequency ranges, particularly in the fingerprint region, strongly indicates that they have the same molecular structures.

6.2.3.3 The Aromatic Region, 910 to 650 cm^{-1}

The IR bands in this region do not necessarily come from the aromatic compounds, but the absence of strong absorption in the 910 to 650 cm^{-1} region usually indicates the lack of aromatic characters. The out-of-plane bending of ring C–H bonds of aromatic and heteroaromatic compounds gives rise to strong IR bands in the range between 910 and 650 cm^{-1}.

6.3 SELECTION RULES IN VIBRATIONAL SPECTROSCOPY

As mentioned earlier, infrared spectroscopy is absorption phenomenon whereas Raman spectroscopy is scattering phenomenon. Both give complementary information but different selection rules are applied. The number of vibrational modes of a linear molecule is $3n - 5$ and for nonlinear it is $3n - 6$ (n = no. of atoms). Considering molecules in the harmonic oscillator gives the allowed transition, $\Delta v = \pm 1$. This is the most prominent and gives an intense band. This is caused by transition of $v = 0$ to $v = 1$, which is known as fundamental vibration. When the molecule is treated more accurately in an inharmonic oscillator, the selection rules relax and allow in principle all vibrational transitions: $\Delta v = \pm 1, \pm 2, \pm 3\ldots$ etc.

A vibration is infrared active, if the vibration is accompanied by a change in electric dipole moment or inactive if there is no change of the dipole moment with changing the bond distances and angles.

A molecular vibration is Raman active if there is a change in polarizabilty during vibrations.

In other words a change in polarizability (induced dipole moment) is required to observe Raman scattering.

In general, homonuclear diatomic molecules such as H_2, Cl_2, and so on are IR inactive since these molecules do not have dipole moment but are Raman active. In general, polar bonds provide intense absorption bands in IR whereas nonpolar bond such as C–C, C=C, C=S, C–S, –S–S–, –N=N–, –C–X, and –CN, which exhibit more polarizability, give strong Raman signals [18]. Usually, if a band is strong in IR it may be weak in Raman and vice versa [19]. In a molecule having a center of symmetry, IR active transitions are Raman inactive and Raman active modes are IR inactive. Vibrations of molecules or ions having an inversion center *i* are either infrared or Raman active. Thus, IR and Raman spectroscopy together allow us to establish the existence of a center of symmetry in a molecule quite easily. Since water has strong dipole moment, the presence of a small quantity of water could interfere with the spectrum. Therefore IR spectroscopy is most commonly applied to the analysis of dried samples and not suitable for *in situ* studies of plants and aqueous samples. On the contrary, water has only week Raman scattering properties and consequently, this technique is more suitable to perform aqueous and *in situ* studies of samples.

6.4 STRUCTURE CHARACTERIZATION BY VIBRATIONAL SPECTROSCOPY

Spectroscopy techniques are indispensable to natural product researchers for structure determination of molecules. More importantly, when the molecule is biologically significant, it is worthy to determine the structure for further understanding of the physiochemical properties of the molecule. Several spectroscopic tools were employed for this purpose such as NMR, mass spectrometry, and IR spectroscopy. Among these techniques, IR is one of the most simple to perform with relatively simple instrumentation and provides structural information in a short time and at a relatively low sample quantity. Several compounds and biological significant molecules have been characterized by IR spectroscopy. Over the years, IR had been used for structure determination and identification of biologically significant molecules such as polysaccharides [20], terpenes [21], and flavanones [22]. Raman spectroscopy gives complementary information to IR but needed an optimized instrumentation. Both techniques were used although infrared technique was paid more attention by organic chemists because of low cost, ready detection of functional groups, and much simpler instrumentation. Uses of IR and Raman spectroscopy as tools for identifying bacteria and characterization of biologically relevant compounds were reported, and a few critical reviews were published in recent years [4,23]. In the following sections we discuss and review the uses of these two important tools for the structure determination of biologically active compounds from marine sources.

6.4.1 Polysaccharides

Polysaccharides are macromolecules made up of many monosaccharides joined together by glycosidic bonds. Polysaccharides differ not only in the nature of their components but also in the length of the chain and in the amount of chain branching. Several polysaccharides are found in marine organisms mainly in brown algae and seaweeds. They have wide value in medicine, food, and cosmetics [24–26].

6.4.1.1 Chitin and Chitosan

Chitin is the second most abundant biopolymer found in nature and most widespread aminopolysaccharides widely distributed in crustaceans (shrimps, crabs, and lobsters). They are isolated from the exoskeleton of crustaceans, mollusks, insects, and certain fungi, but the main commercial sources of chitin are crab and shrimp shells [27]. Potential and usual applications of chitin, chitosan, and their derivatives are estimated to be more than 200 [28]. Chitin and chitosan have shown various biological activities such as having analgesic, antitumor, hemostatic, hypocholesterolemic, antimicrobial, and antioxidant properties [29]. Chitin may be regarded as a cellulose derivative with an acetamido group at carbon 2. It is a linear polymer consisting mainly of β-(1→4)-linked 2-acetamido-2-deoxy-β-D-glucopyranose units (denoted as A residue) and partially of β-(1→4)-linked 2-amino-2-deoxy-β-D-glucopyranose (denoted as D residue) and it is insoluble in water and common organic solvents. Chitosan is the deacylated derivative of chitin that contains high enough amounts of D-units to allow polymers to dissolve in aqueous acidic solutions (pH < 6.0). The structures of real chitin and chitosan as well as ideal chitin and chitosan are shown in Figure 6.1(a). Chitosan is a fully or partially deacylated chitin and can be obtained by alkali hydrolysis of chitin.

Depending on the source, chitin occurs in three allomorphs, namely α-, β-, and γ-forms, and differs in the arrangement of chains in the crystalline regions: α-chitin has a structure of antiparallel chains [30], β-chitin has intrasheet hydrogen-bonding by parallel chains [31], and γ-chitin, being a combination of α- and β-chitin, has both parallel and antiparallel structures [31]. The most abundant and stable form is α-chitin, which is usually obtained from exoskeleton of crustaceans, such as shrimp and crab, β-chitin is obtained from squid pens, and γ-chitin is obtained from fungi and yeast [32].

Several analytical tools such as X-ray [33], Mass [34], and IR including ATIR [35] and DRIFT [36] were used for the analysis of chitin and chitosan. Among these, IR spectroscopy is one of the most important and widely used analytical techniques for the characterization of chitin and chitosan, which is well documented in literature [37]. The α- and β-chitin can be differentiated by infrared spectroscopy from the analysis of the characteristic bands. The description and interpretation of the infrared spectra of these two forms of chitin were published by many scientists [38]. The main interest is that the C=O stretching

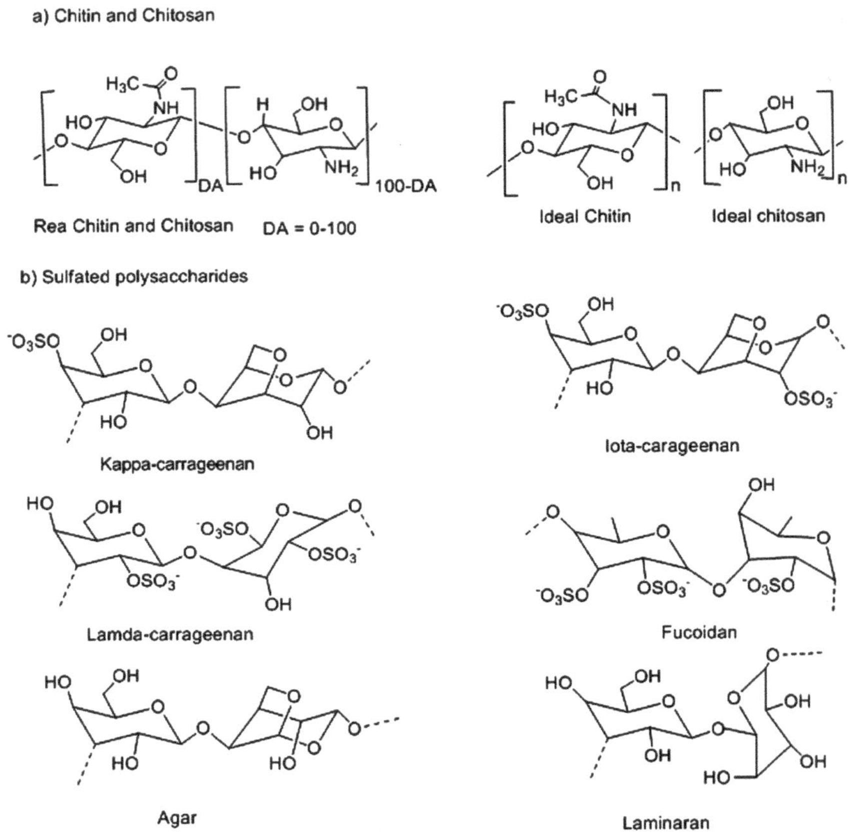

FIGURE 6.1 Chemical structures of chitin, chitosan, and sulfated polysaccharides.

band of amide moiety appeared between 1700 and 1500 cm^{-1}, which shows different signatures for α- and β-chitin. In the case of α-chitin, the amide I band is split into two components at 1660 and 1630 cm^{-1} [39], whereas for β-chitin, it appears as a single band at 1630 cm^{-1}. The amide II band for α-chitin appears at 1558 cm^{-1} whereas for β-chitin, the band appears at a little higher frequency, at 1562 cm^{-1} [32]. The C-H deformations of the β-glycosidic band have also been used as a characteristic marker for these two allomorphs. In β-chitin this band is observed at 890 cm^{-1} and is observed at 895 cm^{-1} in α-chitin. The infrared spectrum of β-chitin usually shows the presence of two additional bands for CH_x deformations at about 1455 and 1374 cm^{-1}, which are not observed in α-chitin [32].

Chitin can be converted to a more soluble and useful derivative chitosan by alkali treatement [40]. The degree of *N*-acetylation is one of the most important chemical parameters capable of influencing the performance of chitosan and chitin in many of their applications [41]. IR spectroscopy is the most convenient, and has drawn a considerable attention among the various analytical techniques developed for determination of the degree of acylation [42]. It is a convenient and relatively quick technique and allows the degree of acylation (DA) values of chitin and chitosan to be determined on the basis of absorption ratios [43].

There is growing interest in the chemical modification of chitin and chitosan to improve their solubility and applications [44]. The most important application of IR spectroscopy in this respect is the structural analysis of chemically modified forms of chitin or chitosan. Typical structural analysis of chitin and chitosan derivatives by FTIR spectroscopy involves the following steps [37]: (1) FTIR analysis of chitin/chitosan, (2) FTIR analysis of the chemical reagent(s) used in the reaction, (3) FTIR analysis of the chitin/chitosan derivative obtained, (4) identification of differences between the spectra, and (5) interpretation of results.

6.4.1.2 Sulfated Polysaccharides

Sulfated polysaccharides comprise a complex group of macromolecules with a wide range of critical biological activities including anticoagulant, antiviral, antitumor, anti-inflammatory, and antioxidant [45–47]. Fucoidan, a sulfated polysaccharide, has been reported from marine algae including *Fucus vesiculosus*, *Laminaria japonica*, and *Undaria pinnatifida*, and it was reported that they had varieties of biological activities such as antioxidant, anti-HIV, and antitumor activities [48]. Although sulfated polysaccharides are found in some other mammals and invertebrates, marine algae are the most important source of sulfated polysaccharides. The chemical structure of the polymers varies according to the algae species [49].

Among the sulfated polysaccharides, carrageenan and agar (Figure 6.1(b)) are the principal sulfated polysaccharides produced by red seaweed. The main difference between the highly sulfated carrageenans from the less sulphated agars (3% sulfated) is the presence of D-galactose and anhydro-D-galactose in carrageenans and of D-galactose, L-galactose, or anhydro-L-galactose in agars. The structure of the various types of carrageenans is defined by the number and position of sulphate groups, the presence of 3,6-anhydro-D-galactose, and conformation of the pyranose ring. FTIR spectroscopy is one widely accepted tool for the characterization of sulfated polysaccharides. The characteristic band for ester sulfate groups were found between 1375 cm^{-1} and 1255 cm^{-1} and the band for the skeleton of galactans and agar appeared in the region around 1072 cm^{-1} and 890 cm^{-1}, respectively [50]. The ethereal bands of 3,6-anhydro-a-L-galactopyranose was usually observed in the region around 931 cm^{-1} and the sulfate groups on the C-4 of galactose was observed at 855 cm^{-1}. The band for 6-sulfate group of D-galactose unit appeared at 820 cm^{-1} [50].

The IR spectroscopy has been used for the ready identification between the two major sulfated polysaccharides, namely agar and carrageenans. In carrageenans the characteristic sulfated ester bands between 1210 and 1260 cm^{-1}

appeared much stronger than in agar [51]. Using FTIR several sulfated polysaccharides were indentified from red algae and brown algae. Pereira *et al.* studied infrared spectral of several seaweeds such as *Gelidium corneum*, *Pterocladiella capillacea*, *Sargassum vulgare*, and *Padina pavonica*, and identified the main sulfated polysaccharide contain in these seaweeds [52]. The IR spectra of brown algae (*Saccorhiza polyschides* and *Undaria pinnatifida*, respectively) exhibited a strong broad band around 1220 to 1260 cm^{-1}, which is a characteristic band for a sulphate ester group (S=O). Similarly, infrared spectra of algae such as *Padina pavonica* and *Sargassum vulgare* exhibited a broad band around 1195 to 1237 cm^{-1} for *Padina* and 1210 to 1280 cm^{-1} for *Sargassum* assignable to an S=O group, which is a characteristic component in fucoidan and other sulfated polysaccharides except alginate in brown seaweeds. Thus, based on infrared spectroscopy, they established that these seaweeds, namely *Padiana pavonica*, *sagassum valgare*, *Saccorhiza polyschides*, and *Undaria pinnatifida*, contain sulfated polysaccharide (mainly fucoidan), while the *Gelidium comeum* and *Ptercladiella capillacea*, which showed relatively weak bands of sulfate ester, contain mainly agar as polysaccharides [52].

6.4.2 Amino Acids and Peptides

Amino acid and peptides constitute an important structural unit of biologically active compounds. They are organic compounds composed of amine ($-NH_2$) and carboxylic (–COOH) functional groups along with side chain "R" specific to each amino acid. There are 20 different naturally occurring amino acids that are the constituents of peptide and protein structures. Marine organisms are rich sources of biologically active peptides such as dipeptide and cyclic peptide [53,54]. Several studies and reviews have been reported on the infrared and Raman spectroscopic regarding the structure of amino acids and peptides [55]. In most of the studies with FTIR spectroscopy, the absorption band of the amino acid C=O group appeared around 1675 to 1680 cm^{-1}, except by L-tyrosine amide, which shows a vibration frequency of the C=O group at 1705 cm^{-1}. The shift of absorption band for C=O groups in L-tyrosine was due to intermolecular hydrogen bonds between the N-atom of the amino acid group and phenolic –OH group [56]. Linder *et al.* [57] have reported IR spectra of five natural amino acids, namely valine, proline, isoleucine, phenylalanine, and leucine. In all the five spectra, the position of the C=O group as well as the –OH group absorption vibration appeared very similar to each other. The amino acid and peptide absorption band appeared in the region 3400 cm^{-1}, which is attributed to O–H and N–H bond stretching vibration. The asymmetric valence vibration of the ammonium (NH_3^+) group appeared in the region of 3030 to 3130 cm^{-1}. The ammonium group deformation vibrations are located together with the absorption characteristic of carboxylate ion in the region of 1500 to 1600 cm^{-1}, whereas the deformation absorption bands associated with carboxylate group were observed in the region of 1610 to 1660 cm^{-1}.

Generally, IR spectra of amino acids contain several bands from the side chains. All the side chain absorption bands are overlapped with the absorption bands of other side chains except the two side chain moieties, SH of Cys (1550–2600 cm^{-1}) and the carbonyl group of protonated carboxyl group (1710–1790 cm^{-1}) for which the absorption is in the spectral region, free from overlapping absorption of the other groups. FT-Raman spectroscopy is suitable to study specific amino acid structure elements such as S-S and S-H groups of cysteine, cystein, aromatic rings of tryptophan, tyrosine, phenylalanine, and imidazole ring of histidine, respectively. Additionally, indentifcation of compounds containing disulfide bonds can be readily obtained by using FT-Raman spectroscopy because the S-S stretching band is prominent in the Raman spectra while the IR spectra is very week due to its nonpolar nature [58]. FT-Raman spectroscopy can also perform a conformational study of the disulfide bridge. The Raman spectra of rice globulin displayed three conformations for the disulfide bond of the cystein [59]. The major conformation was *gauche-gauche-gauche* showing Raman band at 512 cm^{-1}, which is the preferred conformation in many proteins with –S-S-bridge [60]. The weak bands seen at 525 and 540 cm^{-1} were assigned to minor conformation *gauche-gauche-trans* and *trans-gauche-trans*, respectively.

FTIR spectroscopy was also used for the study of DNP-derivative of amino acids (DNP = dinitrophenyl). Friedberg *et al.* [61] reported the finding on the DNP-derivatives of amino acids with IR spectroscopy. All the DNP derivatives examined by them exhibits the band at 1587 cm^{-1} and 1408 cm^{-1}, hence they inferred that these compounds are dipolar ions. Usually, free amino acids have bands at 1449 and 1370 cm^{-1} due to antisymmetric and symmetric CH_3 and CH_2 deformation motions, respectively. All the free amino acids except L-alanine and all DNP-derivatives showed bands at 1333 cm^{-1} due to the CH_2 wagging motion. The bands in the region of 1330 to 1370 cm^{-1} were difficult to interpret since symmetrical aromatic NO_2 stretch bands usually appeared intense in this region. While many free amino acids show a band at 2564 cm^{-1} assigned to the C-H stretching motion [62], this band is missing in almost all DNP-derivatives. On the other hand, the spectra of DNP-derivatives with the exception of DNP-L-tyrosine and DNP-L-arginine possess a band at 1724 cm^{-1}, which is not exhibited by free amino acids. Thus, they have demonstrated that IR spectra analysis could provide a ready characterization of DNP-amino acids and able to distinguish them from the free amino acids.

Moore *et al.* [63] reported an analysis of the vibrational spectra of β polypeptide *viz.* poly (L-alanine) [$(Ala)_n$], poly(L-alanylglycine) [$(Ala\text{-}Gly)_n$]. Both the α- and β-$(Ala)_n$ displayed IR bands at 1096 cm^{-1}, 1050 cm^{-1}, and 909 cm^{-1}, while IR spectra of $(Ala\text{-}Gly)_n$ shows corresponding bands at 1100 cm^{-1}, 1050 cm^{-1}, and 920 cm^{-1}.

6.4.3 Proteins

Proteins are linear biological polymers for which the monomer units are amino acids. Proteins are made up of 20 different amino acids, each distinguished by the identity of R groups. They are an essential nutrient for the human body as an energy source but also as a building agent. Until 1997, approximately 20,000 protein structures were solved primarily using X-ray crystallography, and some of these using newer techniques of multidimensional NMR spectroscopy. Although X-ray crystallography provides the most detailed information concerning position of individual atoms in the protein structure, it is not possible for all protein to form a quality crystal for analysis. NMR is an alternative to study protein in solutions, however, interpretation of NMR spectra for a huge protein is a very cumbersome process. IR spectroscopy is one of the oldest techniques for the analysis of secondary structure of polypeptides and proteins [64,65]. It provides information about the secondary structure content of proteins, unlike X-ray crystallography and NMR spectroscopy, which provide information about the tertiary structure.

Vibrational spectroscopy is used to study the secondary structure of proteins by inspection of the frequencies at which the amide bonds absorb infrared radiation. Information about chemical structure and environment of the amino acid side chain of the molecule can be deduced from the spectral parameters, band position, band width, and absorption coefficient. The band position of a particular amino acid is influenced by hydrogen bonding, protonation state, and coordination of cation. In general, hydrogen bonding lowers stretching vibration but increases the bending vibrations [66]. The polypeptide and protein repeat units give rise to nine characteristics IR absorption bands—Amide A, Amide B, Amide I, Amide II, and Amide III-VII—which are encountered in the secondary structure of proteins (Table 6.2) [67]. Among these absorption bands, three bands (Amide A, Amide I, and Amide II), which characterize N–H stretch, C=O stretch, and N–H bending vibrations, are particularly important for the study of

TABLE 6.2 Characteristic IR Bands of Peptide Bond [67]

Amide Types	Wavenumber (cm^{-1})	Specific Marker Groups
Amide A	3300	N–H stretching
Amide B	3100	N–H stretching
Amide I	1600–1690	C=O stretching (80%)
Amide II	1480–1575	C–N stretching (40%), N–H bending (60%)
Amide III	1229–1301	C–N stretching (40%), NH bending (30%)
Amide IV	625–767	O–C–N bending
Amide V	640–800	N–H bending
Amide VI	537–606	C=O bending
Amide VII	200	Skeleton torsion

peptides and proteins. Since both the C=O and the N–H bonds are involved in the hydrogen bonding that takes place between the different elements of secondary structure, the locations of both the Amide I and Amide II bands are sensitive to the secondary structure content of a protein.

Ambrose and Elliot had demonstrated for the first time that infrared spectroscopy could provide information on the secondary structure of protein [64]. They showed that an empirical correlation existed between the frequency of the so-called Amide I and Amide II absorption of proteins and predominant secondary structure of protein. They have demonstrated that proteins that were predominantly α-helical exhibited strong Amide I and Amide II absorption bands in aqueous solution in the range between 1652 and 1657 cm^{-1} and 1545 and 1551 cm^{-1}, respectively, while protein with β-sheet antiparallel structure exhibited between 1628 and 1635 cm^{-1} and 1521 and 1525 cm^{-1}, respectively. The Amide II band for parallel β-sheet appeared with a little higher frequency at 1530 to 1550 cm^{-1} [68]. Thus, based on IR-spectrum, either a protein structure is α-helical or the β-sheet could be readily determined.

This Amide I band can be easily identified by (H→D) experiment as the amide backbone proton in β-sheet exchange slowly at room temperature, and the difference in frequency between the Amide I and Amide I′ (in D_2O) band is large and shifted to 1647 to 1654 cm^{-1}. The Raman spectrum can also be used for diagnostics of the Amide I band, which is observed as a single strong band at 1670 cm^{-1}. The Amide II band can also be used to distinguish between the parallel and antiparallel β-sheet structures. Peptides and proteins with an antiparallel β-sheet structure have strong Amide II bands between 1510 and 1530 cm^{-1}, while a parallel β-sheet structure is found at somewhat higher frequencies (1530–1550 cm^{-1}) [68]. Protein secondary structure can also be determined based on analysis of the Amide III band. Although Amide III bands are quite weak, they can be identified by deuterium exchange, because the bands shifted to lower frequency at 960 to 1000 cm^{-1} upon deuteration. When using Amide III in combination with Amide I, the assignment may allow a distinction between β-sheet and disordered protein structure, which is not generally possible by Amide I band alone [68]. The unordered or random structure is generally assigned to the band near 1665 cm^{-1} in Raman, while in IR the band appears at 1645 cm^{-1}. Table 6.3 shows a correlation between amide absorption bands and protein structures [69,70]. It should be stressed that the correlation shown in Table 6.3 are guidelines only, because there are proteins and peptides that contain secondary structures that absorb outside the frequency range given in the table [69].

Vibrational spectroscopy had been used for the analysis of natively unfolded proteins. They are proteins that do not fold into compact defined conformation. Raman spectroscopy was used for characterizing the changes in conformation states and for identifying conformational constituents of natively unfolded proteins [71]. Raman spectroscopy has several important advantages in characterizing vibrational spectra of natively unfolded proteins because the spectra can be obtained in dilute solution, and characterization can be applied

TABLE 6.3 Position of the Amide Bands Depending on the Secondary Structure of Protein or Peptide [69]

Conformation	Vibration Frequency, ν (cm^{-1})		
α-helix	Amide A (N–H stretch)	Amide I (C=O stretch)	Amide III (N–H deformation)
β-sheet	3290–3300	1648–1660	1540–1550
β-turns	3280–3300	1625–1640 (strong) 1690 (weak)	1520–1530
Random (unorder)	3250	1662–1660 (deuterated 1640–1648)	1520–1545
3_{10}-Helix		1660–1670	
Aggregated strands		1610–1628	

to proteins in varied states: aqueous solution, amorphous aggregates [72], solid [73], and crystals [74,75]. Since H_2O has weak bands in Raman spectra, this technique obviates the need to use D_2O exchange experiment and reduces the errors inherent in large background solvent substractions. Recently, Anderson *et al.* [71] demonstrated the characterization of natively unfolded protein: α-Synuclein with the help of Raman spectroscopy and characterized the ensemble of secondary structures present in this protein.

6.4.4 Alkaloids

Alkaloids are groups of nitrogen-containing bases and many of them are used for medicinal purposes. They constitute important bioactive natural products obtained from both terrestrial plants and marine organisms. Various spectroscopic techniques had been employed for their identification and structure determination. In spite of the fact that alkaloids contain a broad range of different chemical structures, only a few reports are available on the vibrational spectral measurements of natural substances. It is well known that vibrational spectroscopy could be used for quick fingerprint analysis as well as in the assignment of configuration of certain alkaloids such as quinolizidine and piperidine derivatives [76].

FTIR spectroscopy had been used for identification of amines and assignment of configuration of quinolizidine alkaloids. Until the early 1970s, IR spectroscopy was an important technique for the indentification and assignment of quinolizidine alkaloids. Quinolizidine alkaloids are reported from several marine organisms, particularly from marine sponges. The family of these alkaloids (bisquinolizidine alkaloids) were reported from various marine sponges like *Petrosia Serata* [77], *Oceanapia sp.* [78], *Haliclona exigua* [79], *Petrosia similis*

[80], and *Xestospongia exigua* [81], and exhibited several biological properties including vasodilative [82] and anti-HIV [80]. These alkaloids usually display characteristic bands in the IR spectrum in the region 2800 to 2650 cm^{-1}, known as Bohlmann bands [83]. The bands have been correlated with the stretching of carbon–hydrogen bonds that are anticoplanar to the nitrogen lone pair, and this correlation has been used extensively for the configuration assignment of quinolizidine alkaloids (Figure 6.2(a)).

Bohlmann found that it is possible to distinguish between the *cis* and *trans* configurations of the quinolizidine moiety. The *trans*-fused systems exhibited a characteristic series of bands in the infrared spectrum between 2700 and 2800 cm^{-1}, which were absent in *cis*-fused quinolizidine systems. The Bohlmann band is very useful for ready determination and recognition of the quinolizidine ring [84]. The alkaloids memosamycin (Figure 6.2(b)) and Renieramycines H were isolated from marine sponge *Haliclona cribricutis* [85]. The preliminary structural information of these alkaloids was derived from the IR spectra. The

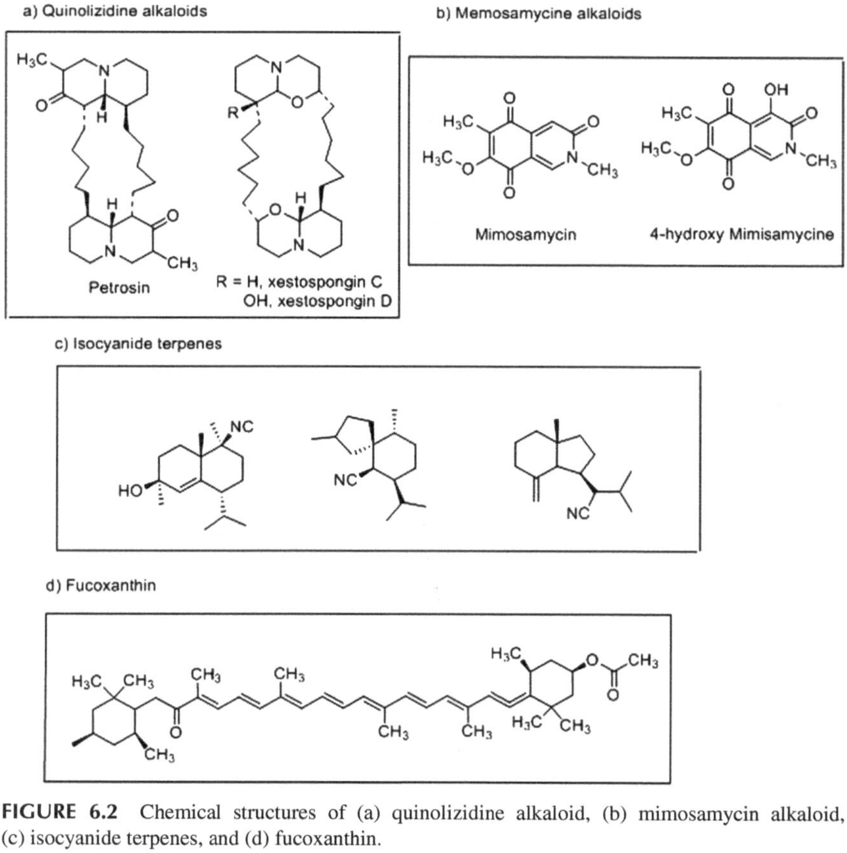

FIGURE 6.2 Chemical structures of (a) quinolizidine alkaloid, (b) mimosamycin alkaloid, (c) isocyanide terpenes, and (d) fucoxanthin.

IR spectra of memosamycin displayed multiple bands in the region 1600 to 1700 cm^{-1}, suggesting the presence of quinone, amide, and C=C groups. The presence of hydroxyl groups was ruled out since the compound had no IR absorption band in the region 3500 cm^{-1}. The Renieramycin H showed IR bands at 3425, 1726, 1690, and 1648 cm^{-1}, which were assigned for the presence of hydroxyl, ester, quinone, and amide groups [85].

6.4.5 Terpenes

Terpenes are a large and diverse class of organic compounds derived from the five carbon isoprene entity. Combination and modification of these isoprene units leads to diverse structures with different chemical and biological properties. Several terpenes have been isolated from marine sources and investigated their biological properties [86]. A majority of terpenes were characterized on the basis of NMR and mass spectral data with information from infrared spectroscopic data analysis. In the recent past vibrational spectroscopy had been employed for characterization of terpenes and their derivatives. It provides rapid characterization of terpenes and also allows recognization of certain structure isomers of monoterpenes such as α-and γ-terpenine. The three groups of monoterpenes (acyclic, monocyclic, and bicyclic) can be noticed based on infrared spectral data though it is difficult to identify individual spectral contribution. A cyclic monoterpene can be recognized readily from the strong stretching vibration of C=C bonds at about 1670 cm^{-1} in Raman spectrum.

Raman spectroscopy can also be used to differentiate between monocylic and bicyclic terpenes. The Raman spectrum of the ring deformation vibration of monocyclic terpenes was observed between 740 and 760 cm^{-1} whereas in bicyclic terpenes it is appeared at 645 and 666 cm^{-1}, allowing a ready differentiation between these two terpenes. FT-Raman and ATR-IR spectra of structure isomers like α-terpenine and γ-terpenine differ significantly, therefore it is possible for the differentiation of these isomers. In the Raman spectrum, the C=C stretching vibration for α-terpenine appeared at 1611 cm^{-1} in contrast to γ-terpenine where it appeared at 1701 cm^{-1}. In ATR-IR spectroscopy, the γ-terpenine can be identified by CH and CH_2-wagging vibrations at 781 and 947 cm^{-1}, whereas α-terpenine displayed only one strong absorption band at 823 cm^{-1}.

Several terpenes isolated from marine sponges and soft corals were investigated by vibrational spectroscopy. The cembrane and briarane-type of diterpenoids isolated from *Sarcophyton stolidotum* and *Ellisella robusta* containing a γ-lactone are readily recognized by IR and UV-vis [87]. The IR spectra of these diterpenoid displayed strong absorption bands at around 1756 to 1770 cm^{-1}, characteristic of γ-lactones. Terpenoids with γ-lactone groups is quite common in marine organisms, and IR spectroscopy is often used to confirm the presence of this group. Thus, the presence γ-lactone in Junceellolide M isolated from gorgonian coral *Junceella gemmacea* and cembranoid diterpenes from *Clavularia koellikeri* and *sinularia inelegans Tixier-durivault* were confirmed by the appearance of IR peak at around 1760 to 1781 cm^{-1} [88,89]. IR spectroscopy was used for the ready confirmation of the presence of certain groups such as isocyanide in secondary metabolites. Thus, the presence of isocyanide group in the terpenes (Figure 6.2(c)) isolated from nudibranch was readily established by the observation of an isocyanide band at 2139 cm^{-1} in the IR spectrum of the terpenoids [21].

Carotenoids, which belong to the category of tetraterpenoids, are classified into the subclasses carotenes (hydrocarbons derivatives) and xanthophyll (oxygented derivative of hydrocarbons). There are nearly 600 carotenoids naturally present, and more than 250 are from marine origin [90]. Among the several carotenoids isolated from marine sources, fucoxanthin is quite common, and contributes more than 10% of total carotenoid production in nature, especially in the marine environment [91]. Fucoxanthin is mainly isolated from the marine brown seaweeds such as *Phaeophyceae* and diatoms (*Bacillariophyta*; Figure 6.2(d)) [92]. It has been reported that fucoxanthin and its metabolites have great antidiabetic, antioxidant, anticancer, and antiphotoaging properties [93]. IR spectral analysis of carotenoids was reported in a few published papers [94]. Rajauria *et al.* [95] reported FTIR analysis of carotenoids isolated from brown seaweed, *Himanthalia elonga.* The IR spectrum shows absorption bands at 3441 cm^{-1}(OH group), 3031 to 2849 cm^{-1} (C–H stretch), 1929 (allene), 1732 cm^{-1} (C=O acetate), 1649 cm^{-1} (C=O conjugated), 1603 to 1450 cm^{-1} (CH_2 stretch), 1335, 1259, and 1246 cm^{-1} (C–O acetate), 1201 to 958 cm^{-1} (*trans*-distributed –C=C–). Thus, based on IR spectrum analysis and comparison with the standard, they demonstrated that the carotenoid isolated from this brown seaweed is fucoxanthin (Figure 6.2(d)).

In recent years, Raman spectroscopy had been used for the analysis of carotenoids. In the Raman spectra of carotenoids, the characteristics bands were observed between 1500 and 1550 cm^{-1} and 1150 and 1170 cm^{-1} ranges due to C=C, C–C and C–CH_3 stretching vibration of the polyene chain, allowing the qualitative identification of carrotenoid present in the sample [96,97]. The Raman spectrum of carotenoids displayed strong bands between 1500 and 1550 cm^{-1} and 1150 and 1170 cm^{-1} ranges due to C=C and C–C stretching vibration of the polyene chain. Schuzl *et al.* [98] have shown the wave number location of these bands is strongly dependent on the length of the carotenoid chain. Generally, carotenoids with 11, 9, 8, 7 conjugated C=C bonds have their characteristic bands at about 1510, 1524, 1530, 1536 cm^{-1}, respectively, which can be used for identification purposes.

6.4.6 Flavonoids

Flavonoids are hydroxyl, methoxyl, and glycoside derivatives of flavone possessing several biological properties. The best known and examined ones include quercetine, rutin, and morin. Flavanoids form complexes with metals ions; the compound such as quercetin, which has the strongest affinities to form a metal complex, exhibits the greatest physiological strength [99]. Flavonoids contain two aromatic rings bonded by a C3 unit (central pyran ring) and can be divided into several classes based on the oxidation state of the pyran ring and on the characteristic color; for example, flavones, flavonols, flavonols glycosides, falvanones, flavanone glycosides, anthocyanins, catechins, and chalcones.

Flavonoids with antioxidants were reported from marine seaweeds [100]. Characterization of the majority of these flavonoids were performed by NMR and mass spectral analysis and little is known for IR spectral analysis of flavonoids. IR spectroscopy analysis of flavonoids appeared in a few limited published papers [100,101]. Systematic studies of vibrational spectroscopy on flavonoids have been conducted since the early 1950s, and most of these studies have been focused on the hydroxyl and carbonyl absorption frequencies [101]. However, with the technical advances in the last two decades, the application of vibrational spectroscopy has become much more relevant in the field of flavonoid analysis, and the techniques have been applied for characterization of flavonoids containing much complexity [102]. Merlin *et al.* [103] reported Raman spectroscopy investigation on phenyl-2-benzopyrillium (anthocyanadins) and discussed some characteristic spectral features on the influence of glycosylation on the benzopyrilium part of the Flavonoids molecule (rings A and C). The spectral range between 500 and 900 cm^{-1} strongly depends on the glycosylation pattern and can be used to determine the position of glycosylation. When the anthocynadin is glycosylated at C-5 position, significant perturabations of spectral features between 640 and 750 cm^{-1} was noticed. For anthocynadin gylcosylated at position 3 of a Carbon ring exhibited a strong resonance Raman signal close to 540 cm^{-1} while 3,5-diglycoside anthocynadins have the strongest feature in the lower frequency range, close to 630 cm^{-1} [104].

Vibrational spectroscopy had been used to study hydroxyl and carbonyl groups of flavonoids. The infrared spectra of carbonyl group bands were observed in the frequencies 1619, 1652, and 1649 cm^{-1}, for fisetin, hesperetin, and flavones, respectively. Brigs and Colebrook studied the shifts of valence vibration bonds of the flavonoids carbonyl group according to the number of OH groups and their position in the pyrone and benzene rings. Raman spectra is much less complex than the IR spectra of same molecule and because of this reason Raman spectroscopy has been gradually taking over IR spectroscopy, although it is common to use both techniques as complements of each other. Vibrational spectroscopy is seldom used alone for the characterization of flavonoids; most studies are accompanied by other spectroscopic techniques such as NMR and mass analysis or quantum chemical computations [105].

6.4.7 Sterols

Recent biological studies have demonstrated that sterols and sterol derivatives possess biological activities. Cytotoxic sterols against various cancer cell lines were isolated from the Formosan brown algae [106] and the phytosterols aringasterol was isolated from several brown algae [107]. The usefulness of IR spectroscopy for the identification and analysis of steroids is limited to the determination of the presence of absorbing functional groups and their stereochemical relationships. The functional groups that contain polar bonds absorbing in the region between 1600 and 400 cm^{-1} are easily identified by IR spectroscopy. Additionally, IR spectroscopy yields information concerning both saturated as well as unsaturated functional groups. Since many steroids exhibit similar UV spectra, IR has been widely used as a complementary tool in structure elucidation. IR spectra of a few functional groups that are common in sterols are discussed next. These absorption data are useful in assignment of position or stereochemistry of IR active functional groups.

6.4.7.1 Functional Group Analysis

Carbonyl stretching of ketones

The carbonyl stretching frequency (in solution) is in the range of 1704 to 1722 $cm^{-1}$for steroids containing saturated ketones at all positions in rings A–C (i.e., C-l to C-12) and the side chain. In a few instances, when keto functionality is located in the 5-membered D-ring, it gives rise to distinctive IR absorption bands in the range of 1738 to 1749 cm^{-1}. Boul's research group showed a correlation between the location of the carbonyls and intensity of methylene absorption in the positional isomers of D-ring monoketones [108]. The classical example includes 15-, 16-, and 17-ketosterol (Figure 6.3(a)).

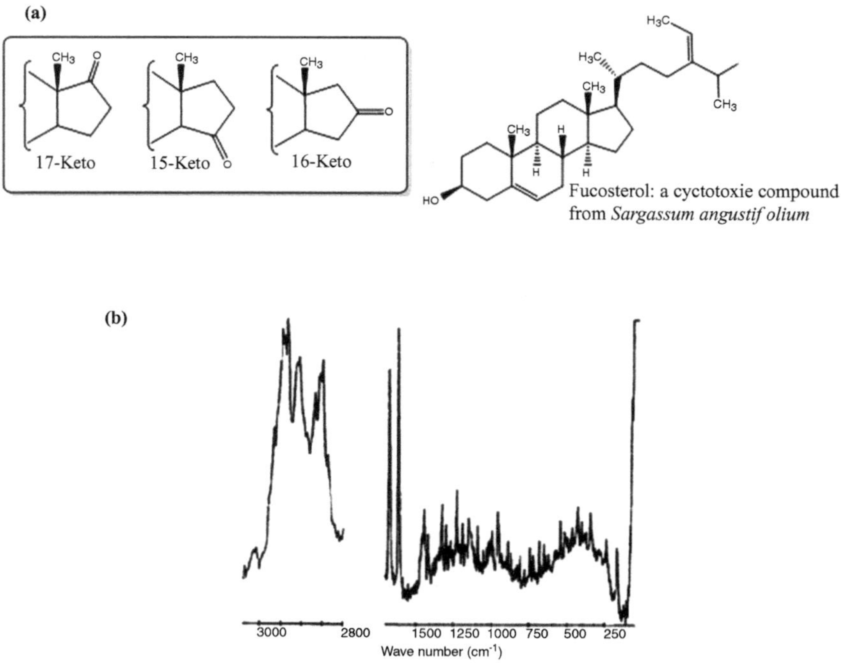

FIGURE 6.3 Chemical structure of (a) Fucosterol and (b) Raman spectrum of Δ^4-Cholesten-3-one.

The C=O frequency of 17- and 15-ketosterol are very close, whereas the intensities of adjacent CH_2 absorption bands are sufficiently different to distinguish between these two isomers. Similarly, the IR spectrum of 16-ketosterol shows a band that is twice as intense as those of 15-ketosterol and 17-ketosterol. Holly *et al.* demonstrated the IR spectrum of pregn-4-ene-3, 20-dione and found that the position and intensity of the C=O stretching frequency of ketones are also altered by the presence of adjacent polar groups of pregn-4-ene-3, 20-dione derivatives [109]. The conjugation of a carbonyl with a double bond (α, β-unsaturated ketones) lowers the frequency of the C=O stretching band and increases its intensity. The IR absorption due to carbonyl stretching of conjugated acyclic and six-membered ring ketones of steroids is in the range of 1660 to 1685 cm^{-1}. The characteristic C=C stretching bands appearing around 1600 cm^{-1} however are also present in the IR spectra of α, β-unsaturated ketones. Hence, due to the overlapping of C=O and C=C absorption in the stretching region often makes IR spectra of α, β-unsaturated keto steroids more complex to interpret. Nevertheless, Jones *et al.* [110] has described the solution to this problem by measuring the individual intensities of overlapping bands, provided that the characteristic intensities of various carbonyl groups in a steroid are known. Recently, Cue *et al.* [111] isolated four sterols from Cowpea (*Vignasinensis K.*) seeds, whereas 7-ketositosterol showed peaks at 3440, 2990, 2850, 1680, 1480, 1395, 1285, and 1080 cm^{-1} in its IR spectrum.

Esters: Carboxylic acid esters and lactones

Carbonyl stretching absorptions of carboxylic acid esters and acetates appear at 1737 to 1742 cm^{-1} and 1733 to 1739 cm^{-1}, respectively. These functional groups are best recognized by the presence of an additional strong, sharp C=O stretching absorption band at 1135 to 1155 cm^{-1} for carboxylic acid esters and at 1200 to 1250 $cm^{-1}$for acetates. Equatorial acetates yield a single band and axial acetates yield doublet or triplet bands in this region. The positions and intensities of the C–O stretching absorptions of esters are affected by their conformations and neighboring unsaturation. Introduction of unsaturation [112], as in enol acetates and phenolic acetates, shifts the C=O stretching frequency to 1748 to 1762 cm^{-1}.

Amides

The absorption due to CO–N stretching appeared in the 1630 to 1680 cm^{-1} region. However, Kastelic-Suhadolc, found that the C=O bands of lactamides of steroidal alkaloids show an unusual but characteristic band at 1730 to 1733 cm^{-1}. Thus, α-hydroxyl groups, when present in steroidal alkaloid structures such as tomatine, shift the amide C=O band from its usually observed frequency (1640 cm^{-1}) to 1730 cm^{-1}, which is a more characteristic peak of esters than amides [113].

Methylene Groups: C=C and C=CH Stretching

Absorptions corresponding to C=C stretching of isolated ethylene groups are observed in the region of 1620 to 1650 cm^{-1}. They are moderately weak in most monounsaturated compounds and have not been widely used to elucidate steroid structures. Conjugation of ethylenic groups displaces these absorptions to 1575 to 1630 cm^{-1}. The C–C stretching absorption is most useful in confirming the presence of an α, β-unsaturated carbonyl. Absorptions due to stretching of olefinic and aromatic C=C–H stretching occur in the region of 3300 to 3100 cm^{-1}. The C–H stretching absorptions of saturated CH, CH_2, and CH_3 appeared near 2850 to 2950 cm^{-1}. These absorptions can be useful if IR spectra are obtained at high resolution [114]. The C–H bending absorptions due to methylene groups adjacent to carbonyl groups can provide guidance for the determination of structures of α, β-unsaturated ketones.

Hydroxyl Absorptions

Polyhydroxy sterols have been found in various groups of marine organisms (e.g., algae, porifera, coelenterata, bryozoa, molluska, echinodermata, arthropoda, tunicata, and chordata); a predominent position is occupied by porifera (sponges). D'Auria *et al.* [115] published a full review of the marine polyhydroxy steroids in 1993. In this paragraph, we had discussed the IR spectrum of hydroxylated sterols. The absorptions corresponding to O–H stretching appear in the region of 3200 to 3650 cm^{-1}. Primary alcohols exhibit this absorption at 3640 cm^{-1} whereas the corresponding absorptions for secondary and tertiary alcohols are at 3630 and 3620 cm^{-1}, respectively. Intramolecularly hydrogen-bonded hydroxyls exhibit sharp bands at 3450 to 3600 cm^{-1} that are not affected by dilution. Solvent utilized in recording IR spectra have a strong effect on absorption frequency. Solvent with hydrogen bonding broadens O–H and usually lowers the frequency of these absorptions. Thus, an intermolecularly hydrogen-bonded hydroxyl gives a broad band at 3200 to 3550 cm^{-1} that disappears upon dilution.

Khanavi *et al.* [116] had identified fucosterol (Figure 6.3(a)) from the hexane fractions of methanolic extract of brown alga *Sargassum angustifolium* as a cytotoxic compound. Fucosterol, the most abundant sterol in brown algae, has been shown to be antioxidant [117], antidiabetic, antifungal, and cytotoxic [118]. The infrared absorbance spectra of cholesterol and fucosterol in CH_2Cl_2 showed the characteristics bands at 3600 cm^{-1}(O–H bond vibration of free hydroxyl group), between 2950 and 2850 cm^{-1} (due the asymmetric vibrations of C–H bonds of CH_2 and CH_3 groups), 1444 cm^{-1} (C–C and C–H vibration), 1372 cm^{-1} (interaction of OH bending and C–O stretching in a CO–H group), and 1049 cm^{-1} that corresponds to secondary C–O vibration in C–O–H group. Recently, Bouzidi *et al.* [119] demonstrated the utility of FTIR spectroscopy for determination of total sterol contained in brown algae *Bifurcaria bifurcata*, *Cladostephu shirsitus*, *Dictyota dichotoma*, and *Cystoseira sedoides*, globally determined as fucosterol, which is the major sterol contained in these algae.

6.4.7.2 Raman Spectra for Steroids

IR and Raman peak in the spectra of a compound is a good evidence for the presence of a center of symmetry. This situation does not apply in the case of a steroid, however, because these molecules have no symmetry at all, except for the identity; and so all molecular vibrations are both IR and Raman active. Moreover, since steroids are relatively large organic molecules, containing 30 or more atoms, there should be at least 80 peaks in each spectrum because of the nonlinearity, a 3n-6 number of vibrational modes can be observed. The numerous peaks that appear in the Raman spectrum of a typical steroid, Δ^4-cholesten-3-one, is shown in Fig. 6.3(b). In 1973, Schrader and Steigner published the results of an exhaustive study of the spectra of about 80 commercially available, crystalline steroids chiefly related to sexual and adrenal hormones [120]. The spectra were recorded using both Hg-arc and Ar^+ ion laser excitation. They examined five major spectral regions: (1) 4000–2800, (2) 2800–1350, (3) 1350–800, (4) 800–200, and (5) 200–20 cm^{-1}. In the 4000–2800 cm^{-1} range, they found that O–H stretching modes had a low Raman intensity as that of the C–H stretching modes of acetylenic groups. The C–H bending modes at 1450 cm^{-1} are of medium intensity. This region is particularly useful in indicating the type of steroid skeleton concerned. The stretching vibrations of the steroid skeleton and the wagging, twisting, and rocking deformations of the C–H bonds appear in the 1350 to 800 cm^{-1} region with varying intensities. The region between 800 and 200 cm^{-1} displays deformation peaks characteristic of the steroid skeleton. A band near 725 cm^{-1} suggests the presence of an aromatic steroid, and a peak at 500 + 5 cm^{-1}, with about half the intensity of the peak near 1450 cm^{-1}, is considered to be indicative of a five-saturated steroid. The very low energy region (200–20 cm^{-1}) is where the translational and rotational modes of the rigid skeleton are expected. Finally, the peaks are particularly intense in the spectra when there are distinctly different electron polarizabilities along the chief axes of inertia. Moreover, a convenient intensity order is observed: aromatic ring >C=C–C=C– (conjugated) >C=C– (unconjugated) > saturated steroids. Schrader and Steigner [120] subjected their accumulated spectral data (peak positions and intensities) to computer analysis in an attempt to locate bands specific to certain steroidal structures, and they proposed an extensive structural correlation scheme shown in Figure 6.4.

6.4.8 Siderophores

Siderophores are iron(III)-chelating compounds produced by microorganisms such as bacteria and fungi that uptake iron ions under iron-deficient conditions and prevent precipitation of iron oxyhydroxides in the natural environment [121]. While the structures of hundreds of terrestrial siderophores are known, relatively few structures of siderophores produced by marine bacteria have been elucidated. Knowledge of the cellular origins of such compounds is desirable, in order to establish whether the sponge or associated micro-organisms are

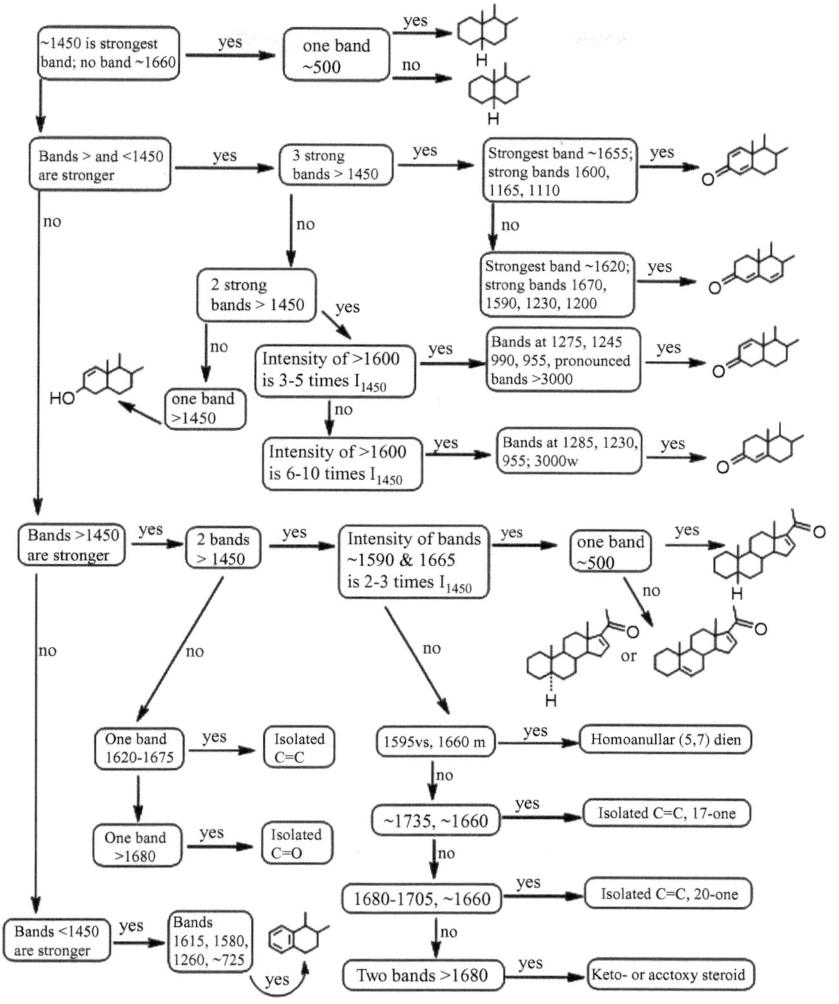

FIGURE 6.4 Correlation Chart of Raman spectral data for elucidation of steroid skeleton (Lit. Schrader and Steigner, 1973 [120]).

producing the compounds of interest. Siderophores are important molecules, which are produced on demand by bacteria in order to obtain sufficient iron necessary for their growth. Siderophores contain three bidentate functional groups (hydroxamate, catecholate, α-hydroxycarboxylate), which form an extremely stable hexadentate, octahedral complex with iron (III). The properties such as coordination chemistry and local bonding environment) of siderophores could be evaluated using vibrational (IR and Raman) and X-ray absorption spectroscopy (XAS).

In general, siderophores are classified by the ligand used to chelate Ferric ion (Fe^{+3}), wherein the major groups are catecholate and hydroxamate types. Since these siderophores possess unique structural features characterized by the presence of a particular organic functional group, the IR spectroscopy is a widely used tool for the characterization of individual types of siderophores. The catecholate, phenolate, hydroxamate, and carboxylate are different in functionality present in each type of siderophores. Two steps are involved in chemical characterization of siderophores. The first step is the determination of the type of siderophore using a qualitative test (assay); for example Csaky's assay [122] for hydroxamate type, Arnow's assay [123] for catecholate type, and so on. The second step involves recording the IR spectra of the active fractions. Some of the case studies for characterization of siderophores are discussed in this section. Supanekar *et al.* had described the identification of catecholate type of siderophores [124] produced by *Klebsiella pneumoniae.* The infrared spectrum analyses of the siderophore showed the presence characteristic peaks for OH, saturated alkanes, a secondary amide linkage, and ethereal linkage at 3351, 2963, 1654, and 1033 cm^{-1}, respectively. Actis *et al.* had isolated Anguibactin, a catecholate type of siderophore [125] from *Vibrio anguillarum 775*, whose IR spectrum showed vibrational modes due to the catecholate moiety at 3400 cm^{-1} for OH stretching, 1463 cm^{-1} corresponding to catechol ring, 1632 cm^{-1} for an amide C=O stretching, and doublet at 3200 cm^{-1} for NH stretching.

Tank *et al.* [126] described biochemical characterization of pyoverdin, a hydroxamate type of siderophore from *Pseudomonas sp.*, using Csaky's reagent and FTIR spectroscopy. The supernatant from culture of *Pseudomonas sp.* was tested with Csaky's reagent wherein the solutions turn pink upon addition of ferric perchlorate indicating the presence of a hydroxamate group of siderophore. Furthermore, FTIR spectrum showed bands at 3189, 2360, 1602, 1566, 780, and 530 cm^{-1} similar to the band shown by standard hydroxamate (i.e., *N*-phenyl benzohydroxamic acid (PBHA)), thus confirming the presence of siderophore with a hydroxamate functional group.

6.4.9 Melanin

Melanins are a diverse group of macromolecules produced in living organisms by oxidative polymerization of various phenolic substances [127]. In nature, they occur widely in plants, animals, and microbial tissues as melanoproteins [128]. Melanins act as photoprotectants against UV and visible light, free-radical scavengers, antioxidants, and so on [129]. Usually alkaline extraction of pigment followed by precipitation with acid produced melanoproteins. However, the term melanin should be reserved solely for alkali insoluble pigment that does not contain protein [130]. Melanin pigments from various strains of marine microorganisms is attractive but still in its infancy compared to terrestrial melanin. Melanin can be extracted as such by alkali or water, but cleavage from protein moiety renders the melanin insoluble in all solvents [130]. They are insoluble in common organic solvents, aqueous acids, and water. Because of their poor nature of solubility in various organic solvents and water, there are limited tools available to study melanin pigment. Among these tools, FTIR is one of the most important and widely used in the study of melanin pigments.

Jones *et al.* [131] described a preliminary investigation of melanin obtained from the fungus *Sclerotinia Sclerotioroum* a couple of decades ago using IR spectroscopy. Following this, Russells *et al.* [132] studied IR spectra of alkali insoluble melanin obtained by acid hydrolysis of melanoprotein extracted from the pigments of ground sclerotia. Melanin from ground sclerotia displayed typical IR absorption bands at 1615 cm^{-1} and 1240 cm^{-1}, weaker bands at 3450, 1700, and 830 cm^{-1}, and a broad unresolved band in the range of 3300 to 2400 cm^{-1}. The spectrum also showed a strong C–H band at 2850 to 2960 cm^{-1} and 1450 cm^{-1}, a strong absorption band at 1700 cm^{-1}, and weak bands of unknown origin at 950, 750, and 700 cm^{-1} [132]. Tarangini and coworkers studied FTIR spectroscopy on melanin pigment isolated from marine *Pseudomonal sp.*, and about the functional group present in the pigment [133]. They found that absorption peak at 3373 cm^{-1} corresponds to the presence of –OH and $-NH_2$ groups. The spectrum also displayed a characteristic absorption peak 1625 cm^{-1} that was attributed to aromatic rings C=C and C=O of a carboxylate group. On the basis of infrared spectra they were able to establish the major functional groups present in these macromolecules.

6.5 CONCLUSION

This chapter presents the basic concepts and application of vibration spectroscopy (IR and Raman) in the characterization of compounds with references to bioactive compounds isolated from marine sources. The chapter compiled the results of the selected published articles on the characterization of biologically significant polysaccharides, alkaloids, terpenes, flavonoids, melanin, sterols, and siderophores. A brief discussion on IR spectral analysis of amino acid, peptide, and proteins were discussed. Usefulness of IR spectroscopy in the identification of different allomorphs of chitin, distinguishing structure isomers of terpenes, and recognition of secondary structures of protein were highlighted. Identification of functional groups present in memosamycin alkaloids by IR spectroscopy and uses of Bohlmann's band for identification and configuration assignment of quinolizidine alkaloids were discussed. The chapter also discussed the IR spectral characterization of bioactive fucoxanthin and flavonoids, isolated from marine sources. We believe that this chapter may be of interest to general readers and researchers working in natural product sciences both from academia and industry. We also acknowledge that several published works on the topic that deserved to be cited may have been excluded due to page limitation.

ACKNOWLEDGEMENTS

Authors are grateful to the CSIR (*OCEAN FINDER*) and DST (SR/FT/CS-001/2010), India as well as the EU (Marex program: FP7-KBBE-2009-3-245137) for financial support. Thank is due to Director, CSIR-NIO for the constant encouragement. MSM thanks to the CSIR-NIO for the award of QHS.

REFERENCES

[1] F. Xu, J. Yu, T. Tesso, F. Dowell, D. Wang, Qualitative and quantitative analysis of lignocellulosic biomass using infrared techniques: A mini review, Appl. Energy 104 (2013) 801.

[2] L.E. R.-Saona, M.E. Allendorf, Use of FTIR for rapid authentication and detection of adulteration of food, Ann. Rev. Food Sci. Tech. 2 (2011) 1.

[3] J. Moros, S. Garrigues, M. de la Guardia, Vibrational spectroscopy provides a green tool for multi-component analysis, Trends in Anal. Chem. 29 (2010) 578.

[4] K. Maquelin, C. Kirschner, L.-P. C.-Smith, N. van den Braak, Ph.H. Endtz, D. Naumann, G.J. Puppels, Identification of medically relevant microorganisms by vibrational spectroscopy, J. Microbiol. Methods 51 (2002) 255.

[5] M. Grube, R. Rutkis, M. Gavare, Z. Lasa, I. Strazdina, N. Galinina, U. Kalnenieks, Application of FT-IR Spectroscopy for Fingerprinting of *Zymomonas mobilis* Respiratory Mutants, Spectroscopy: An International Journal 27 (2012) 581.

[6] P. Rösch, M. Harz, M. Krause, R. Petry, K.-D. Peschke, H. Burkhardt, O. Ronneberger, A. Schüle, G. Schmauz, R. Riesenberg, A. Wuttig, M. Lankers, S. Hofer, H. Thiele, H.W.J.P. Motzkus, Online monitoring and identification of bioaerosol (OMIB), in: J. Popp, M. Strehle (Eds.), Biophotonics, Visions for Better Health Care, Wiley-VCH, Weinheim:, 2006, pp. 596.

[7] K. Virkler, I.K. Lednev, Raman spectroscopic signature of semen and its potential application to forensic body fluid identification, Forensic Sci. Int. 193 (2009) 56–62.

[8] L. Pereira, Identification of phycocolloids by vibrational spectroscopy, in: A.T. Critchley, M. Ohno, D.B. Largo (Eds.), World Seaweed Resources -An authoritative reference system, ETI Information Services Ltd., 2006.

[9] M. Harz, P. Rösch, J. Popp, Vibrational Spectroscopy 22, A Powerful Tool for the Rapid Identification of Microbial Cells at the Single-Cell Level, Cytometry Part A 75A (2009) 104.

[10] J.A. Creighton, in: R.J. Clark, R.E. Hester (Eds.), Spectroscopy of Surface, Wiley, New York, 1988, pp. 21.

[11] R.S. Das, Y.K. Agrawal, Raman spectroscopy: Recent Advancements, techniques and applications, Vibrational Spectroscopy 57 (2011) 163.

[12] T.A. Matitoli, A. Hoffmann, D.G. Sockalingum, Application of near-IR Fourier transform resonance spectroscopy to the study of photosynthetic proteins Spectrochim, Acta A 49 (1993) 785.

[13] B. Schrader, in: B. Schrader (Ed.), Infrared and Raman Spectroscopy. Methods and Applications, VCH, Weinheim, 1995.

[14] N.J. Harrick, Internal Reflection Spectroscopy, John Wiley & Sons Inc., (1967) 342.

[15] M.P. Fuller, P.R. Griffiths, Diffuse reflectance measurements by infrared Fourier transform spectrometry, Anal. Chem. 50 (1978) 1906.

[16] R.G. Messerschmidt, M.A. Harthcock, Infrared microscopy theory and Applications, United States, Mercel Dekker, New York, NY, (1988).

[17] C.N.R. Rao, Chemical application of Infrared spectroscopy, Acedemic press, New York, (1963).

[18] R.J. Dijkstra, F. Ariese, C. Gooijer, U.A. Brinkman, Raman spectroscopy as a detection method for liquid-separation techniques, Trends Anal. Chem. 24 (2005) 304.

[19] J.T. Pelton, L.R. McLean, Spectroscopic Methods for Analysis of Protein Secondary Structure, Anal. Biochem. 277 (2000) 167.

[20] G. Jiao, G. Yu, J. Zhang, H.S. Ewart, Chemical Structures and Bioactivities of Sulfated Polysaccharides from Marine Algae, Mar. Drugs 9 (2011) 196.

[21] E. Manzoo, M.L. Ciavatta, M. Gavagnin, E. Mollo, Y.-W. Guo, G. Cimino, Isocyanide terpene metabolites of Phyllidiella pustulosa, A Nudibranch from the south China Sea, J. Nat. Prod. 67 (2004) 1701.

[22] L. Briggs, L. Colebrook, Infrared spectra of flavonones and flavones. Carbonyl and hydroxyl stretching and CH out-of-plane bending absorption, Spectrochim. Acta 18 (1962) 939.

[23] D. Naumann, H. Labischinski, P. Giesbrecht, The characterization of microorganisms by Fourier-transform infrared spectroscopy (FT-IR), in: W.H. Nelson (Ed.), Modern Techniques for Rapid Microbiological Analysis, VCH Publishers, New York, 1991, pp. 43.

[24] F. Shahidi, J.K.V. Arachchi, Y.-J. Jeon, Food applications of chitin and chitosans, Trends Food Sci. Technol. 10 (1999) 37.

[25] A.D. Martino, M. Sittinger, M.V. Risbud, Chitosan: A versatile biopolymer for orthopaedic tissue-engineering, Biomaterials 26 (2005) 5983.

[26] P.K. Dutta, S. Tripathi, G.K. Mehrotra, J. Dutta, Perspectives for chitosan based antimicrobial films in food applications, Food Chem. 114 (2009) 1173.

[27] M. Rinaudo, Chitin and chitosan: Properties and application, Prog. Polym. Sci. 31 (2006) 603.

[28] M.N.V.R. Kumar, A review of chitin and chitosan applications, React. Funct. Polym. 46 (2000) 1.

[29] M.N.V.R. Kumar, R.A. Muzarelli, C. Muzarelli, H. Sashiwa, A.J. Domb, Chitosan chemistry and pharmaceutical perspectives, Chem. Rev. 104 (2004) 6017.

[30] R. Minke, J. Blackwell, The structure of α-chitin, J. Mol. Biol. 120 (1978) 167.

[31] M.K. Jang, B.G. Kong, Y.I. Jeong, C.H. Lee, J.W. Nah, Physicochemical characterization of α-chitin, β-chitin and γ-chitin separated from natural resources, J. Polym. Sci. Part A Polym. Chem. 42 (2004) 3423.

[32] S.P. C.-Filho, D.D. Britto, E. Curti, F.R. Abreu, M.B. Cardoso, M.V. Battisti, P.C. Sim, R.L. Lavall, Extraction, structures and properties of α- and β-chitin, Quim. Nova 30 (2007) 644.

[33] G.L. Clark, A.F. Smith, X-ray diffraction studies of chitin, chitosan and derivatives, J. Phys. Chem. 40 (1937) 863.

[34] S. Trombotto, C. Ladavière, F. Delolme, A. Domard, Chemical preparation and structural characterization of a homogenous series of chitin/chitosan oligomers, Biomacromolecules 9 (2008) 1731.

[35] W. Sajomsang, U.R. Ruktanonchai, P. Gonil, O. Nuchuchua, Mucoadhesive property and biocompatibility of methylated N-aryl chitosan derivatives, Carbohydr. Polym. 78 (2009) 945.

[36] J.M. Urreaga, M.U. de la Orden, Chemical interactions and yellowing in chitosan-treated cellulose, Eur. Polym. J. 42 (2006) 2606.

[37] J. Kumirska, M. Czerwicka, Z. Kaczyński, A. Bychowska, K. Brzozowski, J. Thöming, P. Stepnowski, Application of Spectroscopic Methods for Structural Analysis of Chitin and Chitosan, Mar. Drugs 8 (2010) 1567.

[38] F.G. Pearson, R.H. Marchessault, C.Y. Liang, Infrared spectra of crystalline polysaccharides V. Chitin, J. Polym. Sci. 13 (1960) 101.

[39] B. Focher, A. Naggi, G. Torri, A. Cosani, M. Terbojevich, Structural differences between chitin polymorphs and their precipitates from solutions-evidence from CP-MAS ^{13}CNMR, FT-IR and FT-Raman spectroscopy, Carbohydr. Polym. 17 (1992) 97.

[40] D. Baskar, T.S.S. Kumar, Effect of deacetylation time on the preparation, properties and swelling behavior of chitosan films, Carbohydr. Polym. 78 (2009) 767.
[41] M. Rinaudo, Chitin and chitosan: Properties and application, Prog. Polym. Sci. 31 (2006) 603.
[42] M.R. Kasaai, Various Methods for Determination of the Degree of *N*-Acetylation of Chitin, Chitosan: A Review, J. Agric. Food Chem. 57 (2009) 1667.
[43] S. Sabnis, L.H. Block, Improved infrared spectroscopic method for the analysis of degree of *N*-deacetylation, Polym. Bull. 39 (1997) 67.
[44] S.S. Silva, S.M.C. Menezes, R.B. Garcia, Synthesis and characterization of polyurethane-g-chitosan, Eur. Polym. J. 39 (2003) 1515.
[45] C.F. Becker, J.A. Guimarães, P.A.S. Mourão, H. Verli, Conformation of sulfated galactan and sulfated fucan in aqueous solutions: implications to their anticoagulant activities, J. Mol. Graph. Mod. 26 (2007) 391.
[46] P.C.S. F.-Tischer, L.B. Talarico, M.D. Noseda, S.M.P.B. Guimarães, M.E.R. Duarte, Chemical structure and antiviral activity of carrageenans from Meristiella gelidium againstherpes simplex and dengue virus, Carbohydrate Polymers 63 (2006) 459.
[47] M.C.R. Souza, C.T. Marques, C.M.G. Dore, F.R.F. Silva, H.A.O. Rocha, E.L. Leite, Antioxidant activities of sulfated polysaccharides from brown and red seaweeds, J. Appl. Phycol. 19 (2007) 153.
[48] S. Koyanagi, N. Tanigawa, H. Nakagawa, S. Soeda, H. Shimeno, Over sulfation of fucoidan enhances its anti-angiogenic and antitumor activities, Biochem. Pharmacol. 65 (2003) 173.
[49] L.S. Costa, G.P. Fidelis, S.L. Cordeiro, R.M. Oliveira, D.A. Sabry, B.G. Camara, L.T.D.B. Nobre, M.S.S.P. Costa, J. A.-Lima, E.H.C. Farias, E.L. Leite, H.A.O. Rocha, Biomed. Pharmacother. 64 (2010) 21.
[50] B.W.S. Souza, M.A. Cerqueira, A.I. Bourbon, A.C. Pinheiro, J.T. Martins, J.A. Teixeira, A. Manuel, B. Coimbra, A. António, Vicente Chemical characterization and antioxidant activity of sulfated polysaccharide from the red seaweed *Gracilaria birdiae*, Food Hydrocolloid 27 (2007) 287.
[51] T. Chopin, B.F. Kerin, R. Mazerolle, Phycocolloid chemistry as a taxonomic indicator of phylogeny in the Gigartinales, Rhodophyceae: a review and current developments using Fourier transform infrared diffuse reflectance spectroscopy, Phycol. Res. 47 (1999) 167.
[52] L. Pereira, S. F. Gheda, P. J. A. R.-Claro, Analysis by vibrational spectroscopy of seaweed polysaccharides with potential use in food, pharmaceutical, and cosmetic industries, Int. J. Carbohy Chem. http://dx.doi.org/10.1155/2013/537202.
[53] A. Aneiros, A. Garateix, Bioactive peptides from marine sources: pharmacological and isolation procedures, J. Chromatgr. B 803 (2004) 41.
[54] S.R.M. Ibrahim, R.A. E.-Ebel, G.A. Mohamed, D.T.A. Youssef, V. Wray, P. Proksch, G. Callyaerin, A new cytotoxic cyclic peptide from the marine sponge Callyspongia aerizusa, ARKIVOC (2008) 164.
[55] A. Barth, The infrared absorption of amino acid side chains, Prog. Biophys Mol. Bio. 74 (2000) 141.
[56] T. Kasai, K. Furukawa, S. Sakamura, Infrared and mass spectra of α- amino acid amides, J. Fac. Agr. Hokkaido Univ. 59 (1979) 279.
[57] R. Linder, M. Nispel, T. Haber, K. Kleinermanns, Gas-phase FT-IR spectra of natural amino acids, Chem. Phys. Lett. 409 (2005) 260.
[58] D. L.-Vien, N.B. Colthup, W.G. Fateley, J.G. Grasselli, The Handbook of Infrared and Raman Characteristic Frequencies of Organic Molecules, Academic Press Inc., San Diego, (1991).
[59] S.W. Ellepola, S.-M. Choi, D.L. Phillips, C.-Y. Ma, Raman spectroscopic study of rice globulin, J. Cereal Sci. 43 (2006) 85.

[60] E.C.Y. L.-Chan, S. Nakai, M. Hirotsuka, in: R.Y. Yada, R.L. Jackson, L.L. Smith (Eds.), Protein Structure-Function Relationships in Foods, Blackie Academic, London, 1994.

[61] F. Friedberg, M.S. O'Dell, Infrared spectra of some DNP-α-amino acids, Can. J. Chem. 37 (1959) 1469.

[62] R.J. Koegel, G.P. Greenstein, M. Winitz, S.M. Birnbaum, R.A. McCallum, Studies on Diastereoisomeric α-Amino Acids and Corresponding α-Hydroxy Acids. V. Infrared Spectra, J. Am. Chem. Soc. 77 (1955) 5708.

[63] W.H. Moore, S. Krimm, Vibrational analysis of peptides, polypeptides and proteins, β-poly(L-alanine) and β-Poly(L-alanylglycine), Biopolymers 15 (1976) 2465.

[64] A. Elliott, E.J. Ambrose, Structure of synthetic polypeptides, Nature 165 (1950) 921.

[65] S. Krimm, J. Bandekar, Vibrational spectroscopy and conformation of peptides, polypeptides, and proteins, Adv. Protein Chem. 38 (1986) 181.

[66] N.B. Colthup, L.H. Daly, S.E. Wiberley, Introduction to Infrared and Raman Spectroscopy, 2nd Edition, Academic Press, New York, (1975).

[67] J. Kong, S. Yu, Fourier transform infrared spectroscopic analysis of protein secondary structures, Acta Biochimica et Biophysica Sinica 39 (2007) 549.

[68] J.T. Pelton, L.R. McLean, Spectroscopic Methods for Analysis of Protein Secondary Structure, Ana. Biochem. 277 (2000) 167.

[69] L.K. Tamm, S.A. Tatulian, Infrared Spectroscopy of Proteins and Peptides in Lipid Bilayers, Q. Rev. Biophys. 30 (1997) 365.

[70] M. Jackson, H.H. Mantsch, The use and misuse of FTIR Spectroscopy in the Determination of Protein Structure, Crit. Rev. Biochem. Mol 30 (1995) 95.

[71] N.C. Maiti, M.M. Apetri, M.G. Zagorski, P.R. Carey, V.E. Anderson, Raman Spectroscopic Characterization of Secondary Structure in Natively Unfolded Proteins: α-Synuclein, J. Am. Chem. Soc. 126 (2004) 2399.

[72] T. Miura, K. Suzuki, N. Kohata, H. Takeuchi, Metal binding modes of Alzheimer's amyloid beta-peptide in insoluble aggregates and soluble complexes, Biochemistry Metal binding modes of Alzheimer's amyloid beta-peptide in insoluble aggregates and soluble complexes, Biochemistry 39 (2000) 7024.

[73] M.C. Chen, R.C. Lord, Laser-excited Raman spectroscopy of biomolecules. VI. Polypeptides as conformational models, J. Am. Chem. Soc. 96 (1974) 4750.

[74] S.U. Sane, S.M. Cramer, T.M. Przybycien, Holistic approach to protein secondary structure characterization using amide I band Raman spectroscopy, Anal. Biochem. 269 (1999) 255.

[75] R.B. Honzatko, R.W. Williams, Raman spectroscopy of avidin: secondary structure, disulfide conformation, and the environment of tyrosine, Biochemistry 21 (1982) 6201.

[76] J.B. Lambert, R.G. Keske, R.E. Charat, A.P. Jovanovich, The conformational rivalry between the nonbonding electron pair and the proton on nitrogen, J. Am. Chem. Soc. 89 (1967) 3761.

[77] J.C. Braekman, P. Daloze, P.M.D. Abreu, C.P. Leopardi, G. Germain, M.V. Meerssche, A Novel type of Bis-quinolizidine alkaloids from the sponge *Petrosia Seriata*, Tetrahedron Lett. 23 (1982) 4277.

[78] K.S. Singh, B. Das, C.G. Naik, Quinolizidines alkaloids: Petrosin and xestospongins from the sponge *Oceanapia* sp., J. Chem. Sci. 123 (2011) 601.

[79] Y. Venkateswaralu, V.R. Reddy, Bis-1-Oxaquinolizidines from the sponge *Haliclona Exigua*, J. Nat. Prod. 57 (1984) 1283.

[80] T.V. Goud, N.R. Ready, T.S. Ram, Y. Venkateswarlu, Anti-HIV Active Petrosins from the Marine Sponge *Petrosia similis*, Biol. Pharm. Bull. 26 (2003) 1498.

[81] K. Y. Orabi, K. A. El Sayed, M. T. Hamann, D. C. Dunbar, Mansour S. A.- Said, T. Higa, M. Kelly, Araguspongines K and L, New Bioactive Bis-1-oxaquinolizidine *N*-Oxide Alkaloids from Red Sea Specimens of *Xestospongia exigua*, J. Nat. prod. 65 (2002) 1782.
[82] M. Kobayashi, K. Kawazoe, I. Kitagawa, Aragupetrosine A. New vasodilative macrocyclic quinolizidine alkaloid from Okinawan marine sponge *Xestospongia sp.*, Tetrahedron Lett. 30 (1989) 4149.
[83] F. Bohlmann, Zur konfigurationsbestimmung von chinolizin Derivaten, Angew Chem. 69 (1957) 641.
[84] D. Tourwe, G. Vanbinst, Carbon-13 Nuclear Magnectic Resonance Spectroscopy of Quinolizidine Derivatives, Heterocycle 2 (1978) 507.
[85] P.S. Parameswaran, C.G. Naik, S.Y. Kamat, Renieramycins H and I, two novel alkaloids from the sponge *Haliclona Cribricutis* Dendy, Ind. J. Chem. 37B (1998) 1258.
[86] H. Gross, G.M. König, Terpenoids from marine organisms: unique structures and their pharmacological potential, Phytochem. Rev. 5 (2005) 115.
[87] P.-J. Sung, M.Y. Chiang, W.-T. Tsai, J.-H. Su, Y.-M. Su, Y.-C. Wu, Chlorinated briarane-type diterpenoids from the gorgonian coral Ellisella robusta (Ellisellidae), Tetrahedron 63 (2007) 12860.
[88] M. Iwashima, Y. Matsumoto, Y. Takenaka, K. Iguchi, T. Yamori, New Marine Diterpenoids from the Okinawan Soft Coral *Clavularia koellikeri*, J. Nat. Prod. 65 (2002) 1441.
[89] K.S. Singh, W. Kaminsky, C. Rodrigues, C.G. Naik, Structural studies and antimicrobial properties of norcembrane diterpenoid from an Indian soft coral Sinularia inelegans Tixier–Durivault, J. Chem. Sci. 121 (2009) 1041.
[90] G. Briton, G. M. Armit, S.Y.M. Lau, A. K. Patel, C. C. Shone, Carotenoproteins. In G. Brition, T. W. Goodwin (Eds.), Carotenoids Chemistry and Biochemistry, 32 (1981) 237.
[91] T. Nomura, M. Kikuchi, A. Kubodera, Y. Kawakami, Proton-donative antioxidant activity of fucoxanthin with 1,1-diphenyl-2-picrylhydrazyl (DPPH), Biochem. Mol. Bio. Intl. 42 (1997) 361.
[92] H. Wang, Z.-M. Fu, C.-C. Han, The Potential Applications of Marine Bioactives against Diabetes and Obesity, Am. J. Mar. Sci. 2 (2014) 1.
[93] J. Peng, J.-P. Yuan, C.-F. Wu, J.-H. Wang, Fucoxanthin, a Marine Carotenoid Present in Brown Seaweeds and Diatoms: Metabolism and Bioactivities Relevant to Human Health, Mar. Drugs 9 (2011) 1806.
[94] A. Khanafari, A. Saberi, M. Azar, Gh. Vosooghi, Sh. Jamili, B. Sabbaghzadeh, Extraction of Astaxanthin esters from shrimp waste by chemical and microbial methods, J. Environ health Sc. Eng. 4 (2007) 93.
[95] G. Rajauria, N. A.-Ghannam, Isolation and partial characterization of bioactive fucoxanthin from Himanthalia elongata brown sea weed: A TLC-based approach, Int. J. Anal. Chem. http://dx.doi.org/10.1155/2013/802573.
[96] V.E. de Oliveira, H.V. Castro, H.G.M. Edwards, L.F.C. de Oliveira, Carotenes and carotenoids in natural biological samples: a Raman spectroscopic analysis, J. Raman Spectrosc. 6 (2010) 642.
[97] R. Withnall, B.Z. Chowdhry, J. Silver, H.G.M. Edwards, L.F.C. de Oliveira, Raman spectra of carotenoids in natural products, Spectrochim, Acta A 10 (2003) 2207.
[98] H. Schulz, M. Baranska, R. Baranski, potential of NIR-FT-Raman spectroscopies in natural carotenoids analysis, Biopolymers 77 (2005) 212.
[99] M. Heneczkowski, M. Kopacz, D. Nowak, A. Kuźnair, Infrared spectrum analysis of some flavonoids, Drug Res. 58 (2001) 415.
[100] G. Kokilam, S. Vasuki, N. Sajitha, Biochemical composition, alginic acid yield and antioxidant activity of brown seaweeds from Mandapam region, Gulf of Mannar, J App. Pharm Sci. 3 (2013) 99.

[101] L. Briggs, L. Colebrook, Infrared spectra of flavanones and flavones. Carbonyl and hydroxyl stretching and CH out-of-plane bending absorption, Spectrochim, Acta 18 (1962) 939.
[102] J.-P. Cornard, J. Barbillat, J.-C. Merlin, Performance of a versatile multichannel microprobe in Raman, fluorescence, and absorption measurements, Microbeam Anal. 3 (1994) 13.
[103] J.-C. Merlin, A. Statoua, J.-P. Cornard, M.S. - Idrissi, R. Brouillard, Resonance Raman spectroscopic studies of anthocyanins and anthocanidins in aqueous solution, Phytochemistry 35 (1994) 227.
[104] J.-C. Merlin, J.P. Cornard, Vibrational analysis of hydroxyflavylium derivatives by IR, Fourier transform Raman and resonace Raman and resonance Raman spectroscopies, Spectrochim, Acta A 50 (1994) 703.
[105] F. Siebert, P. Hildebrandt, Vibrational Spectroscopy in Life Science (1st edition), Wiley VCH, ISBN 978-3527405060, ISA (2007).
[106] J.H. Sheu, G.H. Wang, P.J. Sung, Y.H. Chiu, C.Y. Duh, Cytotoxic Sterols from the Formosan Brown Alga *Turbinariaornata*, Planta Med. 63 (1997) 571.
[107] G.A. Wachter, S.G. Franzblau, G. Montenegro, J.J. Hoffmann, W.M. Maiese, B.N. Timmermann, Inhibition of Mycobacterium Tuberculosis Growth by Saringosterol from *Lessoniani-grescens*, J. Nat. Prod. 64 (2001) 1463.
[108] A.D. Boul, J.W. Blunt, J.W. Browne, V. Kumar, G.D. Meakins, J.T. Pinhey, V.E. Thomas, Microbiological Hydroxylation of Steroids. Part II. Structural information and infrared spectrometry: carbonyl, preturbed methylene and hydroxyl vibrations of Steroidal Ketones and Alcohols, J. Chem. Soc. C (1971) 1130.
[109] S. Holly, G. Jalsovszky, O. Egyed, Infrared Intensities of Pregn-4-ene-3,20-dione Derivatives, J. Mol. Struct. 60 (1980) 139.
[110] R.N. Jones, P. Humphries, F. Herling, K. Dobriner, Steroid Metabolism. XII. Determination of the structure of the side chains of C-21 steroids by Infrared Spectrometry, J. Am. Chem. Soc. 74 (1952) 2820.
[111] E.-J. Cui, H.-J. Park, H. Park, I.-S. Chung, J.-Y. Kim, S.-W. Yeon, N. Baek, Isolation of Sterols from Cowpea (*Vignasinensis*) Seeds and Their Promotion Activity on HO-1, J. Korean Soc. Appl. Biol. Chem. 54 (2011) 362.
[112] G. Aruldhas, G. Jalsovszky, S. Holly, Infrared Absorption Intensities of Steroids, Acta Chim. Acad. Sei. Hung. 104 (1980) 117.
[113] T. K.- Suhadolc, IR-spectra of Amides of Steroidal Alkaloids with Lactic acid, Experientia 39 (1983) 259.
[114] A.M. Smith, C.R. Eddy, High Resolution Infrared Spectra of Steroids in the Carbon-hydrogen Stretching Region, Anal. Chem. 31 (1959) 1539.
[115] M.V. D'Auria, L. Minale, R. Ricco, Polyoxygenated Steroids of Marine Origin, Chem. Rev. 93 (1993) 1839.
[116] M. Khanavi, R. Gheidarloo, N. Sadati, M.S. Ardekani, S.M. Nabavi, S. Tavajohi, S.N. Ostad, Cytotoxicity of Fucosterol Containing Fraction of Marine Algae against Breast and Colon Carcinoma Cell Line, Phcog. Mag. 8 (2012) 60.
[117] S. Lee, Y.S. Lee, S.H. Jung, S.S. Kang, K.H. Shin, Anti-oxidant Activities of Fucosterol from the Marine Algae *Pelvetiasiliqusa*, Arch. Pharm. Res. 26 (2003) 719.
[118] M. Khanavi, M. Nabavi, N. Sadati, M.S. Ardekani, J. Sohrabipour, S.M. Nabavi, Cytotoxic Activity of Some Marine Brown Algae Against Cancer Cell Lines, Biol. Res. 43 (2010) 31.
[119] N. Bouzidi, Y. Daghbouche, M. Hattab, Z. Aliche, G. Culioli, L. Piovetti, S. Garrigues, M. Guardia, Determination of Total Sterols in Brown Algae by Fourier Transform Infrared Spectroscopy, Anal. Chim. Acta 616 (2008) 185.
[120] B. Schrader, E. Steigner, in: E. Heftmann (Ed.), "Modern methods of steroid analysis", Academic Press, New York, 1973 Chapter 9, 231.

[121] R.C. Hider, X. Kong, Chemistry and Biology of Siderophores, Nat. Prod. Rep. 27 (2010) 637.

[122] T. Csaky, On Estimation of Bound Hydroxylamine in Biological Materials, Acta Chem. Scandca. 2 (1948) 450.

[123] E.L. Arnow, Colorimetric Determination of the Component of 3,4-Dihydroxyphenylalanine-tyrosine Mixtures, J. Biol. Chem. 118 (1937) 531.

[124] S.V. Supanekar, A.M. Sorty, A.A. Raut, Catecholate siderophores produced by *Klebsiellapneumoniae* isolated from Rhizosphere of *Saccharumofficinarum L.*, Int. J. Sci. Res. 2 (2013) 423.

[125] A.L. Actis, W. Fish, J.H. Crosa, K. Kellerman, S.R. Ellenberger, F.M. Hauser, J.S. Loehr, Characterization of anguibactin, a novel siderophore from Vibrio anguillarum 775 (pJM1), J. Bact. 167 (1986) 57.

[126] N. Tank, N. Rajendran, B. Patel, M. Saraf, Evaluation and Biochemical Characterization of a Distinctive Pyoverdin from a *Pseudomonas* isolated from Chickpea Rhizosphere, Braz. J. Microbiol. (2012) 639.

[127] S. Sajjan, Purification and physiochemical characterization of melanin pigment from *Klebsiella sp.* GSK,, J. Microbiol. Biotechnol. 20 (2010) 1513.

[128] R.A. Nicolaus, M. Piattelli, Progress in the chemistry of natural black pigments, Rend. Acad. Sc. Fis. Mat. 32 (1965) 83.

[129] J. Geng, P. Yuan, C. Shao, S.B. Yu, B. Zhou, P. Zhou, X. D. Chen Bacterial melanin interacts with double stranded DNA with high affinity and may inhibit cell metabolism in vivo, Arch. Microbiol. 321 (2010) 192.

[130] M.S. Blois, The melanins: their synthesis and structure, Photochem. Photobiol. Rev. 3 (1978) 115.

[131] D. Jones, Ultrastructure and composition of the cell-walls of Sclerotinia sclerotiorum, Trans. Br. Mycol. Soc. 54 (1970) 351.

[132] J.D. Russell, D. Jones, D. Vaughan, A.R. Fraser, A prelimimary study of fungal melanin by infrared spectroscopy, Geoderma 24 (1980) 207.

[133] K. Tarangini, S. Mishra, Production, Characterization and Analysis of Melanin from Isolated Marine *Pseudomonas* sp. using vegetable waste, Res. J. Engineering Sci. 2 (2013) 40.

Chapter 7

Nuclear Magnetic Resonance Spectroscopy for Structural Characterization of Bioactive Compounds

Clementina M.M. Santos[1] and Artur M.S. Silva[2]
[1]*Department of Vegetal Production and Technology, School of Agriculture, Polytechnic Institute of Bragança, Bragança, Portugal*
[2]*Department of Chemistry & QOPNA, University of Aveiro, Aveiro, Portugal*

Contents

7.1 INTRODUCTION

The structural assignment of a new natural product molecule is not only to establish the 3D structure of a compound, but potentially to provide the basis for research in a multitude of disciplines, ultimately generating new therapeutic agents and/or new understanding of disease biology. The development of modern spectroscopic techniques has transformed the structure assignment process, which previously was essentially based on chemical degradation or derivatization followed by partial or total synthesis. Notably, it was only in the specialization era of the spectroscopic structural assignment of natural products that the field of marine natural products chemistry took shape.

Today the processes of marine and terrestrial natural product isolation and structural determination are frequently streamlined and expeditious due to the

Analysis of Marine Samples in Search of Bioactive Compounds, Vol. 65. DOI: 10.1016/B978-0-444-63359-0.00007-0

spectacular advances in chromatographic and spectroscopic technologies as well as chemical synthesis.

The NMR spectroscopy is a powerful tool in structure elucidation because the properties it displays can be related to the molecular structure. The chemical environment of a particular nucleus is associated with the *chemical shift* (δ, ppm), and the *area* of a resonance, usually presented as its relative integral, is related to the number of nuclei giving rise to the NMR signal. The interactions between individual nuclei, mediated by electrons in a chemical bond, determine the *coupling constant* (*J*, Hz). In this chapter we will present the techniques commonly used, basic concepts, and how they are useful for chemists in the structural elucidation of mainly bioactive marine natural products. Its complex planar structure is determined by ^{1}H and ^{13}C NMR analysis strongly supported by other 1D (DEPT) and 2D (COSY, TOCSY, HSQC/HMQC, HMBC) NMR techniques. The stereochemistry is generally based on NOE experiments (NOE difference, NOESY, and ROESY), ^{1}H–^{1}H and ^{1}H–^{13}C coupling constants, chiral derivatizing agents, and also in empirical procedures comparing the chemical shifts of unknown vicinal and proximal centers with libraries of configurationally known stereomodels. However, the most reliable option to assign all the 3D structure of a marine natural product still is their total synthesis.

The use of NMR hyphenated with other chromatographic and spectroscopic techniques and microcoil probes and narrow diameter tube probes for the structural elucidation of bioactive marine natural products, mainly associated with the quantitative NMR determinations, will be also briefly described.

The chapter will finish with a description of the structural characterization of several types of marine natural products using all the referred NMR techniques followed by a small reference to the misassignments that still are very common.

7.2 PROTON RESONANCES ASSIGNMENT

The ^{1}H NMR spectra of marine natural products are generally very complicated since their structures generally possess a great number of aliphatic and/or aromatic carbons. For example, the polyoxy linear carbon chain of prorocentrol **1**, a potential cytotoxic and antidiatom agent, presents 75 proton resonances in a range of 0.89 to 6.52 ppm (aliphatic and olefinic protons) [1] and the *O*-sulfated pyrrole alkaloid baculiferin B **2** presents 12 proton resonances in a small range of the aromatic region 6.35 to 7.85 ppm [2]. However, there is also some simple ^{1}H NMR spectra, due to the structure simplicity (e.g., bromopyrrole sulfonic acid **3** [3]) or symmetry (e.g., chrysazin **4** [4]; Figure 7.1).

In the case of complex NMR spectra, it is necessary to use some ^{1}H–^{1}H correlated 2D NMR techniques to help in the assignment of all the proton resonances. One of these techniques is the ^{1}H–^{1}H COSY (COrrelation SpectroscopY) spectrum, which simply identifies protons that are coupled with each other. In this technique, 1D spectrum is displayed along each axis with a contour projection of this spectrum along the diagonal axis. Off-diagonal peaks represent

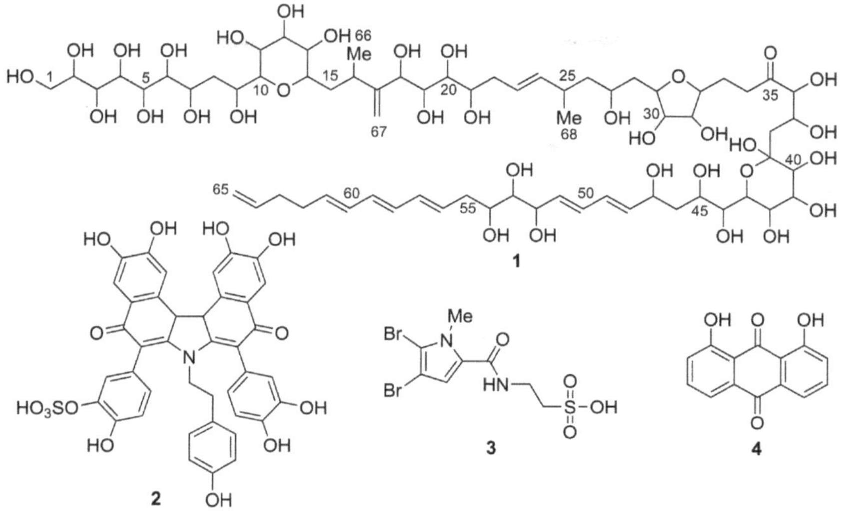

FIGURE 7.1 Structures of compounds **1–4**.

proton shift correlations (or proton couplings). The COSY correlations allow the establishment of some structure moieties, which combined with heteronuclear ^{1}H-^{13}C NMR connectivities, allowed the assignment of all proton resonances and the full structure elucidation. This was the case of the antifungal neomaclafungin A **5** [5] and of the anticholinesterase inhibitor geranylphenazinediol **6** [6] (Figure 7.2). Specifically, the COSY spectrum of geranylphenazinediol **6** showed that protons H-2 (7.17 ppm) and H-3 (7.61 ppm) are coupled with each other (*J* 7.7 Hz) and the long range correlation of H-3 to the methylenic group H-1′ confirm the attachment site of the geranyl substituent. Vicinal and long range correlations of this chain of protons starting from H-1′ allows their complete assignment (Figure 7.2). The protons sequence H-7 to H-9 of the

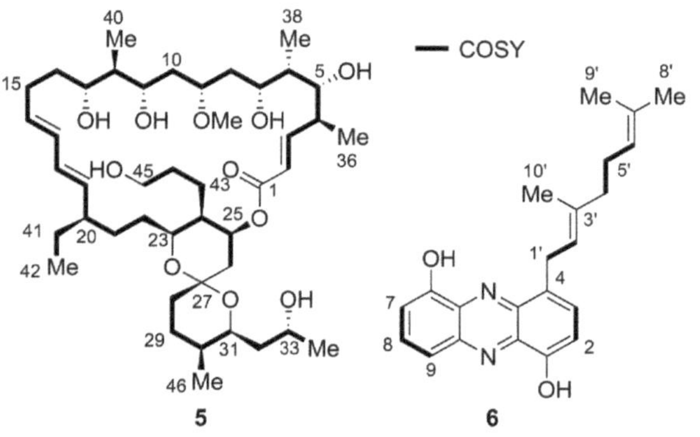

Compound **6**

	δ_H (ppm)	COSY
H-2	7.17, d	3
H-3	7.61, dt	2, 1'
H-7	7.25, dd	8, 9
H-8	7.82, dd	7, 9
H-9	7.73 dd	7, 8
H-1'	4.05, br d	3, 2', 4',10'
H-2'	5.54, tq	1', 4', 10'
H-4'	2.10, m	1', 2', 5'
H-5'	2.13, m	4', 6', 8', 9'
H-6'	5.10, tsept	5', 8', 9'
H-8'	1.59, br s	5', 6'
H-9'	1.55, br s	5', 6'
H-10'	1.80, br s	1', 2'

FIGURE 7.2 COSY correlations of neomaclafungin A **5** and geranylphenazinediol **6**.

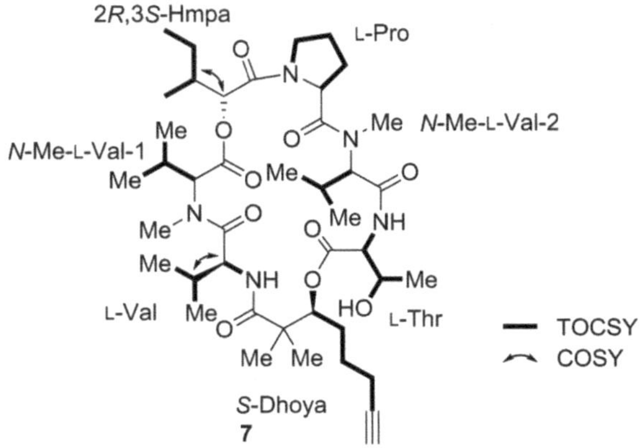

FIGURE 7.3 TOCSY and COSY correlations of viequeamide A **7**.

second aromatic ring of phenazine is also established by the COSY correlations of H-7 (7.25 ppm, dd, *J* 7.7, and 1.0 Hz) with H-8 (7.82, dd, *J* 8.9, and 7.7 Hz) and H-9 (7.73, dd, *J* 8.9 and 1.0 Hz), and also H-8 with H-7 and H-9. Then, the full structure elucidation needed information from the ^{13}C NMR data.

TOCSY (TOtal Correlation SpectroscopY) or HOHAHA (HOmonuclear HArtmann-HAhn) spectrum is another technique used to resolve complex proton NMR spectra. It presents correlations between all the protons within a given spin system, irrespective of whether they are directly coupled or not. For example, the TOCSY experiment of viequeamide A **7**, a cytotoxic agent of H460 human lung cancer cell lines, revealed eight proton spin systems (Figure 7.3). Four of them were assigned to the amino acid residues threonine (L-Thr), proline (L-Pro), and two *N*-methylvaline (*N*-Me-L-Val) units [7]. Two other spin systems were connected by COSY, which suggests that a possible constrained rotation might prevent strong TOCSY correlations within this residue. Heteronuclear correlations confirmed that these two spin systems belong to a valine residue (L-Val) wherein a very small $J_{H\text{-}\alpha,H\text{-}\beta}$ 2.3 Hz was present. The third unassigned and very weak TOCSY spin system was revealed by a combination of homo- and heteronuclear correlations as the hydroxy acid analogue of isoleucine (2*R*,3*S*-Hmpa). The remaining TOCSY spin system was characteristic of a polyketide synthase-derived residue, defined as a derivative of 2,2-dimethyl-3-hydroxy-7-octynoic acid (*S*-Dhoya) based on the long range ^{1}H-^{13}C connectivities. This type of connectivities together with ROESY data (see Section 7.4) allowed the establishment of the sequence of residues of viequeamide A **7**.

A variant of the COSY technique, the phase-sensitive DQF-COSY (Double Quantum Filtered-COrrelated SpectroscopY), is sometimes used to simplify the diagonal of the COSY spectrum and also to eliminate the singlets of this diagonal. The combination of this spectrum with a TOCSY spectrum elucidates partial substructures of prorocentrol **1**: H-1 to H-16, H-18 to H-34, H-36 to H-38, H-40 to H-65 [1] (Figure 7.4).

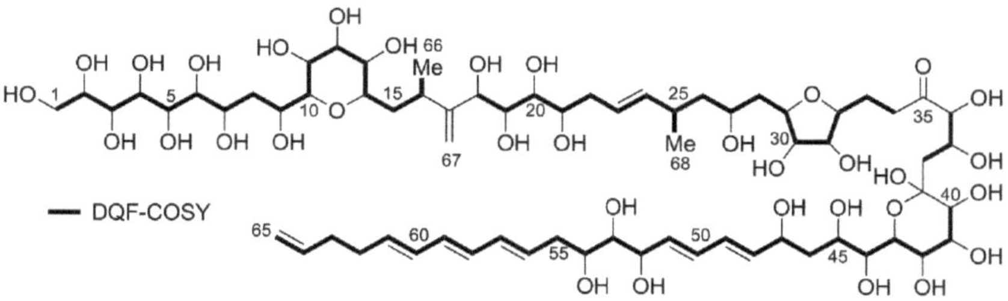

1

FIGURE 7.4 DQF-COSY correlations of prorocentrol **1**.

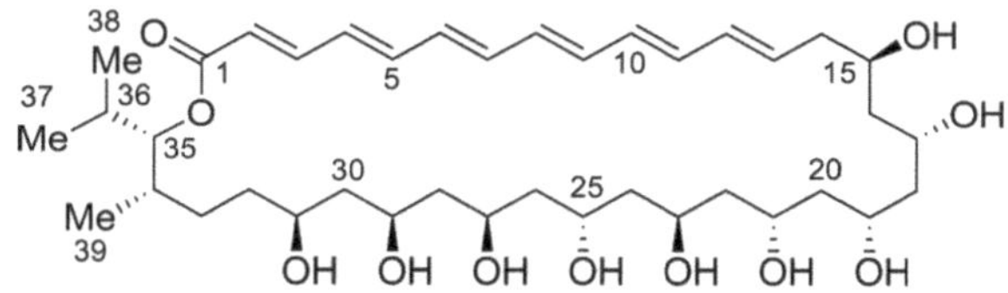

8

FIGURE 7.5 Structure of bahamaolide A **8**.

A general problem that can also appear in the ^{1}H NMR spectra of complex marine products are the overlapping of proton signals with that of the solvent. A possible way to circumvent this problem is to change the NMR solvent [8]. For example, in the ^{1}H NMR spectrum of a pyridine-d_5 solution of the antifungal bahamaolide A **8** (Figure 7.5), a 36-membered macrocyclic lactone, the carbinol resonances are well separated and easily assigned while olefinic signals overlap significantly with the solvent. Changing the solvent for CD_3OD, the alkene protons were well separated and the large vicinal ^{1}H-^{1}H coupling constants (J 14.5–15.5 Hz) allowed assignment of the (E)-configuration for all double bonds [9].

In many samples, OH and NH resonances can be recognized by their characteristic chemical shifts or broadened appearance in the ^{1}H NMR spectrum but they vary depending on the sample preparation, product structure, and ability to form hydrogen bonds. Higher chemical shifts are observed in samples with higher product concentration (increased hydrogen bonding), in H-bonded groups compared to free OH or NH groups, and in H-bonding solvents like DMSO-d_6 or acetone-d_6.

OH and NH resonances are subject to intermolecular exchange processes in some solvents, which may result in broadening or complete loss of coupling with neighboring protons. As example of the last phenomena, the signals of the six hydroxyl groups (δ 9.41–10.10 ppm) in the ^{1}H NMR spectrum of the potential antibiotic agent chrysophaentin A **9** in DMF-d_7 were completely absent in the spectra recorded in $CDCl_3$, CD_3CN, and DMSO-d_6 [10]. The broad singlets attributed to the OH and NH protons in the ^{1}H NMR spectra of the antifungal ambiguine P **10** and the antibacterial ambiguine Q nitrile **11** acquired in DMSO-d_6

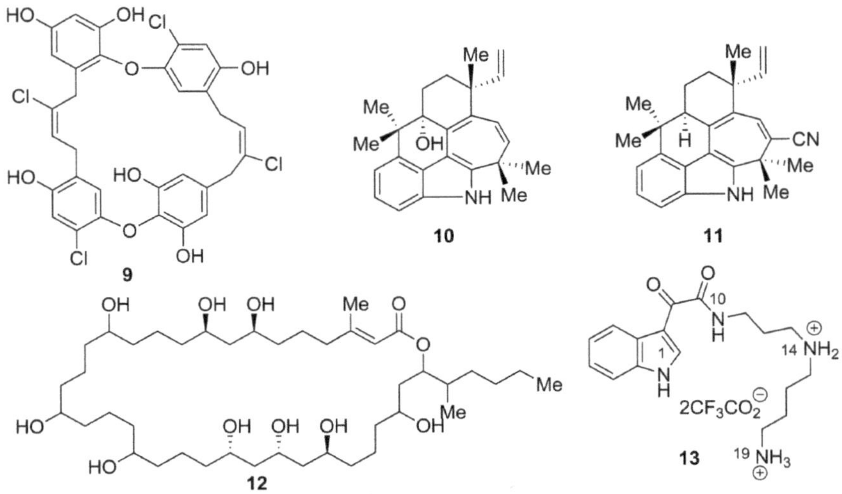

FIGURE 7.6 Structures of compounds **9–13**.

are completely absent in the spectra recorded in CD_3OD at 900 MHz [11]. The ^1H NMR spectra of the macrolactone caylobolide B **12**, a cytotoxic agent against HT29 colorectal adenocarcinoma and HeLa carcinoma cell lines, presented nine OH exchangeable protons at 4.20 to 4.52 ppm [12], and that of the antimicrobial indole spermidine alkaloid didemnidine A **13** exhibited seven exchangeable protons assigned to the NH resonances: indole NH-1 at 12.29 ppm, secondary amine NH-10 at 8.90 ppm, protonated secondary amine NH_2-14 at 8.54 ppm, and finally protonated primary amine NH_3-19 at 7.81 ppm [13] (Figure 7.6).

The rates of exchange are strongly dependent on the temperature, solvent, concentration, and presence of acidic and basic impurities. In $CDCl_3$, the presence of acidic impurities (usually from solvent decomposition) leads to rapid acid-catalyzed exchange between H-bonded groups. In solvents like DMSO-d_6 and acetone-d_6 the strong hydrogen bonds of the H-bonded groups slows down the intermolecular proton exchanges, leading to discrete OH or NH signals with observable coupling to nearby protons.

In some cases the disappearance of all OH and NH signals can be useful. For that, the labile protons can be easily identified by shaking the sample with a drop of D_2O. This process present better effects if the solvent is water immiscible and denser than water ($CDCl_3$, CD_2Cl_2, CCl_4) since the formed DOH is in the drop of water floating at the top of the sample where it is not detected. In water miscible solvents (acetone-d_6, DMSO-d_6, CD_3CN, pyridine-d_5, THF-d_8) the OH and NH signals are converted to OD and ND, but the DOH formed remains in solution and appears in the water region.

Changing solvents can be a problem in terms of structure elucidation, since the use of deuterated protic solvents can give rise to the labeling of some exchangeable protons in the target molecule. Thus changing $CDCl_3$ for CD_3OD,

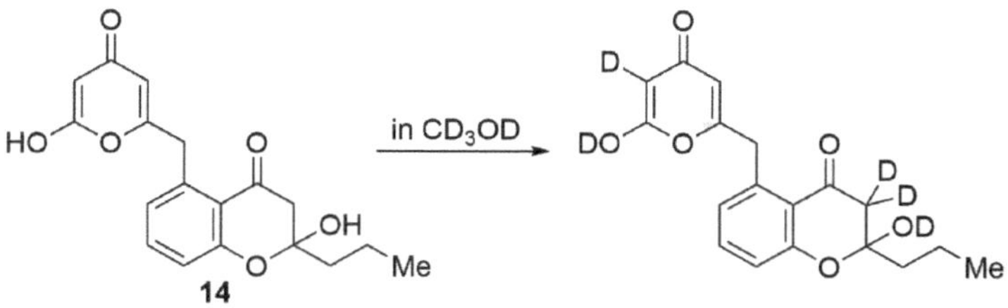

SCHEME 7.1 Deuteration of the antibacterial marine chroman-4-one **14** in CD_3OD.

the ^{1}H NMR spectrum of the antibacterial marine chroman-4-one **14** showed a fast disappearance of three signals corresponding to an olefinic proton, as a singlet at 5.33 ppm, and two methylene protons, as doublets at 2.79 and 2.89 ppm, due to tautomeric equilibria (Scheme 7.1). The incorporation of five deuterium atoms (the previously described plus the OD groups) was confirmed by ESI(+)-MS [14].

7.3 ESTABLISHMENT OF CARBON SKELETON

Similar to ^{1}H NMR spectroscopy, ^{13}C NMR of biologically active marine natural products can also be very complex. For instance, the spectrum of prorocentrol **1** (Figure 7.4), a polyoxy linear compound presents 67 carbon resonances in a range of 21.4 and 136.9 ppm plus the carbonyl carbon at 213.0 ppm. The signals around the acetal and carbonyl groups split in two or three signals, probably due to reversible formation of an acetal ring when measured in CD_3OD or DMF-d_7. This situation was not solved even by varying the NMR conditions such as solvents, temperature, and pH, but in a mixture of pyridine-d_5/D_2O (6:1) at room temperature the spectrum was simplified and improved, being the referred signals observed as a single peak each in the ^{13}C NMR spectrum [1].

The first attempt to assign the signals of complicated ^{13}C NMR spectra is through DEPT (Distortionless Enhancement by Polarization Transfer) experiments, which provide information about the nature of carbon atoms (distinguishing among ^{13}C signals of methyl, methylene, and methine carbons). Three DEPT spectra are required for a full analysis and use a complex series of pulses in both the ^{1}H and ^{13}C ranges: a 45° pulse (DEPT-45) in which all H-bearing carbon peaks are positive, a 90° pulse (DEPT-90) where only methine carbons are seen, and a 135° pulse (DEPT-135) where methyl and methine carbons present positive signals and methylene carbons give negative signals. Quaternary carbons are not seen in DEPT spectra because the technique relies on polarization transfer, which in this case is the transfer of proton magnetization onto the directly bound carbon atoms. In the ^{13}C NMR spectrum of bahamaolide A **8**, where a great number of signals appears in very small ranges of chemical shifts, the DEPT spectra assigned the carbon signals into one quaternary carbon

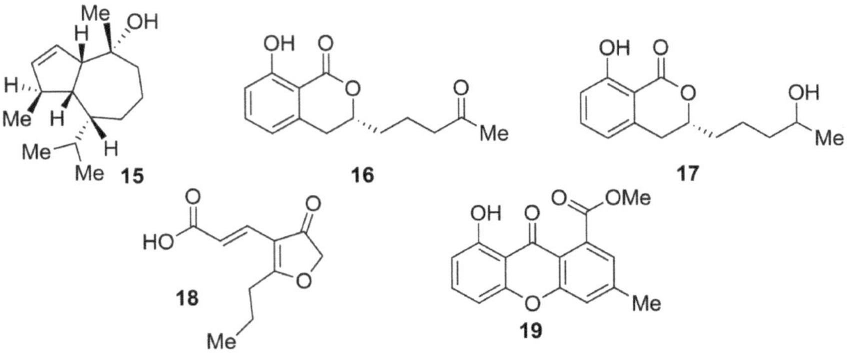

FIGURE 7.7 Structures of compounds **15–19**.

(168.0ppm), 12 olefinic methines (121.1–146.0ppm), 10 oxygenated methines (65.0–79.3ppm), 12 aliphatic methylenes (30.9–48.3ppm), one aliphatic methine (30.2ppm), and three methyls (14.9–20.6ppm) [9]. For a guaiane sesquiterpene derivative **15** where the ranges of some different types of carbons are overlapped, the DEPT spectra were very important to assign their 15 carbon resonances, appearing between 18.8 and 134.5ppm, as two olefinic methines (132.9–134.5ppm), one oxy quaternary carbon (73.4ppm), five aliphatic methines (33.5–49.4ppm), three methylenes (18.8–38.2ppm), and four methyls (19.5–26.3ppm) [15] (Figure 7.7).

The DEPT NMR spectra can also be very useful to assign the type of carbons in molecules bearing aromatic, aliphatic, and/or olefinic moieties. For example, in one of the dihydroisocoumarins isolated from endophytic fungus *Aspergillus* sp. (compound **16**), we can find five quaternary carbon atoms (108.2–208.2ppm), including one lactone carbonyl carbon (169.7ppm) and one ketone carbonyl carbon (208.2ppm), three aromatic methines (116.0–136.1ppm), one aliphatic oxymethine (79.3ppm), four methylenes (18.9–42.7ppm), and one methyl (29.8ppm). The main difference for the second derivative (compound **17**) was the absence of the ketone carbonyl atom and the appearance of an aliphatic oxymethine at 67.3ppm [16] (Figure 7.7). Compounds **16** and **17** presented a weak antibacterial activity against *Staphylococcus aureus* and *Bacillus subtilis*.

Cillifuranone **18**, a natural antibiotic compound produced by a fungal isolate from *Tethya aurantium*, presents 10 signals assigned as three carbonyl or enol carbons (170.9, 196.7, and 201.7ppm), two olefinic methines (118.5 and 133.0ppm), one quaternary olefinic carbon (112.6ppm), one oxymethylene (76.3ppm), two methylenes (20.9 and 31.6ppm), and one methyl (14.0ppm) based on DEPT spectra [17] (Figure 7.7).

In the case of the new antifungal marine xanthone derivative **19** the information given by the DEPT spectra is not very relevant since from their 16 signals, five are aromatic methines, nine are quaternary carbons, and two are methyl groups [18] (Figure 7.7). The structure elucidation of many other marine products involved the use of this 1D NMR technique [3,9,19–21].

For routine work, similar information is obtained from HSQC (Heteronuclear Single Quantum Correlation) and HMQC (Heteronuclear Multiple Quantum Coherence) experiments. These two techniques provide single bond heteronuclear shift correlation, identifying one-bond H–C connectivities within a molecule. The 2D spectrum displays in one dimension ^{1}H chemical shifts and the other ^{13}C chemical shifts and the cross-peaks in the contour plot define to which carbon a particular proton (or group of protons) is directly attached.

HMBC (Heteronuclear Multiple Bond Correlation) is an extremely powerful tool in structure elucidation of marine natural compounds. This experiment establishes the long-range ^{1}H–^{13}C correlations, which typically occur over two and three bonds ($^{2,3}J_{C\text{-}H}$ usually < 10 Hz). Correlations can also be observed to quaternary centers (for example in the assignment of carbonyl resonances) and across heteroatoms other than carbon (for example hydroxyl protons).

These heteronuclear correlation spectra cannot be used alone; they are always combined and analyzed after ^{1}H assignment of a certain molecule. For instance, in the case of trichoderone **20**, a cytotoxic agent isolated from the marine fungus *Trichoderma* sp., the ^{1}H, ^{13}C, and HMQC spectra indicate the presence of two oxygenated methines [δ_C 77.7 (d, C-4) and 81.8 (d, C-5); δ_H 4.66 (^{1}H, br s, H-4) and 4.23 (^{1}H, br s, H-5)], a methyl group [δ_C 10.9 (q, C-7); δ_H 1.20 (3H, t, *J* 7.6 Hz, H-7)], a trisubstituted vinyl group [δ_C 180.1 (s) and 125.6 (d); δ_H 5.99 (^{1}H, s)], a methylene group [δ_C 23.0 (t); δ_H 2.42 and 2.63, (each ^{1}H, m)], as well as a keto carbonyl group [δ_C 202.8 (s)]. All these functionalities were assembled based on the COSY and HMBC spectra [22]. Specifically, in the HMBC spectrum the correlations of H-6 with C-2, C-3, and C-4 and of H-7 with C-3 suggested that C-2, C-4, and C-6 are connected to C-3. Other key correlations, H-2 to C-1 and C-5, revealed that C-1 is directly attached to C-2 and C-5 (Figure 7.8).

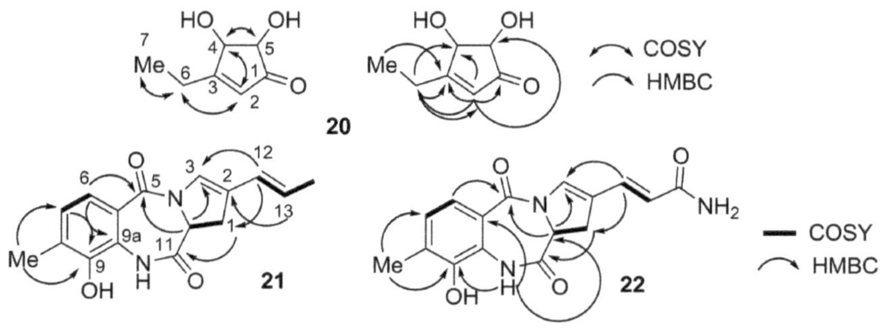

FIGURE 7.8 HMBC and COSY correlations of compounds **20–22**.

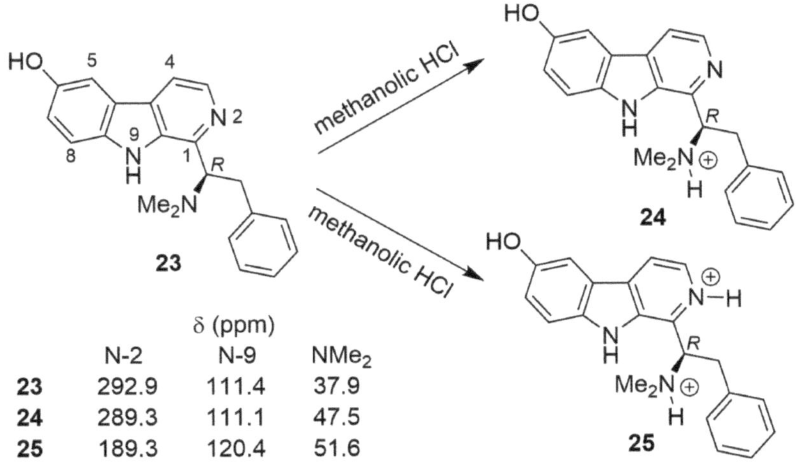

	δ (ppm)		
	N-2	N-9	NMe$_2$
23	292.9	111.4	37.9
24	289.3	111.1	47.5
25	189.3	120.4	51.6

SCHEME 7.2 Formation of dimethylammonium salt **24** and pyridiniumdimethylammonium disalt **25** of the marine alkaloid eudistomin X **23**.

In the case of the neuraminidase inhibitors limazepins G **21** and H **22**, the ^{1}H, ^{13}C NMR spectra together with DEPT experiments and a D_2O exchangeable assay for label protons indicate a pyrrolo[1,4]benzodiazepine skeleton. The combination of COSY, HMQC, and HMBC confirmed this proposal and led to the exact assignment of all ^{1}H and ^{13}C signals. In particular, the key HMBC correlations are: (1) H-6 with C-5 (amide carbonyl) and C-9a and H-7 with methyl and aromatic oxygenated carbons confirmed the substitution pattern of the benzene ring; (2) H-12 with C-1, C-3 and C-14 proves the anchorage of the 1-propenyl group at C-2 [23] (Figure 7.8).

All the referred and described HMBC spectra considered ^{1}H–^{13}C correlations but there are also other possibilities such as ^{1}H–^{15}N. This technique was used in the confirmation of the free-base **23**, dimethylammonium salt **24**, and pyridiniumdimethylammonium disalt **25** of the synthetic antimicrobial marine alkaloid eudistomin X. The ^{15}N NMR chemical shifts were indirectly acquired from the ^{1}H–^{15}N HMBC NMR experiments (Scheme 7.2). The ^{15}N resonance of the dimethylamino group presented only a minor deshielding in the mono salt **24** but the N-2 resonance shifted significantly upon protonation [24].

7.4 STEREOCHEMISTRY

The biological properties attributed to a wide range of chiral natural compounds are closely related to their stereochemistry and therefore, several approaches to determine the absolute configuration have emerged, based on NMR spectroscopy. The establishment of the relative and absolute stereochemistry of stereogenic centers and the stereochemistry of geometric isomers is not an easy task. In some cases, it needs extensive NMR studies, chemical derivatizations together with NMR studies or total synthesis while the presence of isomers mixture led to an increased complexity in the ^{1}H and ^{13}C NMR spectra.

NOE (Nuclear Overhauser Effect) is a tool to establish a spatial relationship between atoms in a molecule. The NOE difference experiment detects and measures NOEs between protons through the space. It involves the application of a radio-frequency field to a single resonance in which the corresponding protons become saturated. Recording of the proton spectrum after the period of saturation may, therefore, show changes in signal intensities for the protons in the vicinity of the saturated proton. Thus, 1D NOE studies present the differences between the proton control spectrum and the irradiated proton spectra. In the 2D NOESY (Nuclear Overhauser Effect SpectroscopY) experiments, only one spectrum is recorded and contains all the information provided by the NOE studies.

A simple example for the utility of NOESY data is the establishment of the configuration of the C2′=C3′ double bond of geranylphenazinediol **6** (Figure 7.2). The correlation between 10′-Me and 1′-CH_2 as well as 2′-CH and 4′-CH_2 and the absence of correlation between 2′-CH and 10′-Me confirmed the (*E*)-configuration of this vinylic system [6]. In another example, the NOESY data was only useful for determining the (*E*)-configuration of the C8=C9 double bond of the tetraene chain of the antibacterial gombapyrone E **26**, since the remaining double bonds geometries were deduced by ^{1}H-^{1}H NMR coupling constants [25] (Figure 7.9).

The relative configuration of salimabromide **27**, a tetracyclic antibiotic compound possessing a brominated benzene ring, a furano lactone residue, and a cyclohexane ring, bridged by a seven-membered cyclic unit, was proved by the NOE connectivities and the (*Z*)-configuration of the C3=C9 double bond was assigned by NOE correlation between H-3 and H-9 and also supported by the ^{1}H-^{1}H NMR coupling constant ($J_{H\text{-}3,H\text{-}9}$ 13.1 Hz) [26] (Figure 7.9).

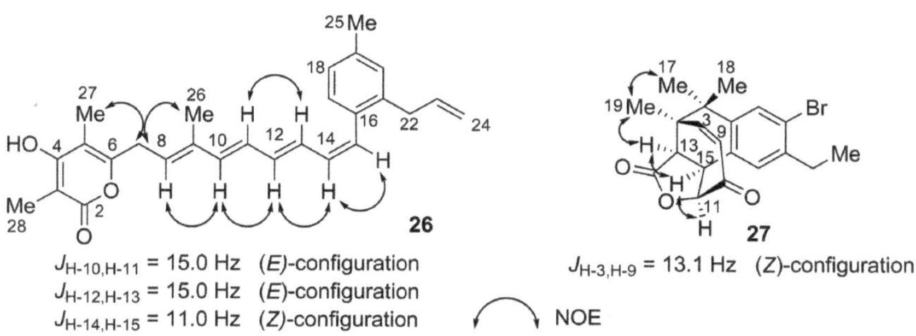

FIGURE 7.9 Stereochemistry of gombapyrone E **26** and salimabromide **27** established by NOEs.

The relative configuration of other marine natural products were elucidated by a combination of NOESY experiments and comparison with the ^{1}H and ^{13}C NMR data of known compounds [27], or using other techniques such as circular dichroism (CD) [28] and overall optical rotation [29].

Contrary to NOESY, ROESY (Rotating-frame Overhauser Effect SpectroscopY) is a quite useful technique for molecules with large molecular sizes (NOEs are too weak to be detectable). In NOESY the cross-relaxation rate constant goes from positive (for small molecules) to negative (for large molecules) as the correlation time increases, or even null, whereas in ROESY the cross-relaxation rate constant is positive for all rotational correlation times. This experiment, in which homonuclear NOE (ROE) effects are measured under spin-locked conditions, presents the signals that arise from proton resonances that are spatially close even if they are not bonded.

Five new potential cytotoxic austalides were isolated from the sponge-associated fungus *Aspergillus* sp. and their relative configuration was elucidated by interpretation of their ROESY spectra. For example, the correlations observed between 24-Me with H-21 and of 31-OMe with H-21 and 24-Me, indicates that rings C and D of austalide M **28** are *cis*-fused. The *trans*-configuration of 24-Me and 27-Me is due to the correlation between 27-Me and H-22 but not with H-21, establishing the (11*S**, 13*R**,14*R**, 20*R**, 21*S**, 22*S**)-relative configuration of **28** [30]. An extensive investigation on the same strain led to the isolation of seven new alkaloids. The correlations observed for both H-12 and 13-CH_2 with H-2, indicated a *cis*-orientation for H-2 H-12, and C-13, and assigns the (2*S**, 3*R**, 12*R**)-configuration of tryptoquivaline K **29** [31]. The relative stereochemistry of the two heterocyclic rings of the antibacterial and antitumor agent fradimycin A **30** was based on the ROESY correlations of H-1′ with H-5′, H-4′ with H-6′, H-7′′ with H-9′′, H-8′′ with H-10′′ and H-9′′ with H-13′′ [32]. The (*E*)-geometry of the two double bonds at C-12 and C-16 of the antibacterial and antitumor agent napyradiomycin A **31** was determined by the ROESY correlations between H-11 with H-12 and 13-CH3-Me and the absence of correlation between H-12 and 13-CH3-Me and also between H-19 with H-16, respectively [33]. ROESY data in combination with *J*-based configurational analysis was used to establish the relative configuration at C-11/C-12 and C-12/C-13 of 13-hydroxyprethyrsenol A **32**, an oxasqualenoid cytotoxic agent isolated from the red alga *Laurencia viridis* [34] (Figure 7.10).

The complexity in the ^{1}H spectrum can be enlarged by the presence of geometrical isomers in solution. Usually, the restricted rotation provided by the carbon–carbon double bond led to the appearance of two sets of signals corresponding to the rapid interconversion of the proton resonances of the *cis*- and *trans*-isomers or more generally, (*Z*)- and (*E*)-isomers. The ^{1}H NMR spectra of some antibacterial butenolides **33a,b** and **34a,b** (Figure 7.11), isolated from korean tunicate *Pseudodistoma antinboja*, showed these two spin signals with an integral ratio of 1:2, the (*Z*)-form being predominant. The difference in the structures of these two isomers is due to the configuration of C4=C5, which can

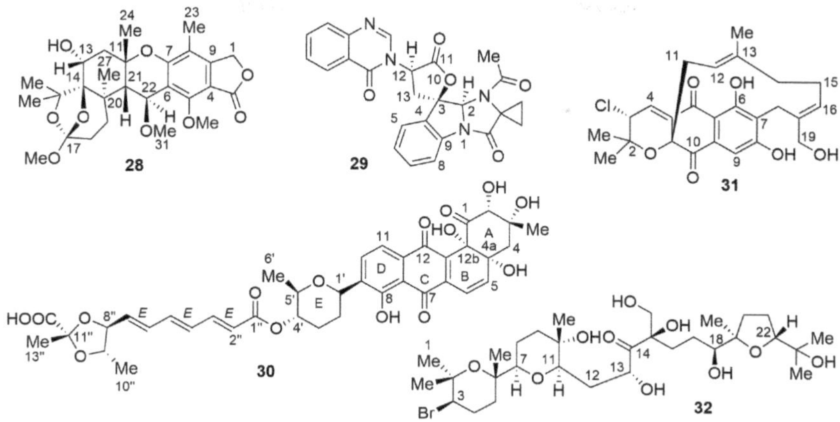

FIGURE 7.10 Structures of compounds **28–32**.

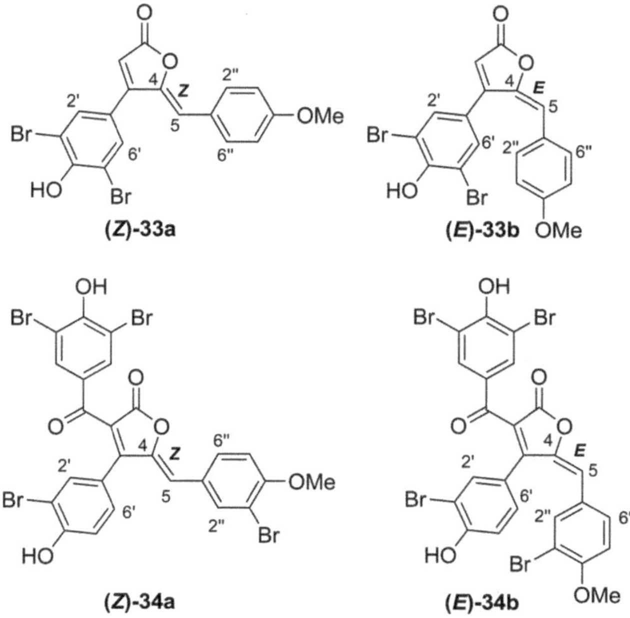

FIGURE 7.11 Geometric isomers of butenolides **33** and **34**.

be assigned by NOE difference experiments. Irradiation of H-5 of **33a** induced enhancements of H-2′,6′ and H-2′′,6′′ signals, indicating a (*Z*)-configuration for this vinylic system. In contrast, irradiation of H-5 of the (*E*)-isomer **33b**, induced NOEs of H-2′′,6′′. Moreover, irradiation of H-2′,6′ induced an enhancement of H-2′′,6′′ indicating a close proximity between the two phenyl rings consistent with the (*E*)-configuration of C4=C5. In the case of isomers **34**, the ROESY spectra unequivocally identified the (*Z*)-isomer by the presence of two ROE cross-peaks of H-5 with H-2′,6′ and H-2′′,6′′. The location of the methoxyl group at C-4′′ was assigned by the ROESY correlations between the methoxyl protons and H-5′′ [35].

As shown earlier, NOE and ROE experiments are very useful in establishing the stereochemistry of vinylic and chiral compounds. However even with these techniques is still very difficult to assign the stereochemistry of highly flexible chains due to the presence of multiple conformers. In these cases, minor populations often originate large NOEs, occasionally leading to contradictory conclusions. Trying to solve this problem, Murata and his group developed a method for the relative configuration analysis of natural products based on ^{13}C-^{1}H spin-coupling constants ($^{2}J_{C,H}$ and $^{3}J_{C,H}$) and ^{1}H-^{1}H spin-coupling constants ($^{3}J_{H,H}$) [36,37]. They conclude that the relative stereochemistry assignment of acyclic systems, where the conformation of adjacent asymmetric centers is represented by staggered rotamers, can only be determined by the combination of these coupling constants, since they allowed the identification of the predominant staggered rotamer out of the six possible from the *threo* and *erythro* configurations, while using $^{3}J_{H,H}$ is inadequate because two H/H-*gauche* rotamers cannot be distinguished. $^{3}J_{C,H}$ follow a Karplus-type equation (Figure 7.12c) and can be used for stereochemistry analysis as is the case of $^{3}J_{H,H}$ (Figure 7.12a). Therefore, when an oxygen functionality on a carbon atom is *gauche* to its geminal proton $^{2}J_{C,H}$ becomes large, and when it is *anti*, the value becomes small (Figure 7.12b).

To assign the stereochemical relationship of a pair of vicinal asymmetric carbons (1,2-methine systems) commonly found in natural products, we have to choose a single conformer with a correct configuration of staggered rotamers possible in *threo*- and *erythro*-diastereomers (Figure 7.13). Four conformers A-1, A-2, B-1, and B-2 can be identified based on $^{3}J_{H,H}$, $^{2}J_{C,H}$, and $^{3}J_{C,H}$ but the other two A-3 and B-3 bearing a H–H *anti*-orientation cannot be distinguished.

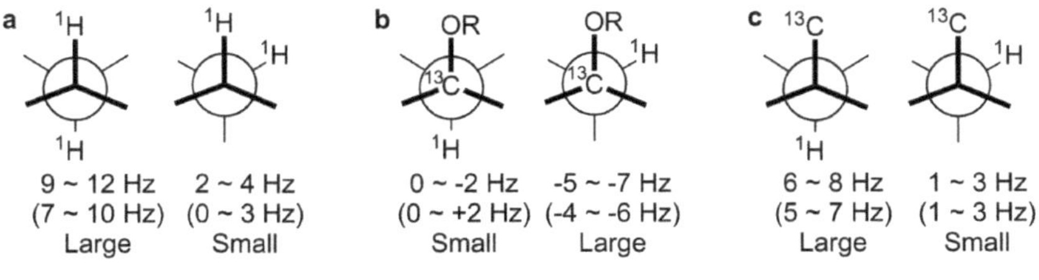

FIGURE 7.12 Dihedral angle dependence of spin-coupling constants: (a) $^{3}J_{H,H}$ for *anti*- and *gauche*-rotamers; (b) $^{2}J_{C,H}$ when dihedral angles between ^{13}C-attached oxygen and the proton are 180° and 60°; (c) $^{3}J_{C,H}$ when the dihedral angles are 180° and 60°. The figures in parentheses represent the values of 1,2-dioxygenated systems. *Adapted from [36].*

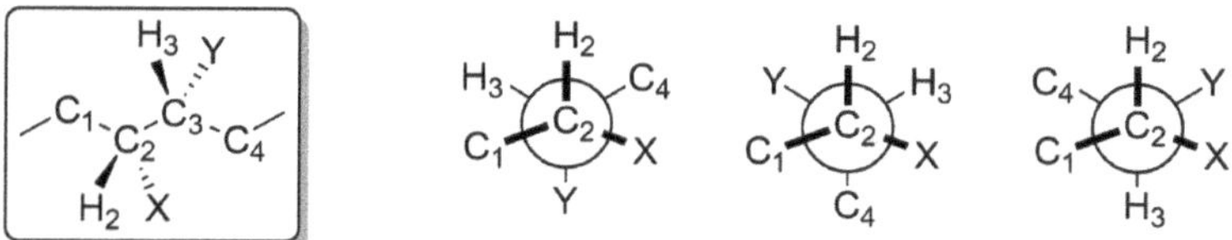

A (*threo*)

		A-1	A-2	A-3
	$^3J_{H\text{-}2,H\text{-}3}$	Small	Small	Large
	$^3J_{H\text{-}2,C\text{-}4}$	Small	Large	Small
	$^3J_{C\text{-}1,H\text{-}3}$	Small	Large	Small
X=Me, Y=OR	$^3J_{C\text{-}x,H\text{-}3}$	Large	Small	Small
	$^2J_{C\text{-}3,H\text{-}2}$	Small	Large	Large
X=OR, Y=OR	$^2J_{C\text{-}2,H\text{-}3}$	Small	Large	Large
	$^2J_{C\text{-}3,H\text{-}2}$	Small	Large	Large

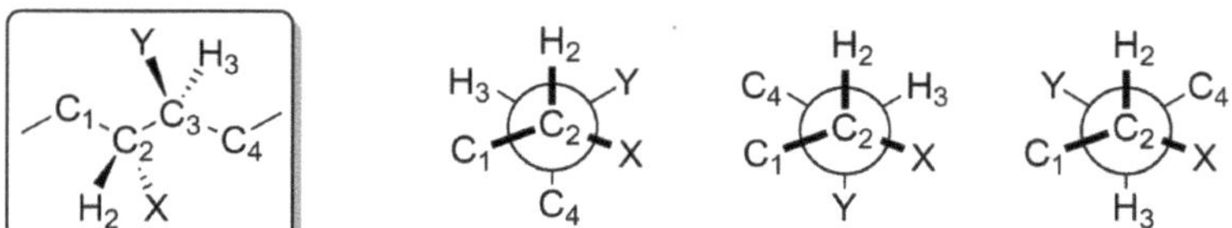

B (*erythro*)

		B-1	B-2	B-3
	$^3J_{H\text{-}2,H\text{-}3}$	Small	Small	Large
	$^3J_{H\text{-}2,C\text{-}4}$	Large	Small	Small
	$^3J_{C\text{-}1,H\text{-}3}$	Small	Large	Small
X=Me, Y=OR	$^3J_{C\text{-}x,H\text{-}3}$	Large	Small	Small
	$^2J_{C\text{-}3,H\text{-}2}$	Large	Small	Large
X=OR, Y=OR	$^2J_{C\text{-}2,H\text{-}3}$	Small	Large	Large
	$^2J_{C\text{-}3,H\text{-}2}$	Large	Small	Large

FIGURE 7.13 Dependence of $^3J_{H,H}$ and $^{2,3}J_{C,H}$ on dihedral angles between vicinal 1,2-methine carbons in butane systems. Occurring pairs of substituents in natural products, Me/OR and OR/OR, are presented. (Me/Me is omitted since it rarely occurs in natural products.) R stands for H, alkyl, and/or acyl groups. *Adapted from [35].*

For these *anti*-conformers, NOE experiments are a practical way to assign their configurations although they are structure dependent [36].

Murata and his group described similar strategies for 1,3- and 1,4-methine systems [36]. This type of comprehensive configuration analysis of complicated natural products based on ^{13}C–^1H coupling constants are scarce in the literature, probably due to the difficulty in determining the $^{2,3}J_{C,H}$ values accurately.

Sato *et al.* used the Murata's method to establish the relative configuration of three substructures (C-4 to C-13, C-31 to C-34, and C-20 to C-42) of the neomaclafungin A **5**, the most abundant compound of the new neomaclafungins

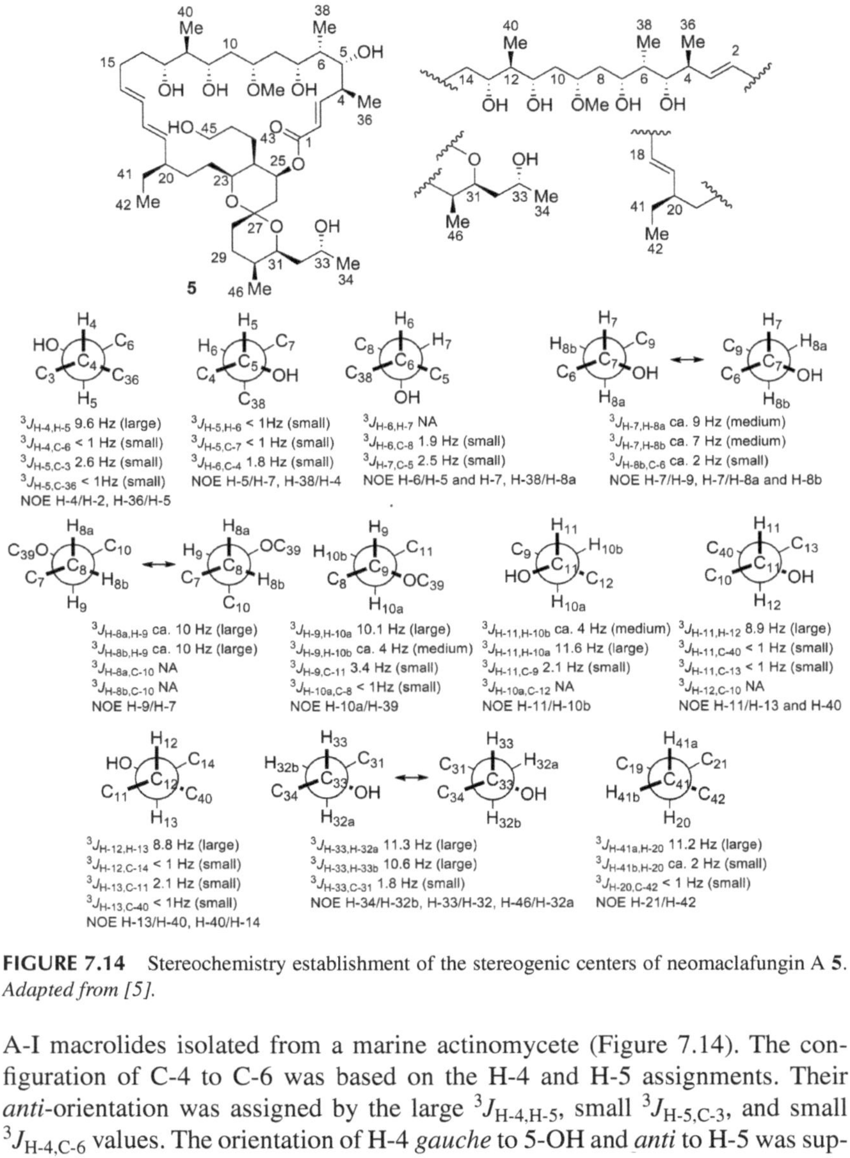

FIGURE 7.14 Stereochemistry establishment of the stereogenic centers of neomaclafungin A **5**. *Adapted from [5].*

A-I macrolides isolated from a marine actinomycete (Figure 7.14). The configuration of C-4 to C-6 was based on the H-4 and H-5 assignments. Their *anti*-orientation was assigned by the large $^3J_{\text{H-4,H-5}}$, small $^3J_{\text{H-5,C-3}}$, and small $^3J_{\text{H-4,C-6}}$ values. The orientation of H-4 *gauche* to 5-OH and *anti* to H-5 was supported by the NOE correlation of H-36 with H-5. The coupling constant values of $^3J_{\text{H-7,H-8a}} \sim 9\,\text{Hz}$, $^3J_{\text{H-7,H-8b}} \sim 7\,\text{Hz}$, and $^3J_{\text{H-8b,C-6}} \sim 2\,\text{Hz}$ led the authors to consider two major rotamers around the C-7 and C-8 carbon bond where H-5 and the two H-8 exhibit both *gauche*- and *anti*-orientations. All the other relative configurations were elucidated by the same method (Figure 7.14) [5].

Another approach for the assignment of relative and absolute configuration of acyclic compounds is based in the Universal NMR Database developed by

Kishi's group [38,39]. It is an empirical procedure that relies on the comparison of the observed chemical shifts of compounds with unknown vicinal or proximal centers with libraries of configurationally known stereomodels. These ^{1}H and ^{13}C NMR databases were first created for the stereochemical assignment of 1,3,5-triols [37,38,40] and then extended for 1,2,3-triols, 1,2,3,4-tetraols, and 1,2,3,4,5-pentaols [41–43]. Using this strategy, the relative configuration of 1,3-diol system at C7/C-9 of caylobolide B **12** was assigned as a *syn*-arrangement since it fits with the carbon chemical shifts of a 1,3-diol model system. The 1,3,5-triol unit was assigned as either *syn*/*anti* or *anti*/*syn* between C-25/C-27 and C-27/C-29, comparing the carbon chemical shifts at C-27 with those of the central carbon of the 1,3,5-triol model system (Figure 7.15). This method cannot differentiate between *syn*/*anti* or *anti*/*syn* orientation. Mosher's method (see later) was also inconclusive on the absolute configuration of **12**, due to the small amount of the sample [12].

The stereochemistry of chiral marine natural products can also be determined based on chiral derivatizing agents (CDA) and ^{1}H NMR experiments. This methodology requires the transformation of a chiral substrate into two different species, by derivatization of one enantiomeric substrate with chiral auxiliary reagents to give two diastereomers or conformers, which can be differentiated by NMR. This can be achieved by two main approaches in which a pure enantiomeric substrate is separately derivatized with two enantiomers of the chiral auxiliary (double derivatization) or with a single enantiomer of the chiral auxiliary (single derivatization). In the former case, the chiral substrate is derivatized with the two enantiomers of the chiral auxiliary (*R* and *S*) and the ^{1}H NMR spectra of the obtained diastereomers are compared between them. In the latter case, the chiral substrate is coupled with only one enantiomer of the chiral auxiliary (either *R* or *S*), and the ^{1}H NMR of the resulting diastereomer is recorded. Then, a conformational change is induced (e.g., temperature change or chelate formation) and a second ^{1}H NMR spectra is taken. The differences between these two spectra and their comparison with the induced conformational changes correlate the configuration of the substrate with that of the reagent. Another option does

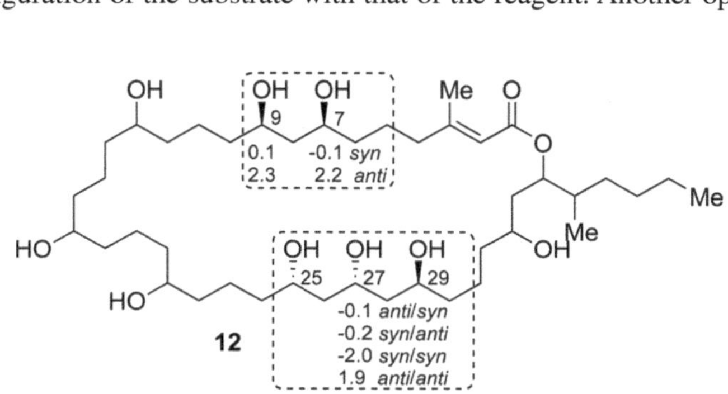

12

FIGURE 7.15 Difference in the ^{13}C NMR chemical shifts of 1,3-diol and 1,3,5-triol of caylobolide B **12** with those of similar model systems.

not involve conformational changes, but a simple comparison of the diastereomer spectrum with that of the substrate prior to derivatization.

As described earlier, two different spectra are always obtained and compared. The sign of the differences in the chemical shifts of the substituents of the asymmetric carbon of the substrate in the two diastereomers (or conformational species) provides information about the configuration of this asymmetric center. For example, in the Mosher's method a chiral secondary alcohol can be derivatized with (*R*)- and (*S*)-α-methoxy-α-phenylacetic acid (MPA) or with α-methoxy-α-trifluoromethyl-α-phenylacetic acid (MTPA) and the differences in the chemical shifts $\Delta\delta^{RS}$ for MPA and $\Delta\delta^{SR}$ for MTPA calculated. $\Delta\delta^{RS}$ is the difference between the chemical shift in the (*R*)-MPA ester derivative and that of the (*S*)-MPA while $\Delta\delta^{SR}$ is the difference between the chemical shift in the (*S*)-MTPA ester derivative and that of the (*R*)-MTPA [44,45].

Since the original Mosher's method [46,47], NMR procedures for the assignment of absolute configuration of chiral compounds have been improved and a large number of CDAs is available for the study of different substrates. The most common one is α-methoxy-α-trifluoromethyl-α-phenylacetic acid (MTPA) known as Mosher's reagent, α-methoxy-α-phenylacetic acid (MPA), α-methoxy-α-(2-naphthyl)acetic acid (2-NMA), α-(9-anthryl)-α-methoxyacetic acid (9-AMA), Boc-phenylglycine (BPG), among others. Over the years, the scope of Mosher's method has been expanded to highly hindered secondary (*sec*) alcohols, primary (*prim*) alcohols, tertiary alcohols, primary and secondary amines, and of *sec*/*sec*, *sec*/*prim*, or *prim*/*sec* 1,n-diols and 1,2-amino alcohols, and carboxylic acids [44,45,48].

Mosher's method can use ^{19}F NMR, particularly the chemical shifts of the CF_3 groups, and ^{1}H and ^{13}C NMR, namely the chemical shifts of the protons and carbons of the chiral alcohol under study, to deduce the absolute configuration of the substrate based on the signs of the $\Delta\delta$ values and using certain empirical models [44]. In this chapter the ^{1}H NMR models will be described since it is the most frequently used in the literature. This model is based on the anisotropic effect that the α-phenyl group of the auxiliary MPA and MTPA exerts on the substituents Z^1 and Z^2 of the alcohol. This effect establishes a correlation of the spatial position of Z^1 and Z^2 with respect to the phenyl of the MPA and MTPA moieties, based on the signs of, respectively, $\Delta\delta^{RS}$ or $\Delta\delta^{SR}$ of the substituents. In these methods it is assumed that the most representative conformation have H-1′, the carbonyl group and the methoxyl group in the same plane in the case of MPA and H-1′, the carbonyl group and the trifluoromethyl group in the same plane in the case of MTPA [44] (Figures 7.16 and 7.17).

The protons of the substituent Z^2 are shielded in the (*R*)-MTPA ester while those of the Z^1 remain unaffected. In the (*S*)-MTPA ester the opposite occurs: Z^1 are shielded while those of the Z^2 remain unaffected. These shieldings are expressed as $\Delta\delta^{SR}(Z^1) = \delta Z^1(S) - \delta Z^1(R)$ and $\Delta\delta^{SR}(Z^2) = \delta Z^2(S) - \delta Z^2(R)$. All the protons shielded in the (*R*)-MTPA derivative will present a positive $\Delta\delta^{SR}$ while those shielded in the (*S*)-MTPA ester will present a negative $\Delta\delta^{SR}$ value (Figure 7.16).

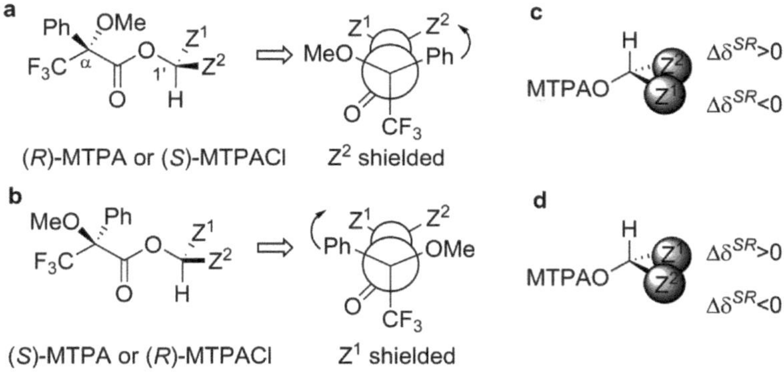

FIGURE 7.16 Model for configurational correlation of MTPA esters: (a, b) assignment by ^{1}H NMR and (c, d) expected sign of $\Delta\delta^{SR}$. *Adapted from [44].*

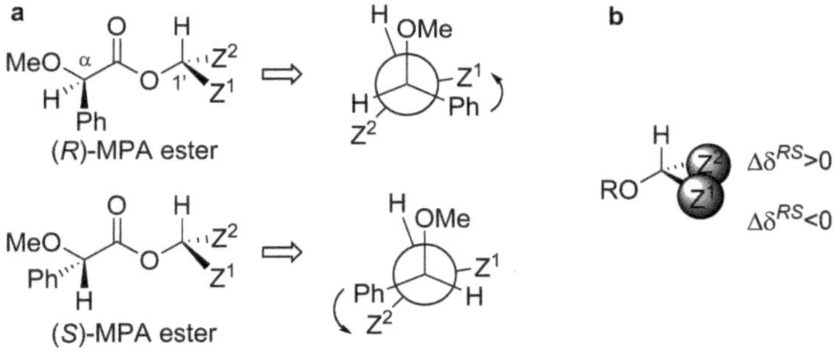

FIGURE 7.17 Model for configurational correlation of MPA esters: (a) assignment by ^{1}H NMR and (b) expected sign of $\Delta\delta^{RS}$. *Adapted from [44].*

In the case of the (*R*)-MPA ester substituent Z^1 is shielded by the phenyl ring whereas Z^2 is unaffected. In the case of the (*S*)-MPA ester substituent Z^2 is shielded whereas Z^1 is unaffected (Figure 7.17) [44].

The absolute configuration of two new and two already known potential antifungal and antitumoral azaphilones, isolated from the endophytic fungus *Chaetomium globosum*, were determined by the Mosher's method [49]. A solution of the new 5′-epichaetoviridin A **35** (Figure 7.18) in dichloromethane was treated with (*R*)- or (*S*)-MPA acids and the mixture was stirred at room temperature for 24 h. The two diastereomeric esters were purified by column chromatography and the analysis of their ^{1}H NMR spectra revealed positive $\Delta\delta^{RS}$ values for H-4′ and 4′-Me (+0.101 and +0.105, respectively) and a negative value for H-6′ (−0.175), that applying the MPA criteria, assigns (*S*)-configuration for C-5′ in compound **35**. A similar procedure was used for the preparation of the (*R*)- and (*S*)-MTPA esters of the known chaetoviridin D **36** (Figure 7.18). Their ^{1}H NMR data revealed negative $\Delta\delta^{SR}$ values for H-10, H-11 and 11-Me

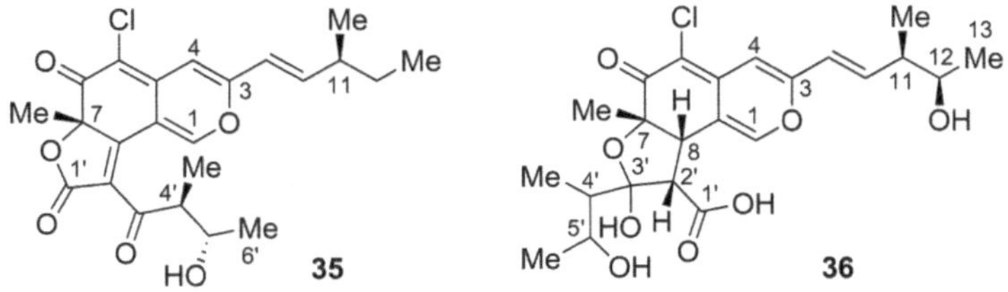

FIGURE 7.18 Structures of 5′-epichaetoviridin A **35** and chaetoviridin D **36**.

(−0.14, −0.062, and −0.086, respectively) and a positive $\Delta\delta^{SR}$ value for H-13 (+0.043) that following the MPTA rules, assigns the (*R*)-configuration of the secondary hydroxyl group at C-12. It was not possible to determine the C-5′ configuration by this method because it was not possible to derivatize the secondary alcohol linked to this carbon. Comparative analysis of the ^{1}H NMR data of chaetoviridin D **36** with chaetoviridin A already reported [50] suggested (*S*)-configuration at both C-7 and C-11. NOE experiments also confirmed this disposition.

Advanced Mosher's method was also quite useful in elucidating the absolute configuration of the fijiolides A **37** and B **38**, inhibitors of TNF-α-NF-kB activation obtained from a marine-derived bacterium of the genus *Nocardiopsis*, that possess in their structures an aminosugar, an NH group, and an OH group [51].

The absolute configuration of the ribopyranose moiety was determined by the calculation of $\Delta\delta^{SR}$ values of the bis-MPTA esters of fijiolide A **37a,b** (Figure 7.19) that indicate (*R*)-configuration at C-2′, allowing the establishment of the stereochemistry of the remaining chiral centers of the ribopyranose moiety as (1′*S*, 2′*R*, 3′*R*, 4′*S*) based on the vicinal coupling constants ($^{1}J_{C\text{-}1',H\text{-}1'}$ 161 Hz indicates a β-configuration for the anomeric carbon, $^{3}J_{H\text{-}1',H\text{-}2'}$ 8 Hz indicates an axial configuration for H-2′) and NOESY spectroscopy data (NOE of H-2′ with H-3′ and of H-3′ with H-4′ establish that H-2′ and H-3′ are in the axial and equatorial positions, respectively).

The absolute configuration at C-17 was elucidated by the ^{1}H NMR data of MTPA amides of fijiolide B **38a,b** (Figure 7.19) that presented positive $\Delta\delta^{SR}$ values for H-19 and H-23 and negative $\Delta\delta^{SR}$ values for H-16, assigning the (*S*)-configuration at this carbon. To assign the absolute configuration at C-13, the 1,2-diol groups first were protected and then the MTPA derivatization was performed. Calculation of $\Delta\delta^{SR}$ values of the MPTA esters of fijiolide A **37** indicates an (*R*)-configuration at C-13. The absolute configuration of the remaining C-8 and C-9 were established by the correlations presented in the NOESY spectra [51].

The absolute configuration of eunicidiol **39**, a potent anti-inflammatory agent, was unambiguously assigned by the Mosher's method. Comparing the ^{1}H NMR data of the (*S*)-MPA ester **41** of the eunicidiol **39** with that of the corresponding (*R*)-MPA ester **40**, H-10, H-17, and H-18 of the former were strongly shielded by the phenyl ring while H-16a,b were less shielded. Thus, the $\Delta\delta^{RS}$

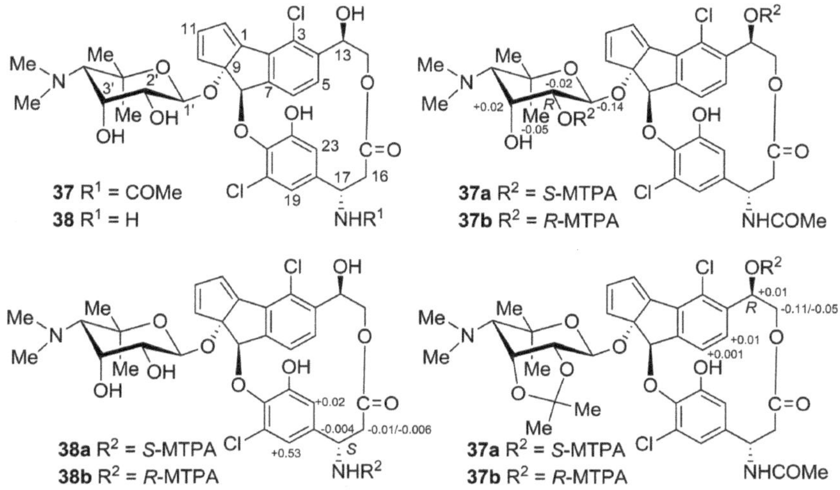

FIGURE 7.19 Chiral derivatization products of fijiolide A **37** and B **38**.

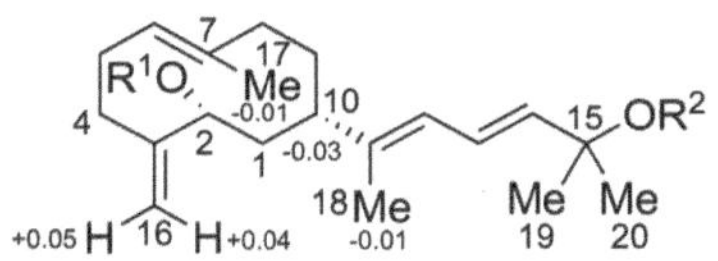

39 R^1 = R^2 = H
40 R^1 = (*R*)-MPA, R^2 = Me
41 R^1 = (*S*)-MPA, R^2 = Me

FIGURE 7.20 Structures of eunicidiol **39** and corresponding (*R*)- and (*S*)-MPA esters **40** and **41** and the $\Delta\delta^{RS}$ data for the MPA esters **40** and **41**.

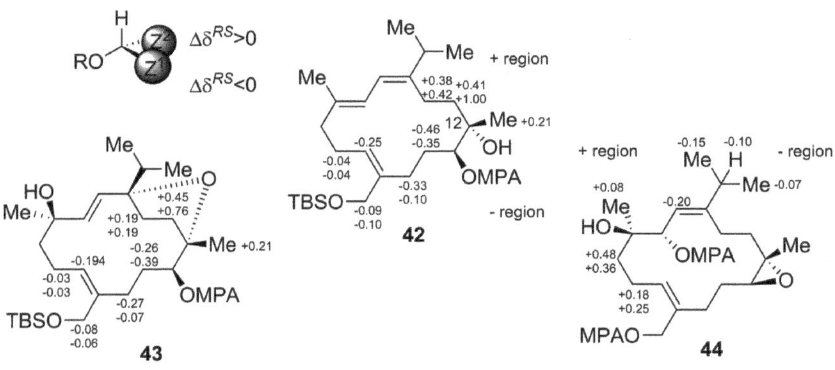

FIGURE 7.21 The $\Delta\delta^{RS}$ data for the MPA esters of sinulariols A **42**, D **43**, and L **44**.

values indicate the (2*R*)-configuration, which combined with the NOESY NMR data, provided the absolute configuration of **39** as (2*R*,10*S*) [52] (Figure 7.20). A similar procedure was applied to determine the absolute configuration of C-12 of the antimicrobial sinulariols A **42**, D **43**, and L **44** [53] (Figure 7.21).

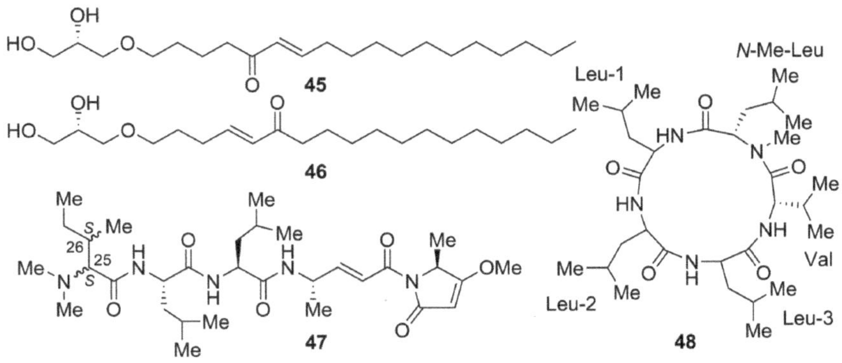

FIGURE 7.22 Structures of compounds **45–48**.

The total synthesis of marine natural products and the comparison of the spectra from the synthetic samples with those of the natural ones assigns the stereochemistry of the natural compounds. For example, the structures of the androgen receptor antagonists niphatenones A **45** and B **46** were assigned by NMR and MS data and confirmed by total synthesis [54]. A comparison of the NMR data of the isolated antimalarial agent gallinamide A **47** and of the four synthetic diastereomers assigned the absolute configuration as (25*S*,26*S*) at the *N*,*N*-dimethylated isoleucyl residue (Figure 7.22). Besides that, it helped to demonstrate that no tautomerization and epimerization reactions took place in the preparation of the analogues and the retention of the stereochemical integrity in the key steps [55].

The planar structure of the antimicrobial lajollamide A **48** (Figure 7.22), a cyclopentapeptide isolated from marine fungus *Asteromyces cruciatus*, was determined by NMR spectroscopy in combination with mass spectrometry but the absolute configuration was unambiguously solved by total synthesis [56]. This synthetic sequence provided three additional unnatural diastereomers and showed that the unexpected acid-mediated epimerization of leucine units in the lajollamide framework can be a danger in misassigning natural products structure when elucidated by spectroscopic methods alone.

In the case of the thrombin and Factor VIIa inhibitor dysinosin A **49**, the structure was determined by detailed NMR studies and its absolute stereochemistry deduced from an X-ray structure of a complex with thrombin [57]. Therefore, the total enantioselective synthesis of dysinosin A **49** provides definitive evidence for its structural and configurational identity [58] (Figure 7.23).

The structure of elisabethin A **50**, which was also named elisabethane, was established by spectroscopic methods and confirmed by an X-ray diffraction experiment, which also yielded its relative stereochemistry [59]. However, the absolute configuration of this molecule was not determined by the X-ray experiment, being accomplished only by their total synthesis [60]. The same occurred

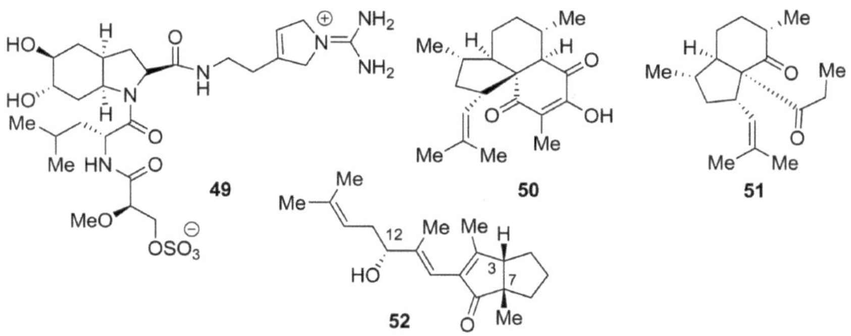

FIGURE 7.23 Structures of compounds **49–52**.

with elisabethin C **51**, a *bisnor*-elisabethane skeleton possessing antituberculosis activity [61] (Figure 7.23).

The structure of xestenone **52**, a marine norditerpenoid found in the northeastern Pacific sponge *Xestospongia vanilla*, was established by spectroscopic methods and the relative configuration of their C-3 and C-7 stereogenic centers determined by NOESY spectral analysis. However, the relative configuration of C-12 and the absolute configuration of this compound were only successfully determined by their total synthesis [62] (Figure 7.23).

7.5 HYPHENATED NMR TECHNIQUES

In the last decades, a large variety of publications have emerged involving NMR spectroscopy hyphenated to other techniques to provide useful information for the structural elucidation on the metabolites prior to isolation.

HPLC-NMR (High Pressure Liquid Chromatography coupled to Nuclear Magnetic Resonance) spectroscopy has been successfully used in the identification of metabolites of crude extracts of marine natural products. An advantage of this technique includes recovery of all components *in situ* for further analytical or biological assays and is a nondestructive experiment. The LC-NMR experiments can be performed by on-flow and stop-flow approaches. In the on-flow mode of operation, each fraction of the crude extract is subjected to the LC analysis and subsequent online ^{1}H NMR analysis, a valuable insight for the identification of major components. The limited sensitivity of this approach was increased by the introduction of the stop-flow mode. In the stop-flow measurements, selected chromatographic peaks are trapped in the LC-NMR flow cell and retained for the required period to extend the NMR acquisition times, allowing sensitivity and resolution of the 1D and 2D NMR spectra obtained.

The first reports using this hyphenated technique for the structure elucidation of marine natural products was conducted by Bobzin *et al.* in 2000, taking

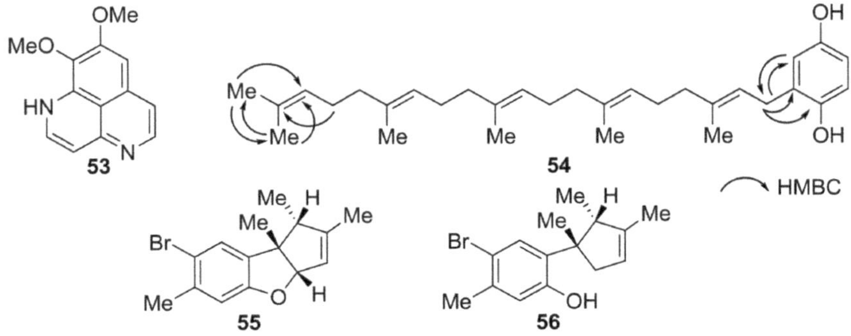

FIGURE 7.24 Structures of compounds **53–56**.

advantage of the new probe designs that allowed the use of gradient pulse sequences to provide efficient and specific suppression of the NMR signals due to the HPLC solvents. LC-NMR has been applied to identify the marine alkaloid aaptamine **53** as the active component (glutamine:fructose-6-phosphate amidotransferase inhibitor) in the crude dichloromethane extract of the sponge *Aaptos* sp. [63,64] (Figure 7.24). Since then, other bioactive compounds have been isolated and characterized (^{1}H, ^{13}C, COSY, ROESY, and one-bond and long-range heteronuclear correlations) from fungal isolates derived from marine sponges using this online stop-flow approach [65,66].

Later, Urban and her group dedicated their more recent works in the identification of several Australian marine natural products using on-flow and stop-flow protocols with a WET solvent suppression sequence [67]. They identified the antitumoral pentaprenylated *p*-quinol **54** as the major constituent in the dichloromethane extract of the marine sponge *Dactylospongia* sp. involving 1D (^{1}H, ^{13}C, and DEPT spectra) and 2D (COSY, HSQC, and HMBC spectra) NMR experiments [68]. Similar studies on the marine alga *Laurencia filiformis* led to the identification of the new potential antimicrobial compounds cycloisoallolaurinterol **55** and isoallolaurinterol **56** from their crude extract as well as the *in situ* monitoring of the chemical conversion between them [69] (Figure 7.24).

Using these strategies, in the crude extract of *Laurencia elata* it was possible to identify, by 1D (^{1}H, ^{13}C, DEPT, and NOE experiments) and 2D (COSY, HSQC, and HMBC spectra), NMR techniques, two novel potential antitumoral C_{16} chamigrenes, cycloelatanene A **57** and B **58**, and several previously reported compounds [70]. In the red alga *Plocamium angustum*, the identification of the cytotoxic agent plocamenone **59** and the unstable double bound isomer isoplocamenone **60** was achieved (Figure 7.25). Conventional offline chemical investigations allowed the structural revision of the former compound as well as the collection of sufficient amount for the biological assays [71]. More recently, seven resorcinol and two polyene secondary

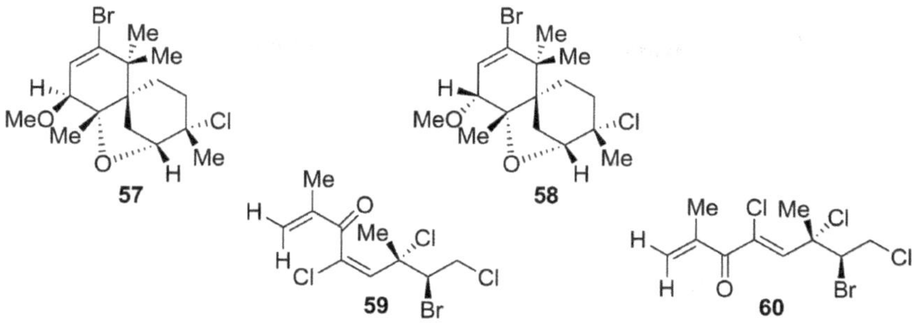

FIGURE 7.25 Structures of compounds **57–60**.

metabolites were obtained from the crude dichloromethane extract of the marine brown alga *Cystophora tondosa* [72]. In most cases, for completion of the structural elucidation, the compounds were isolated by preparative HPLC and subjected to NMR and MS (Mass Spectrometry) measurements. Although the online LC-NMR-MS technique gained an increased importance for a rapid isolation and identification of complex mixtures in clinical, biological, and natural products research [73], to the best of our knowledge the benefits of this hyphenated protocol has not been applied in the field of marine natural products investigation.

SPE (Solid Phase Extraction) coupled to the LC-NMR system emerged in recent years in order to increase the amount of analyte in the flow cell. As the SPE cartridge is placed after the HPLC separation, this approach presents several advantages that reduces drastically the analysis costs: (1) increased amount of sample; (2) HPLC separation without deuterated solvents; (3) wide range of solvents and gradients in the HPLC separation. After the collection of the analyte in the cooled nitrogen SPE cartridge, it is dried with a stream of nitrogen and finally washed with a deuterated solvent prior to injection in the NMR flow cell. A final advantage of this LC-SPE-NMR technique is the use of low-volume probes to maximize sample concentration and NMR data quality [67]. In this field, several reports have appeared, providing comprehensive information about novel products obtained directly from plant crude or prepurified extracts [67,74]. However, the application of this methodology to marine natural products is quite scarce. A single example uses a HPLC-SPE-NMR setup coupled to a diode array detector (DAD) monitoring HPLC separation and triggering SPE trappings. A HPLC-DAD-SPE-NMR system was used to resolve complex natural isolates of marine-derived *Streptomyces* spp. in which four known and 12 new rare reveromycin classes (Figure 7.26 shows two examples of these compounds **61** and **62** and the corresponding key NMR correlations) of antiproliferative polyketide spiroketals were identified. These results were supported by further mass spectrometry analysis using a separate HPLC-DAD-MS installation [75].

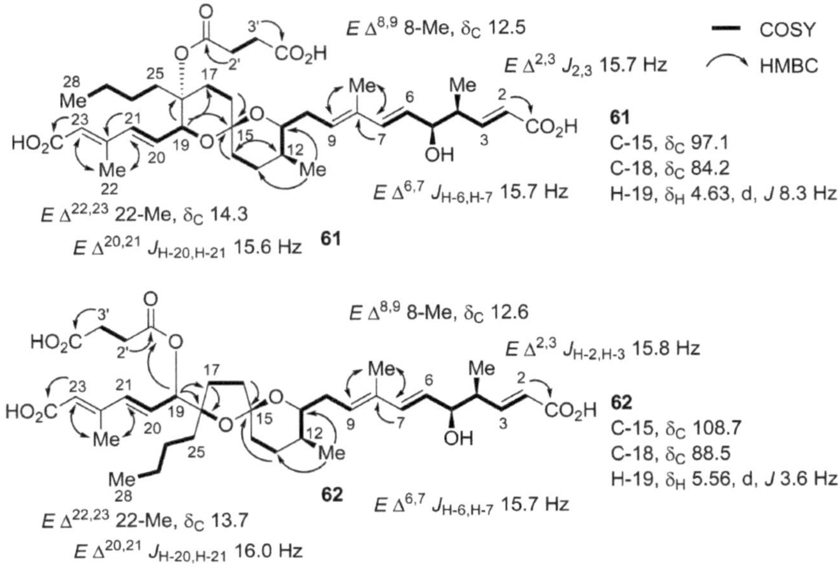

FIGURE 7.26 Key 2D NMR correlations of the 6,6-spiroketal **61** and 5,6-spiroketal **62**.

7.6 QUANTITATIVE NMR

High-resolution NMR spectroscopy is an almost unique tool in the quantitative evaluation of mixtures: it is nondestructive and with a few precautions in terms of parameter choice [76] the integral intensity of a given signal is directly proportional to the number of nuclei contributing to that signal; that is, intensities directly reflect molar ratios. Unlike optical spectroscopy, no calibration is required, and it is not necessary to make assumptions about the identity of the compounds to be determined. Further, compared to quantification based on chromatographic separation, there is no risk (except for enantiomers, since they present the same NMR spectrum) of not being able to resolve the different components. However, a number of factors can distort signal intensities [77].

1D and 2D proton quantitative NMR (qHNMR) has proved to be highly suitable for the simultaneous selective recognition and quantitative determination of metabolites in complex biological matrixes, such as quantitation of single natural products in plant extracts, dietary materials, and materials representing different metabolic stages of (micro)organisms. As a general rule, qHNMR provides an opportunity for the routine determination of constituents at the 1% level with 1% error and is well capable of analyzing possibly down to the 0.1% level and below with errors as low as 0.1%, depending on the nature of the sample [78].

Quantitative NMR possesses the required accuracy and precision to become a routine quantitative tool in many analytical laboratories. Chemical referencing

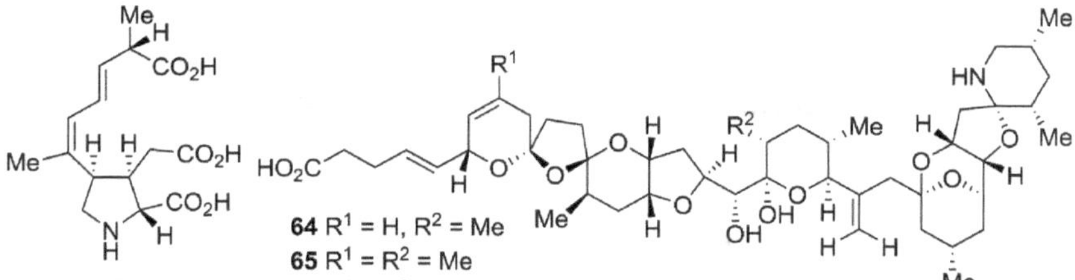

FIGURE 7.27 Structures of compounds **63–66**.

with internal standards can provide accurate quantification, but this technique contaminates the sample. As a consequence the use of external references or the use of ^{13}C satellites as an internal standard in 1H NMR spectroscopy allows this contamination problem to be circumvented. It is noteworthy to notice the ability of this method to function accurately with external calibration, requiring only nonidentical calibrants, and simultaneously quantitating several compounds [79].

Although both internal and external referencing methods have been used, several electronic referencing approaches have also been reported in an attempt to provide accurate and precise quantitative results. A very recent publication makes a comparison of quantitative nuclear magnetic resonance methods using internal, external, and electronic referencing [80].

Burton *et al.* [81,82] discussed the application of qHNMR with external standards (e.g., caffeine) for monitoring phycotoxins (e.g., domoic acid **63**; Figure 7.27), a challenging task because their isolation from marine matrixes such as shellfish or plankton is tedious and generally presents a total yield of low milligram to submilligram range. They concluded that qHNMR has the potential for accurate determination of molar phycotoxin concentrations in crude extracts of marine organisms. This method was used to accurately certify the concentration of certified calibration solutions of the potent marine biotoxins azaspiracid-1 **64**, -2 **65,** and -3 **66** [83] (Figure 7.27).

qHNMR was used to quantify pyrrole-2-carbaldehyde derivatives in the extracts of wild and cultured sponges, using 4-hydroxy-3-methoxybenzaldehyde (vanillin) as an internal standard [84] and also to determine the concentration of stevensine **67** (Figure 7.28) in butanol partitions of crude extracts from the sponge *Axinella corrugate*, using cyclooctatetraene as an internal standard [85]. Compound **67** showed deterred feeding in laboratory aquarium assays.

The Molinski group developed a simple and accurate method for quantitation of small amounts of natural products, taking advantages of the developed technologies in microcoil probes and narrow diameter tube microprobes [86]. They used the solvent ^{13}C-satellites (QSCS) method for the quantitation, which uses the integration of ^{13}C satellite peaks of deuterated solvents, in particular

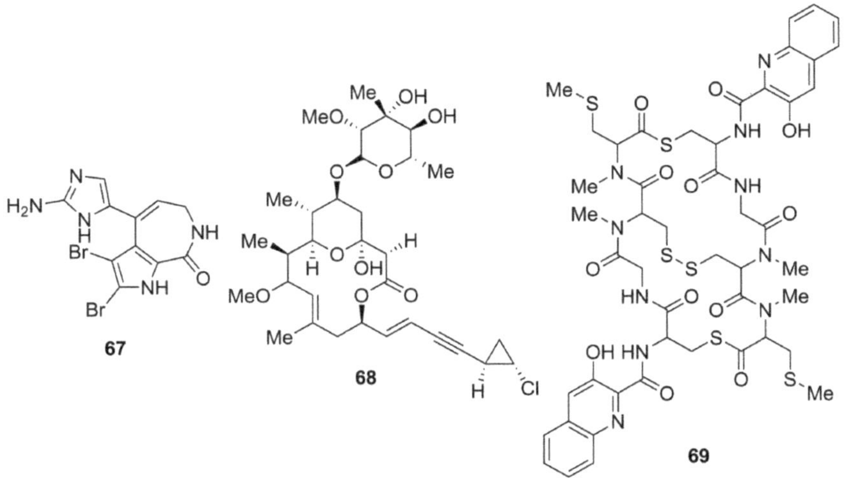

FIGURE 7.28 Structures of compounds **67–69**.

$CDCl_3$, that have favorable intensities for measurements of samples in NMR microcoils and microprobe tubes in the 1 to 200 nanomole range. Highly sensitive cryoprobes allow ready observation of the so-called ^{13}C-satellites of the $^{13}CHCl_3$ signal ($^{1}J_{C\text{-}H}$ 209 Hz), even after only one scan. Although the main $^{12}CHCl_3$ signal intensity is generally too large for accurate quantitation of nanomole samples due to dynamic range considerations, the $^{13}CHCl_3$ solvent satellite integrals are comparable in magnitude to those of typical methine CH signals in the natural product sample being measured. They concluded that the number of moles of a sample, *m*, in a $CHCl_3$ solution is related to the ratio of a signal integral A_{CH} to satellite integrals A_{sat} by a proportionality constant, *c*. This constant may be obtained from the slope of a standard curve of the ^{1}H NMR integral ratio for methine signals of a standard sample, A_{CH}, to residual $^{13}CHCl_3$ integral, A_{sat}, against the amount of a standard in a fixed volume of sample $CDCl_3$ solution.

$$\begin{aligned} m &= (A_{CH} / A_{sat}) / c \\ A_{sat} &= A_L + A_R \text{ (}A_L\text{ being the left-hand and }A_R\text{ right-hand} \\ &\quad \text{components of the }^{1}\text{H-}^{13}\text{C couplet of residual }^{13}\text{CHCl}_3\text{)} \\ A_{CH} &= \text{integral of a methine signal} \end{aligned} \tag{7.1}$$

After establishing the proportionality constant, *c*, in a specific spectrometer, we can apply Eq. (7.1) to determine the amount of a specific metabolite in a $^{13}CHCl_3$ solution. The method was illustrated for the structure elucidation and the quantitation of a new potential cytotoxic natural product phorbaside F **68** from the marine sponge *Phorbas* sp. (Figure 7.28).

QSCS expands the window of discovery and characterization of new natural products, especially those that can be procured only in vanishingly small amounts [87,88].

qHNMR is very useful tool for determining metabolite concentration, but sometimes severe signal overlapping of complex metabolite mixtures hinders the accurate quantification. Markley *et al.* [89] have developed a 2D extrapolated time-zero ^{1}H-^{13}C HSQC ($HSQC_0$) to simultaneously quantify and identify individual chemicals in metabolite mixtures. In this method, the extension of 1D ^{1}H to 2D ^{1}H-^{13}C HSQC leads to the dispersion of peaks along the ^{13}C dimension and greatly alleviates peak overlapping. In $HSQC_0$ the cross-peaks intensities is proportional to concentration of individual metabolites, which can be converted to absolute concentrations by reference to an internal standard of known concentration. This technique has been applied to quantify individual compounds in complex natural mixtures, without isolation or purification. Thiocoraline **69** is a scarce example of a marine product that was selective quantified at a level of 1% (w/w) in an extract from a *Verrucosispora* sp. isolated from the sponge *Chondrilla caribensis* f. *caribensis* [90] (Figure 7.28). This compound has shown potent cytotoxicity against several cancer cell lines [91].

7.7 COMBINED NMR TECHNIQUES AND DIFFERENT FAMILIES OF COMPOUNDS

In this chapter we have described the most important NMR techniques used in the structural characterization of natural compounds from marine sources, always exemplified with the structural elucidation of specific bioactive marine products. Since these compounds present complex and not repetitive structures, making it not possible to compile their NMR characteristics by families, we will describe the full characterization of some derivatives using a combination of the described NMR techniques.

El-Amraoui *et al.* reported the structure elucidation of the antifungal haliscosamine **70** (Figure 7.29), a new sphisgosine derivative, isolated from the Moroccan marine sponge *Haliclona viscosa* [92]. The ^{1}H NMR spectrum revealed the presence of 45 protons in the range of 0.9 to 5.4 ppm, while the ^{13}C APT spectrum (the Attached Proton Test experiment is a common way to assign C-H multiplicities in ^{13}C NMR; the phase of CH and CH_3 peaks is 180° different from C and CH_2 peaks) combined with the HSQC spectrum identified 22 carbons (1 methyl, 15 methylene, and 6 methines) bounded to 39 protons. From

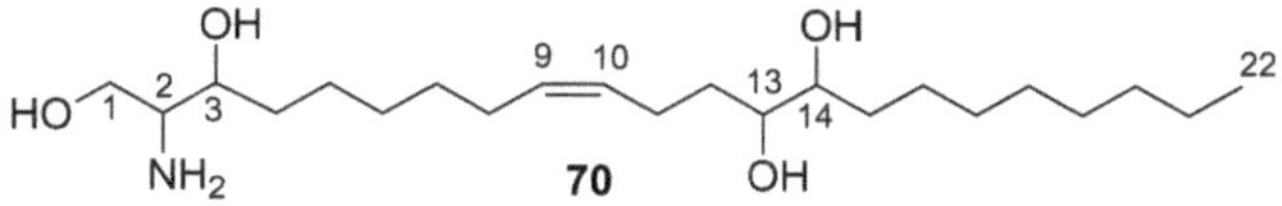

70

FIGURE 7.29 Structure of haliscosamine **70**.

these they have assigned the resonances of oxygenated carbons [59.1 (C-1), 67.7 (C-3), 73.4 (C-13), and 73.8 (C-14) ppm], a probably nitrogenated one [57.7 (C-2) ppm] and two double bounded carbons [129.1 (C-9) and 129.8 (C-10) ppm] (Table 7.1).

TABLE 7.1 NMR Spectroscopic Data of Haliscosamine **70** [92]

Position	δ ^{1}H ppm *J*(Hz)	δ ^{13}C ppm	HMBC	COSY	TOCSY
1a	3.67, dd, *J* 11.7, 4.0	59.1	2, 3	1b, 2, 3	2,3
1b	3.79, dd, *J* 11.7, 4.0		2, 3	1a, 2, 3	2, 3
2	3.07, m	57.7	1, 3, 4	1	1, 3
3	3.69, m	67.7	1, 2, 4, 5	1,4	1, 2, 4, 6
4a	1.47,m	33.5	2, 3, 5		
4b	1.57, m		2, 3, 5		
5a	1.40, m	24.9	3		
5b	1.57,m		3		
6	1.37, m	28.94			
7	1.37, m	29.45	8	8	9
8	2.09, m	26.7	7, 9	7, 9, 10	
9	5.39, m	129.1	8	8, 10	7, 8
10	5.40, m	129.8	11	8, 9, 11	11, 13
11a	2.13, m	23.3	10, 12, 13	10, 12	10, 13
11b	2.24, m		10, 12, 13	10, 12	10, 13
12a	1.54, m	32.6	11, 13		13
12b	1.59, m		11, 13		13
13	3.42	73.4	11, 12, 14, 15		10, 11, 12, 15
14	3.40	73.8	13, 15, 16	15	
15a	1.44, m	32.7	13, 14, 16	14	13
15b	1.54, m		13, 14, 16	14	13
16a	1.37, m	25.6	14, 15		
16b	1.52, m		14, 15		
17	1.37, m	29.17			
18	1.37, m	29.18			
19	1.37, m	29.20			
20	1.33, m	31.7	21, 22		
21	1.36, t, *J* 6.5	22.3	20, 22		
22	0.92	13.1	20, 21		

The referred data together with the high resolution mass spectrum indicate that the remaining six protons must be due to the presence of one NH_2 and four OH groups (this information could be confirmed running the ^1H NMR spectrum with some drops of D_2O).

The information given by the COSY, HMQC, HMBC, and TOCSY spectra allowed to assign three units of the structures C-1 to C-5, C-8 to C-16, and C-20 to C-22, but the connectivities between them were not possible due to the strong overlapping in the ^1H NMR spectrum. In order to solve this problem and to simplify the analysis of the spectrum, the authors performed an extensive study of ^1H–^1H decoupling experiments. (This experiment belongs to a class of experiments known as double-resonance experiments and offers a simple and effective means to identify coupled protons. The idea is to selectively saturate one multiplet in the spectrum with a radio-frequency during acquisition. This causes a loss of all couplings with the saturated proton and, hence, the multiplet structure of its partners will change.) These series of experiments allowed assembly of the entire structure of haliscosamine **70** (Figure 7.29).

Finally the (*Z*)-geometry of the double was established from a ^1H NMR of **70** acquired in CD_3OD with some drops of benzene, which increased the resolution in the vinylic part of the spectrum and allowed to determine the characteristic coupling constant value of 10 Hz. The small difference between the carbon resonances of C-8 (26.8 ppm) and C-11 (23.3 ppm) also confirms the stereochemistry of this double bond.

Jaspers *et al.* [93] described the discovery and structure elucidation of the cytotoxic laurefurenynes A-F **71-76** (Figure 7.30), cyclic ether acetogenins from the aqueous extract of the marine red alga *Laurencia* sp. The analysis of the ^1H and ^{13}C NMR spectra in $CDCl_3$ of laurefurenyne A **71** (Tables 7.2 and 7.3) revealed the presence of a C_{15} acetogenin, bearing one methyl group (belonging to an ethyl group), four methylenes, six oxygenated-sp^3 methines, two sp^2 methines, one sp methine, and a quaternary carbon. Based on the coupling constants and COSY and HSQC spectra it was possible to establish a *cis*-ene-yne moiety [6.06 (H-4, dt, *J* 10.9 and 7.9 Hz), 5.57 (H-3, dd, *J* 10.9 and 2.0 Hz), 3.10 (H-1, d, *J* 2.0 Hz); 139.9 (C-4), 111.3 (C-3), 80.0 (C-2), and 82.4 (C-1) ppm].

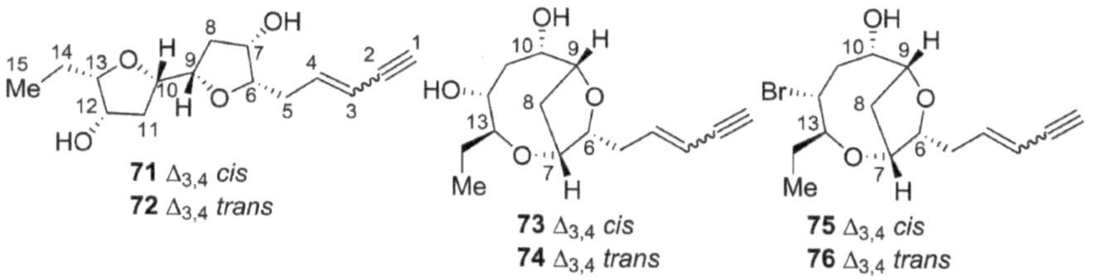

FIGURE 7.30 Structures of laurefurenynes A-F **71–76**.

TABLE 7.2 ^{1}H NMR Spectroscopic Data of Laurefurenynes A-F 71-76 [93]

Position	71 δ ^{1}H ppm *J*(Hz)	72 δ ^{1}H ppm *J*(Hz)	73 δ ^{1}H ppm *J*(Hz)	74 δ ^{1}H ppm *J*(Hz)	75 δ ^{1}H ppm *J*(Hz)	76 δ ^{1}H ppm *J*(Hz)
1	3.10, d, *J* 2.0	2.83, d, *J* 2.1	3.09, d, *J* 2.1	2.76, d, *J* 1.6	3.07, d, *J* 2.0	2.79, d, *J* 1.7
2	—	—	—	—	—	—
3	5.57, dd, *J* 10.9, 2.0	5.58, dd, *J* 15.5, 2.1	5.55, dd, *J* 11.2, 2.1	5.59, dd, *J* 16.0, 1.6	5.51, dd, *J* 10.9, 2.0	5.55, dd, *J* 15.9, 1.7
4	6.05, dt, *J* 10.9, 7.9	6.22, dt, *J* 15.5, 7.6	6.08, dt, *J* 11.2, 8.0	6.26, dt, *J* 16.0, 7.5	6.01, dt, *J* 10.9, 7.8	6.18, dt, *J* 15.9, 7.5
5a	2.61, m	2.44, m	2.85, m	2.54, m	2.79, m	2.56, m
5b		2.37, m	2.69, m		2.68, m	2.45, m
6	3.90, ddd, *J* 10.1, 6,5, 3.8	3.90, ddd, *J* 10.2, 6.6, 3.7	3.87, m	3.73, m	3.72, ddd, *J* 8.8, 7.1, 2.0	3.67, ddd, *J* 8.6, 7.3, 2.1
7	4.12, m	4.10, m	4.02, m	3.99, dd, *J* 3.6, 2.0	3.94, m	3.94, m
8a	1.86, m	1.85, m	2.98,d, *J* 15.2	3.02, d, *J* 16.2	2.24, m	2.24, d, *J* 16.1
8b	1.69, m	1.68, m	1.96, m	1.93, m	1.92, m	1.92, m
9	4.37, m	4.40, m	4.30, m	4.29, dd, *J* 9.8, 3.9	4.10,dd, *J* 9.6, 4.0	4.10, dd, *J* 9.6, 3.9
10	4.14, m	4.16, m	4.09, m	4.07, m	3.96, m	3.96, m
10-OH	—	—	3.34, br s	3.37, br s	—	—
11a	2.19, m	2.20, m	2.21, m	2.20, m	3.01, dddd, *J* 15.3, 11.3, 9.7, 1.8	2.98, dddd, *J* 15.4, 11.2, 9.8, 1.8
11b	1.81, dd, *J* 14.5, 3.1	1.79, dd, *J* 13.9, 3.0			2.28, dd, *J* 15.2, 5.1	2.28, dd, *J* 15.1, 5.0
12	3.98, dd, *J* 5.1, 2.3	4.01, dd, *J* 5.5, 2.3	4.29, m	4.30, m	4.55, t, *J* 9.6	4.54, t, *J* 9.6
12-OH	—	—	2.72, br s	2.80, br s		
13	3.49, ddd, *J* 13.6, 6.8, 2.4	3.52, ddd, *J* 9.4, 6.9, 2.4	3.81, m	3.76, m	3.86, ddd, *J* 11.8, 9.6, 2.8	3.84, ddd, *J* 13.5, 9.6, 2.4
14	1.66, m	1.69, m	1.74, m, 1.57, m	1.77, m, 1.57, m	1.93, m, 1.66, m	1.96, m, 1.65, m
15	0.94, t, *J* 7.5	0.97, t, *J* 7.5	1.01, t, *J* 7.5	0.98, t, *J* 7.5	1.02, t, *J* 7.2	1.01, t, *J* 7.2

TABLE 7.3 ^{13}C NMR Spectroscopic Data of Laurefurenynes A–F 71–76 [93]

Position	71 δ ^{13}C ppm	72 δ ^{13}C ppm	73 δ ^{13}C ppm	74 δ ^{13}C ppm	75 δ ^{13}C ppm	76 δ ^{13}C ppm
1	82.4	76.7	81.9	76.2	81.7	76.6
2	80.0	81.7	—	82.2	80.7	81.8
3	111.3	112.0	110.0	110.8	110.2	111.3
4	139.9	140.7	141.8	142.5	141.5	142.1
5	34.1	37.1	30.4	32.8	30.1	32.8
6	85.7	85.4	83.0	82.9	83.2	83.4
7	74.8	74.9	71.0	70.2	70.2	70.1
8	37.1	37.3	32.0	31.7	33.1	33.2
9	79.1	79.1	79.5	79.7	78.6	78.4
10	78.2	78.1	73.3	73.4	70.5	70.5
11	34.5	34.5	30.8	29.7	38.8	38.9
12	70.8	70.8	72.8	73.1	51.8	51.8
13	85.6	85.7	78.9	79.3	83.2	83.5
14	21.8	21.8	23.1	22.3	23.1	23.2
15	10.5	10.8	11.3	11.4	11.8	11.9

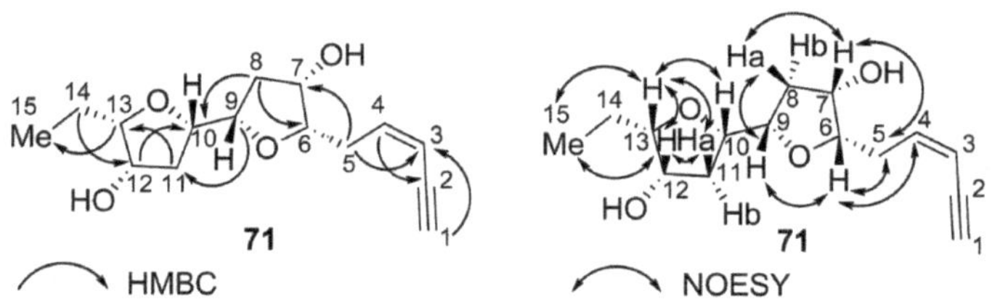

FIGURE 7.31 Main HMBC and NOESY correlations of laurefurenyne A **71**.

The combined analysis of the COSY, selective 1D TOCSY and HSQC spectra established a continuous chain of carbon atoms (C-1 to C-15) of the C_{15} acetogenin, which was confirmed by the HMBC correlations (Figure 7.31). In order to circumvent some doubts around the two ether bridges of the structure due to an extensive overlapping of signals, the authors acquired the ^1H NMR spectrum of laurefurenyne A **71** in DMSO-d_6, which combined with a new COSY spectrum allowed the location of the two hydroxyl groups at C-7 and C-12.

Since there were three possible carbon frameworks, a deep analysis of the carbon chemical shifts led to the proposal of a central 2,2′-bifuran skeleton. The stereochemistry around the two tetrahydrofuran rings was established by a 2D NOESY spectrum (Figure 7.31). The relative stereochemistry of the two tetrahydrofuran rings was difficult and needed some modeling, which confirmed the strong NOE correlations of H-9 with both H-10 and H-11. Thus the relative stereochemistry of laurefurenyne A **71** was assigned as (6*S**,7*S**,9*S**,10*S**,12*S**,13*S**).

The ^{1}H and ^{13}C NMR spectra of laurefurenyne A **71** and B **72** were very similar (Tables 7.2 and 7.3), except for the coupling constant of the vinylic system (*J* 15.5 Hz) of **72**, which is consistent with a *trans*-configuration consistent with a *trans*-ene-yne moiety.

The presence of a *cis*-ene-yne moiety of laurefurenyne C **73** was easily detected as was laurefurenyne A **71** (Table 7.2). The C1-C15 contiguous spin system was inferred from a deep analysis of the COSY and HSQC spectra of laurefurenyne C **73** acquired in both $CDCl_3$ and DMSO-d_6 due to strong signal overlapping, but the two hydroxyl groups in this case were located at C-10 and C-12. This was confirmed by a selective 1D TOCSY experiment as well as by HMBC correlations, which allows the establishment of the new location of one of the ether linkages to C7-C13 (H-7 → C-13) and leaving the other ether bridge in the C6-C9 carbons. Thus the carbon skeleton of laurefurenyne C **73** was identified as a 2,8-dioxabicyclo[5.2.1]decane. Once again, the stereochemistry of the stereogenic centers of this laurefurenyne C **73** was established by 2D NOESY experiments (NOE cross-peaks of H-8 with H-6, H-7, H-9; of H-8a with H-7, H-10, H-11b, H-12, H-14; and of H-7 with H-15) and assigned as (6*R**,7*R**,9*R**,10*S**,12*R**,13*S**).

The structural difference of laurefurenyne D **74** relatively to the C derivative was only the stereochemistry of their vinylic system, which was identified by the characteristic coupling constant (*J* 16.0 Hz) of a *trans*-configuration. All the other spectroscopic data (Tables 7.2 and 7.3) and 2D NOESY correlations of laurefurenyne D **74** were very similar to those of laurefurenyne C **73**.

The high resolution mass spectra (HRMS) of laurefurenynes E **75** and F **76** revealed that a bromine replaced an hydroxyl group in the structure of the corresponding laurefurenynes C **73** and D **74**, while the ^{1}H and ^{13}C NMR indicate that it was the 12α-hydroxyl group replaced by a 12α-bromine substituent (Tables 7.2 and 7.3).

Four antimicrobial fatty acids, ieodomycins A-D **77-80** (Figure 7.32) [94], have been isolated from the ethyl acetate extract of a culture of a marine *Bacillus* sp. The UV spectrum of **77** indicated the presence of a conjugated diene, while the IR revealed the presence of hydroxyl and ester carbonyl groups. The ^{1}H, ^{13}C NMR, and HSQC spectra of ieodomycin A **77** showed four vinylic carbons (115.1–140.1 ppm), including a terminal vinylic methylene (115.1 ppm), two oxygenated methines, two methyls, one of them oxygenated, and an ester

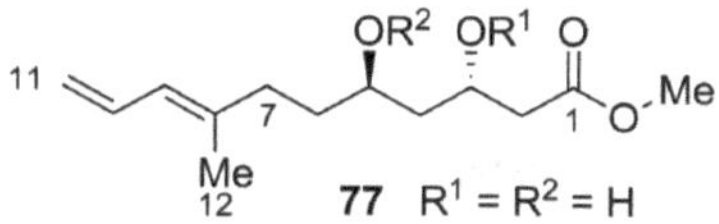

77 $R^1 = R^2 = H$
77a $R^1 = R^2 =$ (*S*)-MTPA
77b $R^1 = R^2 =$ (*R*)-MTPA

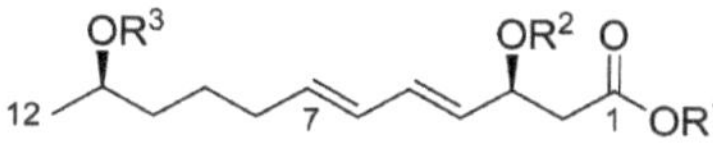

79 $R^1 = R^2 = R^3 = H$
79a R^1 = Me, $R^2 = R^3 = H$
79b R^1 = Me, $R^2 = R^3 =$ (*S*)-MTPA
79c R^1 = Me, $R^2 = R^3 =$ (*R*)-MTPA

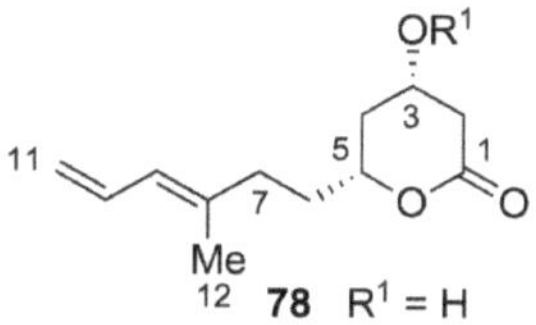

78 $R^1 = H$
78a $R^1 =$ (*S*)-MTPA
78b $R^1 =$ (*R*)-MTPA

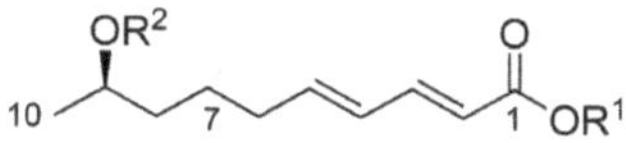

80 $R^1 = R^2 = H$
80a R^1 = Me, $R^2 = H$
80b R^1 = Me, $R^2 =$ (*S*)-MTPA
80c R^1 = Me, $R^2 =$ (*R*)-MTPA

FIGURE 7.32 Structures of ieodomycins A–D **77–80** and their corresponding MTPA esters.

carbonyl carbon (Tables 7.4 and 7.5). The COSY spectrum indicated two isolated spin systems (H-2 to H-7 and H-9 to H-11), their connectivity being established through the HMBC correlations (H-7 → C-8 and C-9) (Figure 7.33). The other HMBC correlations confirmed all the other connectivities of ieodomycin A **77**. The ROESY correlations of H-10 with H-12 and of H-9 with H-11b establishes the (*E*)-stereochemistry of the C8=C9 double bond (Figure 7.33). The carbon resonances of C-3 and C-5 are consistent with an *anti*-arrangement of the 1,3-diol model system, according to the Kishi's Universal NMR Database [41,42] (Figure 7.34). Moreover the absolute configuration was established by the modified Mosher's method, preparing the esters (*S*)-MTPA-**77a** and (*R*)-MTPA-**77b** starting from ieodomycin A **77** (Figure 7.34). The negative values of $\Delta\delta^{SR}$ of H-3 (–0.11 ppm), H-4 (–0.12/-0.13 ppm), and H-5 (–0.17 ppm) correspond to an *anti*-1,3-diols [95], indicating a (3*S*,5*R*) configuration. Thus the structure of ieodomycin A **77** was established as (3*S*,5*R*,8*E*)-methyl 3,5-dihydroxy-8-methylundeca-8,10-dienoate.

The spectroscopic data (^{1}H, ^{13}C NMR, and HRMS) of ieodomycin B **78** were very similar to those of ieodomycin A **76**, but having one less carbon, four hydrogens, and one oxygen atom. The comparison of ^{1}H and ^{13}C NMR spectra of **77** and **78** revealed that the latter does not have the methoxyl group and its C-1 was attached to the oxygen at C-5, forming a δ-lactone ring (Figure 7.32; Tables 7.4 and 7.5). The configuration of ieodomycin B **78** was assigned based on the ^{1}H–^{1}H coupling constants and ROESY correlations, which suggested a chair conformation for the δ-lactone ring (Figure 7.33). The absolute configuration was also established by application of the modified Mosher's method, which indicate a (3*S*)-configuration (Figure 7.34). The (5*R*)-configuration was established based on the ROE cross-peaks observed in the ROESY spectrum that indicate an *anti*-relationship between C-3 and C-5. All the spectroscopic data support the structure of ieodomycin B **78** as

TABLE 7.4 ^{1}H NMR Spectroscopic Data of Ieodomycins A–D 77–80 [94]

Position	77 δ ^{1}H ppm *J*(Hz)	78 δ ^{1}H ppm *J*(Hz)	79 δ ^{1}H ppm *J*(Hz)	80 δ ^{1}H ppm *J*(Hz)
1				
2	2.47, dd, *J* 10.0, 5.0	2.36, dd, *J* 16.8, 7.3 (H-2ax)	2.45, d, *J* 6.0	5.78, d, *J* 15.3
		2.86, dd, *J* 16.8, 1.3 (H-2eq)		
3	4.25, m	4.19, dddd, *J* 10.4, 7.3, 3.5, 1.3 (H-3ax)	4.50, m	7.21, dd, *J* 15.3, 10.5
4	1.52, m	2.24, ddd, *J* 13.5, 10.4, 7.3 (H-4ax)	5.60, dd, *J* 15.3, 6.5	6.24, dd, *J* 15.3, 10.5
		2.27, ddd, *J* 13.5, 4.2, 3.5 (H-4eq)		
5	3.76, m	4.24, dddd, *J* 10.4, 7.3, 4.2, 3.5 (H-5ax)	6.22, dd, *J* 15.3, 10.5	6.16, m
6	1.56, m	1.80, m	6.03, dd, *J* 15.3, 10.5	2.20, m
7	2.10, m	2.18, m	5.70, dt, *J* 15.3, 7.0	1.45, m
	2.19, m	2.25, m		1.56, m
8			2.10, m	1.43, m
9	5.86, d, *J* 10.5	5.88, d, *J* 10.6	1.38, m	3.72, m
10	6.57, ddd, *J* 16.7, 10.5, 10.5	6.58, ddd, *J* 16.6, 10.6, 10.6	1.50, m	1.14, d, *J* 6.0
11	4.93, d, *J* 10.5 (H-11b)	4.96, dd, *J* 10.6, 2.2, (H-11b)	1.42, m	
	5.04, d, *J* 16.7 (H-11a)	5.07, dd, *J* 16.6, 2.2 (H-11a)	3.70, m	
12	1.75, s	1.77, s	1.13, d, *J* 6.0	
OMe	3.67, s			

TABLE 7.5 ^{13}C NMR Spectroscopic Data of Ieodomycins A–D 77–80 [94]

Position	77 δ ^{13}C ppm	78 δ ^{13}C ppm	79 δ ^{13}C ppm	80 δ ^{13}C ppm
1	174.0	174.1	175.6	171.5
2	43.9	40.2	43.7	121.3
3	66.6	64.4	70.2	146.5
4	45.3	38.7	133.6	130.0
5	68.8	78.5	132.1	145.3
6	37.5	34.9	131.0	34.0
7	37.0	36.1	136.1	26.2
8	140.1	139.0	33.7	39.7
9	127.0	127.6	26.7	68.4
10	134.7	134.5	39.8	23.6
11	115.1	115.6	68.5	
12	16.8	16.6	23.6	
OMe	52.1			

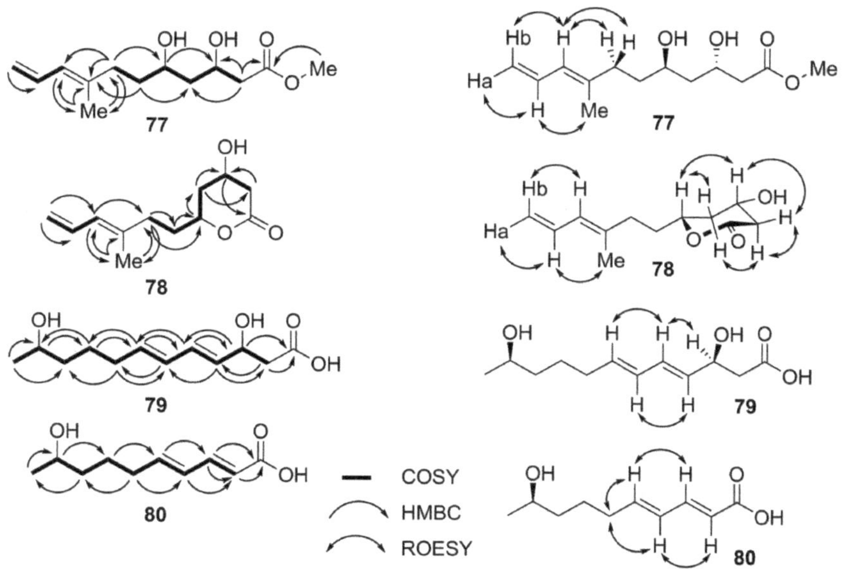

FIGURE 7.33 Main COSY, HMBC, and ROESY correlations of ieodomycins A–D **77–80**.

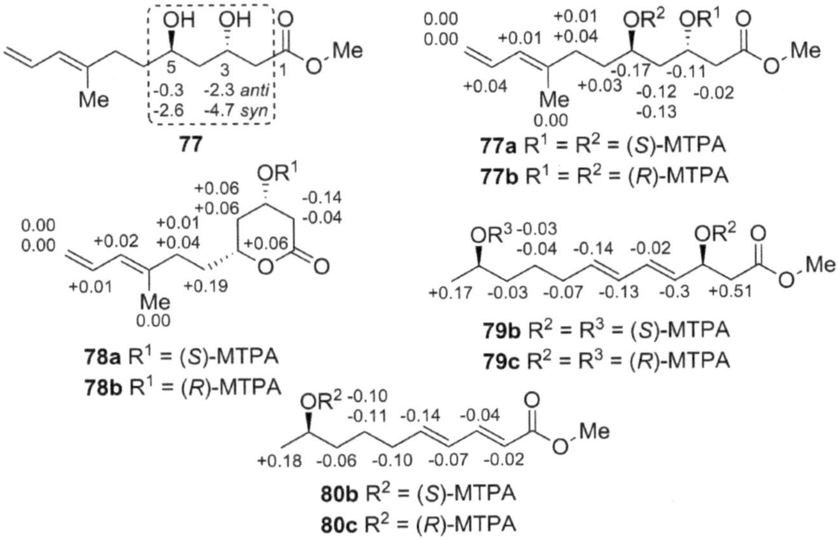

FIGURE 7.34 The $\Delta\delta^{SR}$ data for the MTPA esters of ieodomycins A–D **77–80**.

(3*S*,5*R*,8*E*)-3-hydroxy-5-(8-methylhexa-8,10-dienyl)tetrahydro-1*H*-pyran-1-one (Figures 7.33 and 7.34).

The COSY spectrum of ieodomycin C **79** indicates a single spin system from H-2 to H-12, while the (*E*)-configuration of both vinylic systems was based on the characteristic scalar coupling constants (both *J* 15.3 Hz), although also confirmed by the ROESY correlations (H-4 → H-6 and H-5 → H-7). All the proton and carbon resonances were then assigned through the analysis of their ^{1}H, ^{13}C, HSQC, and HMBC spectra (Figure 7.33; Tables 7.4 and 7.5). The synthesis of the modified Mosher's diesters and the analysis of the $\Delta\delta^{SR}$ values assigns the configuration of C-3 and C-11 as (*S*) and (*R*), respectively (Figure 7.34). Thus ieodomycin C **79** was identified as (3*S*,4*E*,6*E*,11*R*)-3,11-dihydroxydodeca-4,6-dienoic acid.

The analysis of the spectroscopic data (^{1}H, ^{13}C NMR, and HRMS) of ieodomycin D **80** shows that this structure has two carbons, four hydrogens, and one oxygen less than ieodomycin C **79** (Tables 7.4 and 7.5). Like in the case of **79**, the COSY spectrum of ieodomycin D **80** revealed the presence of only one spin system, the (*E*)-configurations of their vinylic systems were based on the large ^{1}H–^{1}H coupling constants (both *J* 15.3 Hz) and confirmed by ROESY correlations, the carbon resonances were assigned based on the HSQC correlations, while the carbon skeleton was established by the HMBC connectivities (Figure 7.33). The (9*S*)-absolute configuration was assigned by the modified Mosher's method (Figure 7.34). Thus the structure of ieodomycin D **80** was unambiguously established as (2*E*,4*E*,9*R*)-9-hydroxydeca-2,4-dienoic acid.

7.8 REVISED STRUCTURES

Despite all the development in the spectroscopic techniques, including those of the NMR experiments described in this chapter and the detailed modern spectral analyses, the sad truth is that structures of natural products and also those of marine sources originally proposed based on these techniques and analyses often include misassignments. There were two articles on natural products' structural revisions covering the periods 1990 to 2005 [96] and 2005 to 2009 [97] where they focused the application of chemical (particularly total) synthesis to structure elucidation, including terrestrial and marine-derived metabolites.

In a more recent review, McPhail *et al.* [98] surveyed marine natural products structural revision based on spectroscopic and chemical synthesis. This covers the period of 2005 to 2010 and identifies the techniques used in the misassignment of all the structures revised during this period, most notably that the NOE experiments (NOE difference, NOESY, and ROESY) and the NMR comparison are responsible for around half of these misassignments. The high frequency of NOE associated errors implies that configurational misassignments predominate over constitutional errors. The source of misassignments of marine natural products can be divided into the subcategories constitution errors and configuration errors, described as follows.

Constitution errors include cyclization (e.g., five- vs six-membered cycles; differentiation of fused six-membered from two five-membered rings linked by a single bond), regioisomerism (e.g., placement of substituents on cycles), substituent identity (e.g., identity of heteronuclear substituents, based on MS techniques and subject to errors depending on the substituent stability and the ionization technique used), hydrocarbon chains (e.g., a challenge topic of the MS technique based on the molecular ion and fragmentation patterns), and symmetry of dimerization (based on incorrect formula).

Configuration errors are those such as absolute and relative configuration. Generally, in the case of absolute configuration it is necessary to perform chemical degradation and/or derivatization followed by comparison of chromatographic retention and spectroscopic properties with a standard of model compounds. Since the isolation quantity of marine natural products is generally very small, the spectroscopic comparison of the new metabolites with known assigned compounds to extrapolate absolute configuration is common and can lead to propagation of errors. Mosher's analyses rely on the application of empirically derived principles and consequently errors can be sustained and propagated. Taking into account all these errors and their origin, with a sufficient sample, X-ray crystallography can be considered a gold standard for assignment of absolute configuration, but when there is a very small amount of sample, the errors can almost exclusively await detection and revision by total synthesis. The NOE data are responsible for the majority of the relative configuration misassignments of marine natural products. This situation highlights the need for a careful evaluation of all the molecular models that are consistent with NOE constraints, which support the use of computational modeling for this analysis.

Considering the sources of misassignments leads to the question of how these errors are detected and which techniques are used to revise the corresponding structure. Nondestructive and relatively sensitive NMR methods, although the greatest source of misassignments, still provide the major contribution to marine natural products revision, immediately followed by total and partial synthesis.

The increasingly high-field magnets and sensitive probes do not necessarily attenuate the rate of structural misassignments, since they are still to be reported [e.g., 99,100], but they permit the attempted structure elucidation of increasingly limited quantities of minor components from natural products extracts, as well as larger molecules of greater structural complexity. Although total syntheses of marine natural products require extremely high-level work in organic chemistry, it will surely continue to be central to the confirmation of natural product structure assignment [e.g., 101,102], as well as providing material for biological testing toward pharmaceutical development, and investigations of biosynthetic pathways.

ACKNOWLEDGEMENTS

The authors acknowledge University of Aveiro, Fundação para a Ciência e Tecnologia (FCT, Portugal), European Union, QREN, FEDER, and COMPETE for funding the QOPNA Research Unit.

REFERENCES

[1] K. Sugahara, Y. Kitamura, M. Murata, M. Satake, K. Tachibana, J. Org. Chem. 76 (2011) 3131–3138.
[2] G. Fan, Z. Li, S. Shen, Y. Zeng, Y. Yang, M. Xu, T. Bruhn, H. Bruhn, J. Morschhäuser, G. Bringmann, W. Lin, Bioorg. Med. Chem. 18 (2010) 5466–5474.
[3] T. Hertiani, R. Edrada-Ebel, S. Ortlepp, R.W.M. van Soest, N.J. de Voogd, V. Wray, U. Hentschel, S. Kozytska, W.E.G. Müller, P. Proksch, Bioorg. Med. Chem. 18 (2010) 1297–1311.
[4] H. Yamazaki, H. Rotinsulu, T. Kaneko, K. Murakami, H. Fujiwara, K. Ukai, M. Namikoshi, Mar. Drugs 10 (2012) 2691–2697.
[5] S. Sato, F. Iwata, S. Yamada, M. Katayama, J. Nat. Prod. 75 (2012) 1974–1982.
[6] B. Ohlendorf, D. Schulz, A. Erhard, K. Nagel, J.F. Imhoff, J. Nat. Prod. 75 (2012) 1400–1404.
[7] P.D. Boudreau, T. Bryum, W.-T. Liu, P.C. Dorrestein, W.H. Gerwick, J. Nat. Prod. 75 (2012) 1560–1570.
[8] W.M. Abdel-Mageed, B.F. Milne, M. Wagner, M. Schumacher, P. Sandor, W. Pathom-aree, M. Goodfellow, A.T. Bull, K. Horikoshi, R. Ebel, M. Diederich, H.-P. Fiedler, M. Jaspars, Org. Biomol. Chem. 8 (2010) 2352–2362.
[9] D.-G. Kim, K. Moon, S.-H. Kim, S.-H. Park, S. Park, S.K. Lee, K.-B. Oh, J. Shin, D.-C. Oh, J. Nat. Prod. 75 (2012) 959–967.
[10] A. Plaza, J.L. Keffer, G. Bifulco, J.R. Lloyd, C.A. Bewley, J. Am. Chem. Soc. 132 (2010) 9069–9077.

[11] S. Mo, A. Krunic, B.D. Santarsiero, S.G. Franzblau, J. Orjala, Phytochemistry 71 (2010) 2116–2123.
[12] L.A. Salvador, V.J. Paul, H. Luesch, J. Nat. Prod. 73 (2010) 1606–1609.
[13] R. Finlayson, A.N. Pearce, M.J. Page, M. Kaiser, M.-L. Bourguet-Kondracki, J.L. Harper, V.L. Webb, B.R. Copp, J. Nat. Prod. 74 (2011) 888–892.
[14] I. Djinni, A. Defant, M. Kecha, I. Mancini, Mar. Drugs 11 (2013) 124–135.
[15] K. Chakraborty, A.P. Lipton, R. Paulraj, R.D. Chakraborty, Eur. J. Med. Chem. 45 (2010) 2237–2244.
[16] S. Li, M. Wei, G. Chen, Y. Lin, Chem. Nat. Comp. 48 (2012) 371–373.
[17] J. Wiese, B. Ohlendorf, M. Blümel, R. Schmaljohann, J.F. Imhoff, Mar. Drugs 9 (2011) 561–585.
[18] C. Li, J. Zhang, C. Shao, W. Ding, Z. She, Y. Lin, Chem. Nat. Comp. 47 (2011) 382–384.
[19] Y. Chen, J. Pan, F. Xu, F. Liu, J. Yang, C. Huang, C. Xu, Y. Lu, X. Cai, Z. She, Y. Lin, Chem. Nat. Comp. 46 (2010) 230–232.
[20] F. Alihosseini, J. Lango, K.-S. Ju, B.D. Hammock, G. Sun, Biotechnol. Prog. 26 (2010) 352–360.
[21] F. Song, H. Dai, Y. Tong, B. Ren, C. Chen, N. Sun, X. Liu, J. Bian, M. Liu, H. Gao, H. Liu, X. Chen, L. Zhang, J. Nat. Prod. 73 (2010) 806–810.
[22] J. You, H. Dai, Z. Chen, G. Liu, Z. He, F. Song, X. Yang, H. Fu, L. Zhang, X. Chen, J. Ind. Microbiol. Biotechnol. 37 (2010) 245–252.
[23] R.H. Jiao, H. Xu, J.T. Cui, H.M. Ge, R.X. Tan, J. Appl. Microbiol. 114 (2013) 1046–1053.
[24] R. Finlayson, A. Brackovic, A. Simon-Levert, B. Banaigs, R.F. O'Toole, C.H. Miller, B.R. Copp, Tetrahedron Lett. 52 (2011) 837–840.
[25] H.B. Park, H.O. Yang, K.R. Lee, H.C. Kwon, Molecules 16 (2011) 3519–3529.
[26] S. Felder, S. Dreisigacker, S. Kehraus, E. Neu, G. Bierbaum, P.R. Wright, D. Menche, T.F. Schäberle, G.M. König, Chem. Eur. J. 19 (2013) 9319–9324.
[27] N. Liu, F. Shang, L. Xi, Y. Huang, Mar. Drugs 11 (2013) 1524–1533.
[28] F.-Z. Wang, Z. Huang, X.-F. Shi, Y.-C. Chen, W.-M. Zhang, X.-P. Tian, J. Li, S. Zhang, Bioorg. Med. Chem. Lett. 22 (2012) 7265–7267.
[29] K. Tidgewell, N. Engene, T. Byrum, J. Media, T. Doi, F.A. Valeriote, W.H. Gerwick, ChemBioChem 11 (2010) 1458–1466.
[30] Y. Zhou, A. Mándi, A. Debbab, V. Wray, B. Schulz, W.E.G. Müller, W. Lin, P. Proksch, T. Kurtán, A.H. Aly, Eur. J. Org. Chem. (2011) 6009–6019.
[31] Y. Zhou, A. Debbab, A. Mándi, V. Wray, B. Schulz, W.E.G. Müller, M. Kassack, W. Lin, T. Kurtán, P. Proksch, A.H. Aly, Eur. J. Org. Chem. (2013) 894–906.
[32] W. Xin, X. Ye, S. Yu, X.-Y. Lian, Z. Zhang, Mar. Drugs 10 (2012) 2388–2402.
[33] Y.-B. Cheng, P.R. Jensen, W. Fenical, Eur. J. Org. Chem. (2013) 3751–3757.
[34] F. Cen-Pacheco, J.A. Villa-Pulgarin, F. Mollinedo, M. Norte, A.H. Daranas, J.J. Fernández, Eur. J. Med. Chem. 46 (2011) 3302–3308.
[35] W. Wang, H. Kim, S.-J. Nam, B.J. Rho, H. Kang, J. Nat. Prod. 75 (2012) 2049–2054.
[36] N. Matsumori, D. Kaneno, M. Murata, H. Nakamura, K. Tachibana, J. Org. Chem. 64 (1999) 866–876.
[37] D. Menche, Nat. Prod. Rep. 25 (2008) 905–918.
[38] Y. Kobayashi, J. Lee, K. Tezuka, Y. Kishi, Org. Lett. 1 (1999) 2177–2180.
[39] Y. Kobayashi, J. Lee, K. Tezuka, Y. Kishi, Org. Lett. 1 (1999) 2181–2184.
[40] Y. Kobayashi, C.-H. Tan, Y. Kishi, Helv. Chim. Acta 83 (2000) 2562–2571.
[41] Y. Kobayashi, C.-H. Tan, Y. Kishi, J. Am. Chem. Soc. 123 (2001) 2076–2078.

[42] S. Higabayashi, W. Czechtizky, Y. Kobayashi, Y. Kishi, J. Am. Chem. Soc. 125 (2003) 14379–14393.
[43] E. Fleury, M.-I. Lannou, O. Bistri, F. Sautel, G. Massiot, A. Pancrazi, J. Ardisson, Eur. J. Org. Chem. (2009) 4992–5001.
[44] J.M. Seco, E. Quiñoá, R. Riguera, Chem. Rev. 104 (2004) 17–117.
[45] J.M. Seco, E. Quiñoá, R. Riguera, Chem. Rev. 112 (2012) 4603–4641.
[46] J.A. Dale, H.S. Mosher, J. Am. Chem. Soc. 95 (1973) 512–519.
[47] G.R. Sullivan, J.A. Dale, H.S. Mosher, J. Org. Chem. 38 (1973) 2143–2147.
[48] N. Harada, Chirality 20 (2008) 691–723.
[49] W.S. Borges, G. Mancilla, D.O. Guimarães, R. Durán-Patrón, I.G. Collado, M.T. Pupo, J. Nat. Prod. 74 (2011) 1182–1187.
[50] M. Takahashi, K. Koyama, S. Natori, Chem. Pharm. Bull. 38 (1990) 625–628.
[51] S.J. Nam, S.P. Gaudêncio, C.A. Kauffman, P.R. Jensen, T.P. Kondratyuk, L.E. Marler, J.M. Pezzuto, W. Fennical, J. Nat. Prod. 73 (2010) 1080–1086.
[52] D.H. Marchbank, F. Berrue, R.G. Kerr, J. Nat. Prod. 75 (2012) 1289–1293.
[53] D. Lai, Y. Li, M. Xu, Z. Deng, L. van Ofwegen, P. Qian, P. Proksch, W. Lin, Tetrahedron 67 (2011) 6018–6029.
[54] L.G. Meimetis, D.E. Williams, N.R. Mawji, C.A. Banuelos, A.A. Lal, J.J. Park, A.H. Tien, J.G. Fernandez, N.J. de Voogd, M.D. Sadar, R.J. Andersen, J. Med. Chem. 55 (2012) 503–514.
[55] T. Conroy, J.T. Guo, R.G. Linington, N.H. Hunt, R.J. Payne, Chem. Eur. J. 17 (2011) 13544–13552.
[56] T.A.M. Gulder, H. Hong, J. Correa, E. Egereva, J. Wiese, J.F. Imhoff, H. Gross, Mar. Drugs 10 (2012) 2912–2935.
[57] A.R. Carroll, G. Pierens, G. Fechner, P.A. Leone, A. Ngo, M. Simpson, J.N.A. Hooper, S.-L. Boström, D. Musil, R.J. Quinn, J. Am. Chem. Soc. 124 (2002) 13340–13341.
[58] S. Hanessian, R. Margarita, A. Hall, S. Johnstone, M. Tremblay, L. Parlanti, J. Am. Chem. Soc. 124 (2002) 13342–13343.
[59] A.D. Rodríguez, E. González, S.D. Huang, J. Org. Chem. 63 (1998) 7083–7091.
[60] T.J. Heckrodt, J. Mulzer, J. Am. Chem. Soc. 125 (2003) 4680–4681.
[61] H. Miyaoka, D. Honda, H. Mitome, Y. Yamada, Tetrahedron Lett. 43 (2002) 7773–7775.
[62] K. Ota, T. Kurokawa, E. Kawashima, H. Miyaoka, Mar. Drugs 7 (2009) 654–671.
[63] S.C. Bobzin, S. Yang, T.P. Kasten, J. Chromatogr. B 748 (2000) 259–267.
[64] S.C. Bobzin, S. Yang, T.P. Kasten, J. Ind. Microbiol. Biotecnhol. 25 (2000) 342–345.
[65] R.A. Edrada, M. Heubes, G. Brauers, V. Wray, A. Berg, U. Gräfe, M. Wohlfarth, J. Mühlbacher, K. Schaumann, Sudarsono, G. Bringmann, P. Proksch, J. Nat. Prod. 65 (2002) 1598–1604.
[66] G. Bringmann, G. Lang, S. Steffens, E. Günther, K. Schaumann, Phytochemistry 63 (2003) 437–443.
[67] R. Brkljača, S. Urban, J. Liq. Chromatogr. Rel. Technol. 34 (2011) 1063–1076.
[68] D.A. Dias, S. Urban, J. Sep. Sci. 32 (2009) 542–548.
[69] D.A. Dias, J.M. White, S. Urban, Nat. Prod. Commun. 4 (2009) 157–172.
[70] D.A. Dias, S. Urban, Phytochemistry 72 (2011) 2081–2089.
[71] M.A. Timmers, D.A. Dias, S. Urban, Mar. Drugs 10 (2012) 2089–2102.
[72] S. Urban, M.A. Timmers, Nat. Prod. Commun. 8 (2013) 715–719.
[73] Z. Yang, J. Pharm. Biomed. Anal. 40 (2006) 516–527.
[74] J.W. Jaroszewski, Planta Med. 71 (2005) 795–802.
[75] L. Fremlin, M. Farrugia, A.M. Piggott, Z. Khalil, E. Lacey, R.J. Capon, Org. Biomol. Chem. 9 (2011) 1201–1211.

[76] M.L. Martin, J.-J. Delpuech, G.J. Martin, Practical NMR Spectroscopy, Heyden, London, (1980).
[77] C. Szántay Jr., Trends Anal. Chem. 11 (1992) 332–344.
[78] G.F. Pauli, B.U. Jaki, D.C. Lankin, J. Nat. Prod. 68 (2005) 133–149.
[79] G.F. Pauli, T. Gödecke, B.U. Jaki, D.C. Lankin, J. Nat. Prod. 75 (2012) 834–851.
[80] C.H. Cullen, G.J. Ray, C.M. Szabo, Magn. Reson. Chem. 51 (2013) 705–713.
[81] I.W. Burton, M.A. Quilliam and J.A. Walter, 42nd Experimental Nuclear Magnetic Resonance Spectroscopy Conference (ENC): Orlando, Florida, March 11-16, 2001.
[82] I.W. Burton, M.A. Quilliam, J.A. Walter, Anal. Chem. 77 (2005) 3123–3131.
[83] R. Perez, N. Rehmann, S. Crain, P. LeBlanc, C. Craft, S. MacKinnon, K. Reeves, I.W. Burton, J.A. Walter, P. Hess, M.A. Quiliam, J.E. Melanson, Anal. Bioanal. Chem. 398 (2010) 2243–2252.
[84] J.L. Carballo, B. Yañez, E. Zubía, M.J. Ortega, C. Vega, Mar. Biotechnol. 12 (2010) 516–525.
[85] D.M. Wilson, M. Puyana, W. Fenical, J.R. Pawlik, J. Chem. Ecol. 25 (1999) 2811–2823.
[86] T.F. Molinski, Curr. Opin. Biotechnol. 21 (2010) 819–826.
[87] D.S. Dalisay, T.F. Molinski, J. Nat. Prod. 72 (2009) 739–744.
[88] T.F. Molinski, Nat. Prod. Rep. 27 (2010) 321–329.
[89] K. Hu, W. Westler, J.L. Markley, J. Am. Chem. Soc. 133 (2011) 1662–1665.
[90] K. Hu, T.P. Wyche, T.S. Bugni, J.L. Markley, J. Nat. Prod. 74 (2011) 2295–2298.
[91] A. Negri, E. Marco, V. García-Hernández, A. Domingo, A.L. Llamas-Saiz, S. Porto-Sandá, R. Riguera, W. Laine, M.-H. David-Cordonnier, C. Bailly, L.F. García-Fernández, J.J. Vaquero, F. Gago, J. Med. Chem. 50 (2007) 3322–3333.
[92] B. El-Amraoui, J.-F. Biard, A. Fassouane, SpringerPlus 2 (2013) 252–271.
[93] W.M. Abdel-Mageed, R. Ebel, F.A. Valeriote, M. Jaspars, Tetrahedron 66 (2010) 2855–2862.
[94] M.A.M. Mondol, J.H. Kim, M. ah Lee, F.S. Tareq, H.-S. Lee, Y.-J. Lee, H.J. Shin, J. Nat. Prod. 74 (2011) 1606–1612.
[95] F. Freire, J.M. Seco, Q. Emilio, R. Riguera, J. Org. Chem. 70 (2005) 3778–3790.
[96] K.C. Nicolaou, S.A. Snyder, Angew. Chem. Int. 44 (2005) 1012–1044.
[97] M.E. Maier, Nat. Prod. Rep. 26 (2009) 1105–1124.
[98] T.L. Suyama, W.H. Gerwick, K.L. McPhail, Bioorg. Med. Chem. 19 (2011) 6675–6701.
[99] F. Zhu, G.Y. Chen, J. Wu, J. Pan, Nat. Prod. Res. 27 (2013) 1960–1964.
[100] J.W. Cha, J.-S. Park, T. Sim, S.-J. Nam, H.C. Kwon, J.R. Del Valle, W. Fenical, J. Nat. Prod. 75 (2012) 1648–1651.
[101] S. Das, R.K. Goswami, J. Org. Chem. 78 (2013) 7274–7280.
[102] M. Ebine, M. Kanemoto, Y. Manabe, Y. Konno, K. Sakai, N. Matsumori, M. Murata, T. Oishi, Org. Lett. 15 (2013) 2846–2849.

Chapter 8

Mass Spectrometry for Determination of Bioactive Compounds

Supriya Tilvi, Mahesh S. Majik and Keisham S. Singh
Bio-Organic Chemistry Laboratory, CSIR-National Institute of Oceanography, Dona Paula Goa, India

Contents

8.1 INTRODUCTION

Natural products research is primarily concerned with identifying useful compounds from natural sources, including plants, fungi, bacteria, and marine organisms. The power and utility of mass spectrometry (MS) in solving structure elucidation in natural product chemistry has been demonstrated on numerous occasions over the past 25 years. Since the development of devices that could first measure the masses of molecules, the technology of MS has continued to enjoy exponential growth. In the last 5 years much of this growth has been in the development of new ionization methods, primarily serving the needs of

Analysis of Marine Samples in Search of Bioactive Compounds, Vol. 65. DOI: 10.1016/B978-0-444-63359-0.00008-2

forensic, proteomic, and metabolomic investigations. A major problem in the discovery of new biologically active compounds from natural products is the reisolation of known compounds. Such reisolation wastes time and resources, distracting chemists from more promising leads. To address this problem, dereplication strategies are needed that enable crude extracts to be screened for the presence of known compounds before isolation efforts are initiated. The current approaches to discover new bioactive natural products is to screen extracts or partially purified fractions on therapeutic targets for the presence of active compounds and subsequent investigation of their biological activities. A quicker, simpler, and more sensitive method to screen natural product extracts that would allow significant time savings is needed. The MS allows quick identification of the molecular mass of the natural products. The chemist then applies the screening to identify active compounds in natural product extracts. Notable successes were achieved in correlating spectra with structure, particularly with natural steroids, terpenes, and alkaloids.

This chapter on application of MS in the determination of bioactive compounds will be of interest to the biochemical, biomedical, and pharmaceutical communities, as well as those working in other branches of analytical bioscience. Here, a brief introduction of the principles of mass spectrometry and instrumentation currently used is given. As a part of the research program on marine natural product chemistry at CSIR-National Institute of Oceanography, Goa, Tilvi *et al.* have identified several metabolites based on mass spectrometric technique, which has been presented as a case study for selected bioactive compounds that includes peptides, alkaloids, and lipids.

8.2 MASS SPECTROMETRY (MS)

Mass spectrometers can be divided into three fundamental parts, the ionization source, the analyzer, and the detector. The sample has to be introduced into the ionization source of the instrument. Once inside the ionization source, the sample molecules are ionized and extracted into the analyzer region of the mass spectrometer where they are separated according to their mass-to-charge (m/z) ratios. The separated ions are detected and this signal is sent to a data system where the m/z ratios are stored together with their relative abundance for presentation in the format of an m/z spectrum. The analyzer and detector of the mass spectrometer, and often the ionization source too, are maintained under a high vacuum to give the ions a reasonable chance of travelling from one end of the instrument to the other without any hindrance from air molecules. The entire operation of the mass spectrometer, and often the sample introduction process also, is under complete data system control on modern mass spectrometers.

Matrix assisted laser desorption ionization (MALDI) or electrospray ionization (ESI) are currently the most common used ionization sources. Because MALDI and ESI are soft ionization methods they allow routine MS on biologically relevant molecules and have become the standard in the life sciences.

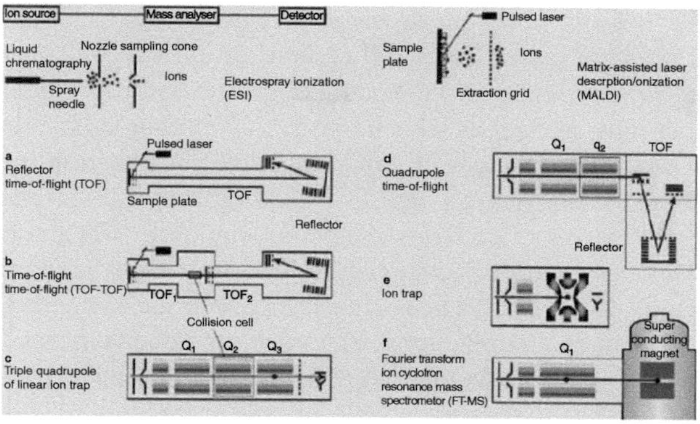

The left and right upper panels depict the Ionization and sample Introduction process In electrospray Ionization (ESI) and matrix-assisted laser desorption/ionization (MALDI). The different Instrumental configurations (**a-f**) are shown with their typical ion source. **a**, In reflector time-of-flight (TOF) instruments. the ions are accelerated to high kinetic energy and are separated along a flight tube as a result of their different velocities. The ions are turned around in a reflector. which compensates for slight differences In kinetic energy, and then impinge on a detector that amplifies and counts arriving ions. **b**, The TOF-TOF instrument Incorporates a collision cell between two TOF sections. Ions of one m/z ratio are selected in the first TOF section. fragmented in the collision cell, and the masses of the fragments are separated in the second TOF section. **c**, Quadrupole mass spectrometers select by time-varying electric fields between four rods, which permit a stable trajectory only for ions of a particular desired m/z. Again, ions of a particular m/z are selected in a first section (Q_1). fragmented in a collision cell (q_2), and the fragments separated in Q_3. In the linear ion trap, ions are captured in a quadruple section, depicted by the red dot in Q_3. They are then exited via resonant electric field and the fragments are scanned out, creating the tandem mass spectrum. **d**, The quadrupole TOF instrument combines the front part of a triple quadruple instrument with a reflector TOF section for measuring the mass of the ions. **e**, The (three-dimensional) ion trap captures the ions as in the case of the linear ion trap, fragments ions of a particular m/z, and then scans out the fragments to generate the tandem mass spectrum. **f**, The FT-MS instrument also traps the ions, but does so with the help of strong magnetic fields. The figure shows the combination of FT-MS with the linear ion trap for efficient isolation, fragmentation and fragment detection in the FT-MS section.

FIGURE 8.1 Schematic representation of mass spectrometer.

It was therefore no surprise that the development of soft desorption ionization methods was recognized with the Nobel Prize in 2002 [1,2]. However, since the development of MALDI and ESI, other modes of ionization and improvements to existing techniques are providing a rich suite of complementary methods. These new methods enable MS to be used in a variety of new and innovative applications. Many of these newer methods are well suited to analyzing surfaces, are tolerant to impurities, and do not require extensive sample preparations. Examples include low-temperature plasma ionization (LTP), direct analysis in real time (DART), electrospray desorption/ionization (ELDI), desorption electrospray ionization (DESI), and many others [3–6]. Collectively, these approaches are called ambient ionization methods, a term first introduced by Graham Cooks's laboratory in 2004 with the introduction of DESI [5]. There are a number of mass analyzers currently available, the better known of which include quadrupoles, time-of-flight (TOF) analyzers, magnetic sectors, and both Fourier transform and quadrupole ion traps (Figure 8.1) [7].

8.3 TANDEM MASS SPECTROMETRY (MS^n)

Tandem mass spectrometry is at the heart of most modern mass spectrometric investigations of complex mixtures [8–10]. Tandem MS is considered a mass spectrum of a mass spectrum raised to the power n, (MS^n) where n is the number of MS/MS spectra, obtained per molecular ion. It has more than one analyzer, in practice usually two. The two analyzers are separated by a collision cell into which an inert gas (e.g., helium, nitrogen, or argon) is introduced. Collision induced dissociation (CID) involves activation of a precursor ion via collisions with a target gas and may produce charged and neutral fragments (product ions). The nature of the product ions, as well as their intensities, are often indicative of the structure of the precursor ion and thus can yield useful information for the identification of unknown analytes, as well as providing a useful screening technique for different classes of analytes. The combination ESI-MS and MALDI-MS with collision-induced dissociation (CID) is what gives tandem MS its power in the analysis of mixtures. The analyzers can be of the same or of different types, the most common combinations being quadrupole–quadrupole, magnetic sector–quadrupole, magnetic sector–magnetic sector, and quadrupole–time-of-flight (QTOF). Both analyzers are capable of producing ions of low energy, and hence have the power to generate ions from biological macromolecules of molecular masses in excess of 100,000, and in many cases these masses can be measured with accuracies better than 0.01%. The basic modes of data acquisition for tandem mass spectrometry experiments are as follows.

8.3.1 Product or Daughter Ion Scanning

The first analyzer is used to select user-specified sample ions arising from a particular component, usually molecular-related (i.e., $[M+H]^+$ or $[M-H]^-$ ions). These chosen ions pass into the collision cell, are bombarded by gas molecules, which cause fragment ions to be formed, and these fragment ions are analyzed (i.e., separated according to their *m/z* ratios) by the second analyzer. All the fragment ions arise directly from the precursor ions specified in the experiment, and thus produce a fingerprint pattern specific to the compound under investigation. This type of experiment is particularly useful for providing structural information concerning small organic molecules and for generating peptide sequence information.

8.3.2 Precursor or Parent Ion Scanning

The first analyzer allows the transmission of all sample ions, whereas the second analyzer is set to monitor specific fragment ions generated by bombardment of the sample ions with the collision gas in the collision cell. This type of experiment is particularly useful for monitoring groups of compounds contained within a mixture that fragment to produce common fragment ions; for example, glycosylated peptides in a tryptic digest mixture, aliphatic hydrocarbons in an oil sample, or glucuronide conjugates in urine.

8.3.3 Constant Neutral Loss Scanning

This involves both analyzers scanning, or collecting data, across the whole *m/z* range, but the two are offset so that the second analyzer allows only those ions that differ by a certain number of mass units (equivalent to a neutral fragment) from the ions transmitted through the first analyzer. This type of experiment could be used to monitor all the carboxylic acids in a mixture. Carboxylic acids tend to fragment by losing a (neutral) molecule of carbon dioxide, CO_2, which is equivalent to a loss of 44 Da or atomic mass units. All ions pass through the first analyzer into the collision cell. The ions detected from the collision cell are those from which 44 Da have been lost.

8.3.4 Selected/Multiple Reaction Monitoring

Both of the analyzers are static in this case as user-selected specific ions are transmitted through the first analyzer and user-selected specific fragments arising from these ions are measured by the second analyzer. The compound under scrutiny must be known and have been well characterized previously before this type of experiment is undertaken. This methodology is used to confirm unambiguously the presence of a compound in a matrix (e.g., drug testing with blood or urine samples). It is not only a highly specific method but also has very high sensitivity.

For known compounds, mass spectra can be used much like fingerprints. A match is extremely strong evidence that the compounds are identical. For unknowns, results from a mass spectrum often provide significant information that can help elucidate the structure of an organic compound. Analysis of a mass spectrum can provide a formula and considerable information about structure. As fragment ions are produced, the challenge is to put the pieces of a puzzle together to form the whole structure. Applications include the sequencing and analysis of peptides and proteins, studies of noncovalent complexes and immunological molecules, DNA sequencing, and the analysis of intact viruses [11,12].

8.4 LIQUID CHROMATOGRAPHY–MASS SPECTROMETRY (LC-MS)

LC-MS is a chemistry technique that combines the physical separation capabilities of liquid chromatography (or HPLC) with the mass analysis capabilities of MS. LC-MS is a powerful technique used for many applications, which has very high sensitivity and selectivity. Generally its application is oriented toward the general detection and potential identification of chemicals in the presence of other chemicals (in a complex mixture). A preparative LC-MS system can be used for fast and mass-directed purification of natural-product extracts and new molecular entities important to food, pharmaceutical, agrochemical, and other industries. LC-MS is the key analytical technique on which the emerging "-omics" technologies of proteomics, metabolomics, and lipidomics are based. It provides both structural and quantitative data and can be used in a global or targeted manner, allowing on the one hand the identification of thousands of proteins from a tissue, or on the other the detection of biologically active metabolites at levels of a few parts-per-billion. It can be expected that the continued incremental developments in LC-MS along with data-handling routines will soon bear fruit in the quest for a greater understanding of human disease, leading to new drug targets and therapies.

8.5 APPLICATION OF MS

MS has played a pivotal role in a variety of scientific disciplines including organic and inorganic chemistry (especially natural product chemistry), forensic science, biotechnology, and many other areas of natural product chemistry, environmental analysis, forensic analysis, clinical research, proteomics and genomics, generation of physico-chemical data, and so on. Tandem mass spectrometry (MS^n) has wide applications in sequencing peptide and proteomics and is becoming a method of choice for identification of new metabolites in complex biological samples; some that are not easily determined by nuclear magnetic resonance (NMR) analysis are readily available from MS analysis. It has wide applications for the qualitative and quantitative analysis.

In this section, the application of MS in characterization of selected families of bioactive molecules such as peptides, alkaloids, and lipids is discussed. A case study with a representative example of each class of above core is also presented.

8.6 PEPTIDES

With the advent of soft ionization techniques such as ESI [13–15] and MALDI [16,17], which efficiently ionizes by protonation a wide variety of peptides and proteins, tandem MS [18,19] has become a method of significant importance for the sequencing of peptides through CID studies. When a peptide is analyzed by mass spectrometry, the obtained MS spectrum provides the peptide mass. When tandem MS is performed, such a peptide is fragmented into daughter ions (product ions), which provide information regarding the amino acid sequence of the peptide.

Several parameters influence this fragmentation process, including amino acid composition, size of the peptide, excitation method, time scale of the instrument used, ion charge state, among others [20]. The nomenclature of the daughter ions generated by CID was first established by Roepstorff and Fohlmann [21] and later reviewed by Biemann [22], where the b-series ions extend from the N-terminal and the y-series ions extend from the C-terminal (Figure 8.2). The calculation of the mass difference between consecutive daughter ions belonging to the same ion series (for example b or y ion-series) allows the determination of peptide's primary sequence. A major mode of fragmentation in many cases involves cleavage of an amide bond in the protonated peptides. When the charge is retained by the C-terminus fragment migration of a labile hydrogen from the N-terminus, a neutral fragment occurs to form a protonated amino acids (y_1) or peptide (y_n) [23–25]. When the charge remains on the *N*-terminus fragment, a neutral amino acid or peptide is eliminated and b_n ions are formed.

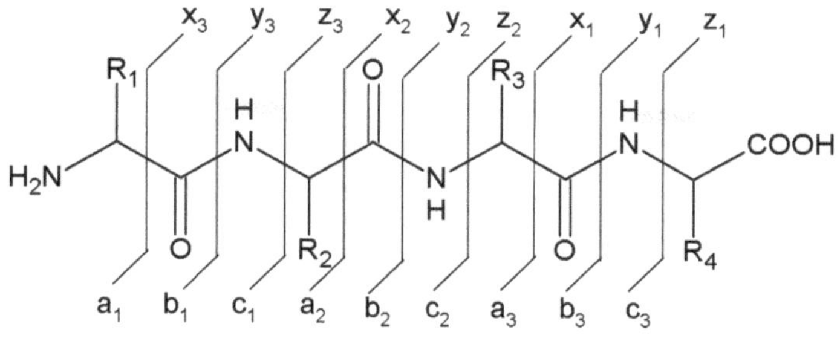

FIGURE 8.2 Peptide fragmentation.

Presently, there are several fragmentation processes available in commercial mass spectrometers, namely CID, electron capture dissociation (ECD), electron transfer dissociation (ETD), and such. The appearance of peptides CID spectra, especially those of $[M+H]^+$ ions, are highly dependant on the collision energy. The CID spectra of peptides recorded at low collision energy (<100 eV) tend to show abundant fragment ions formed by cleavage of the peptide bonds. In low energy CID a peptide carrying a positive charge fragments mainly along its backbone, generating predominantly a, b, and y ions. In addition, peaks due to loss of ammonia (−17 Da), denoted a*, b*, and y*, and water (−18 Da), denoted a^0, b^0, and y^0, are also observed. While high-energy CID (KeV) spectra often contain fragment ions formed by other backbones x, c, z and also side-chain cleavages are observed [26,27]. In case of the cyclic depsipeptides, the fragmentation pattern is different. For an ideal fragmentation process the sequence of a peptide could be simply determined by converting the mass differences of consecutive ions in a spectrum to the corresponding amino acids. This ideal situation would occur if the fragmentation process could be controlled so that each peptide was cleaved between every two consecutive amino acids [28,29].

The application of this technique has been widely used in the identification of several peptides and proteins. This section of the chapter highlights the identification of cyclodepsipeptides, kahalalides as a case study to understand the fragmentation pattern in the peptide sequencing.

8.6.1 Case Study: Cyclodepsipeptides, Kahalalides

In the investigation, spectra obtained using ESI-QTOF MS (QSTAR XL Applied Biosystem, Canada) of the mollusk *Elysia grandifolia* showed a cluster of molecular ion peaks centered at a molecular mass of 1478 Da (kahalalide F, an anticancer agent) [30–36]. Two new molecules, kahalalide R (*m/z* 1464) and S (*m/z* 1492) (Figure 8.3) were characterized using tandem MS [37]. The mass differences of 14 Da suggested that they are homologous molecules. The amino acid sequences of kahalalide R and S were proposed using CID experiments of singly and doubly charged molecular ions and by comparison with the amino acid sequence of kahalalide F. The pathway is presented for the loss of amino acid residues in kahalalide F. It is observed that there is sequential loss of amino acids in the linear

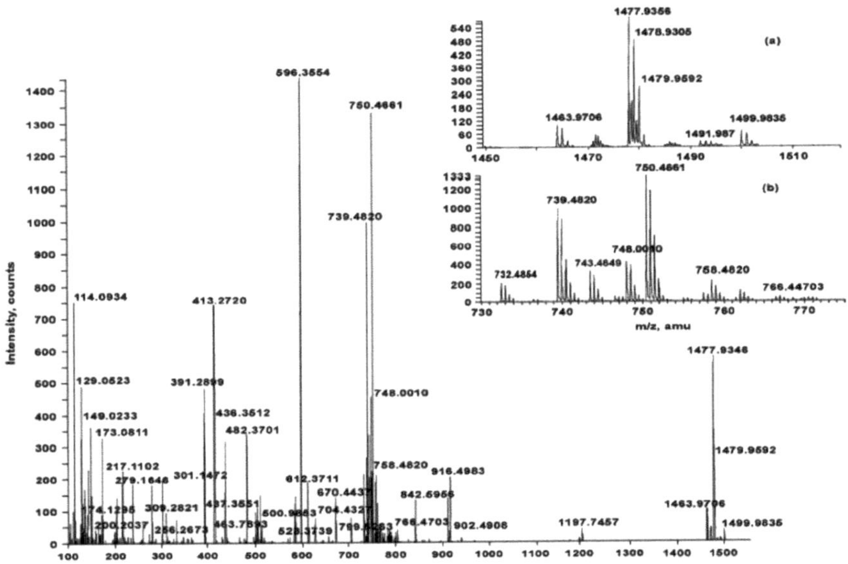

FIGURE 8.3 Full-scan ESI-MS spectrum in the positive ion mode of the peptide-rich fraction of *Elysia grandifolia*: (a) singly charged molecular ion peaks $[M+H]^+$; (b) doubly charged molecular ion peaks $[M+2H]^{2+}$.

peptide chain, but in the cyclic part the ring opens at the amide bond rather than at the lactone linkage, and the loss of amino acid residues is not sequential. The CID experiment of the alkali-metal-cationized molecular ions shows that the sodium and potassium ions coordinate to the amide nitrogen/oxygen in the linear peptide chain of the molecule and not to the lactone oxygen of the lactone.

The fragmentation is illustrated by giving example of kahalalide R. The MS/MS at *m/z* 1464 $[M+H]^+$, 732 $[M+2H]^{2+}$ and 743 $[M+H+Na]^{2+}$ were studied (Figure 8.4). Molecular ion $[M+H]^+$ showed peaks at *m/z* 212 (b_1), 313 (b_2), 412 (b_3), 511 (b_4), and 609 (b_5) identical with those observed in case of kahalalide F (KF). The corresponding y ions were observed at 1253 (y_{12}), 1152 (y_{11}), 1053 (y_{10}), 953.6054 (y_9), and 855(y_8). The fragment ion y_6 at *m/z* 643 (y_6) is also observed, which reveals a cyclic part identical to that of KF. This indicated that the part of the amino acid sequence in the linear region is common with the 5-Me-Hex-D-Val-L-Thr-L-Val-D-Val- peptide sequence. Since the macrolidic region is also identical the change in the peptide sequence should be in the remaining part of the linear region of the molecule. (Please note that in the case of the *m/z* values all decimal points are changed to a round figure and some of the product ion peaks are observed in MS/MS spectra of doubly charged ions.)

As we know the addition of b and the corresponding y ion gives the total molecular ion. The same theory is applied here to get the total sequence of the molecule. The peak at *m/z* 821 when added to 643 (y_6) gives the total 1464. Thus, peak at *m/z* 821 is confirmed to be b_7. The difference between the 609 (b_5) and

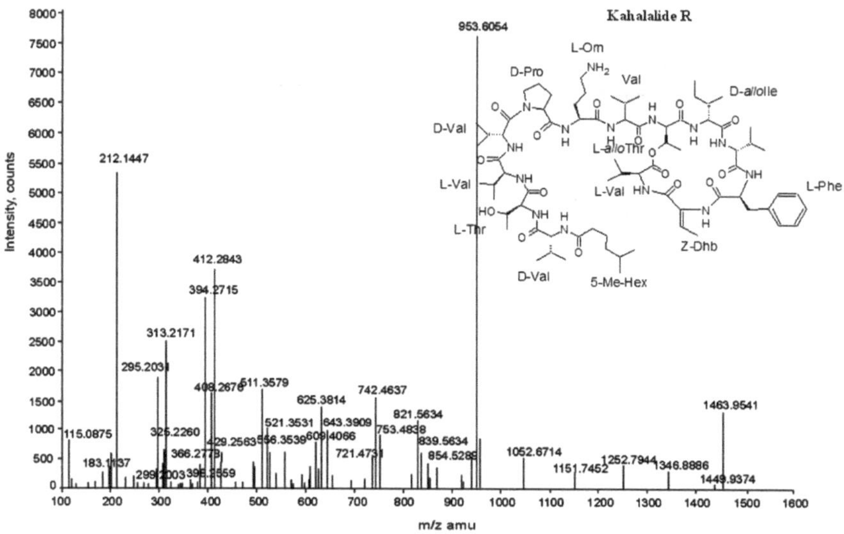

FIGURE 8.4 MS/MS spectrum at *m/z* 1464 $[M+H]^+$ of kahalalide R.

723 is 114 and the difference between the 855 (y_8) and 742 is 114, which corresponds to the ornithine residue. Thus, the peak at *m/z* 723 and 742 was fixed as b_6 and y_7 fragment ions, respectively. The mass difference between the 821 (b_6) and 723 (b_7) is 99, which could account for the amino acid valine instead of D-*allo*Ile. Also, the difference between y_7 and y_6 is exactly 99 amu, confirming the presence of valine residue. For any cyclic peptides the first dissociation step involves opening the cyclopeptide ring by cleavage of either the lactone group or one of the amide bonds. In most cases the basic site gets protonated first, resulting in the weakening of the C–N or C–O bond.

In the case of a cyclic part, the ring opening is at the lactone oxygen with the first cleavage of L-Val, which results in the formation of a fragment ion at *m/z* 542, which is not observed in the spectra. Similarly the sequential loss of other amino acid residues D-Zhb, Phe, Val, and Thr will result in the fragment ions at *m/z* 558, 311, 212, and 199, respectively, and are also not observed in the spectra. This indicates that the ring opening at the lactone oxygen is not favorable, but instead is favored with the initial loss L-Phe giving peak at *m/z* 494 $[y_6\text{-L-Phe}]^+$. This loss can be attributed to the N–C ring opening. The second loss of D-Val gave peak at *m/z* 394. This complex undergoes internal cyclization and the loss of L-Val gives a very intense peak at *m/z* 295. The peak at *m/z* 197 is due to the cleavage of an ester bond of L-*allo*-Thr and Z-Dhb. The electron donating aromatic ring makes the phenylalanine more basic residue, which gets protonated before lactone oxygen and is the reason for opening at amide bond. Thus, the pathway involves loss of amino acid residues that is not sequential. Similarly, following sequence was observed for kahalalide S, given in Table 8.1

TABLE 8.1 Conclusion of Fragmentation and Sequence of the Linear Chain Deduced from +1 and +2 Charged Molecular Ion Peaks

Parent Fragment	Found Peaks and Assigned Sequence	
	b type fragments (*m/z*)	y type fragments (*m/z*)
Kahalalide F 1478 $C_{75}H_{124}N_{14}O_{16}$	b_1 b_2 b_3 b_4 b_5 b_6 b_7 5MeHex-Val- Thr- Val- Val- Pro- Orn- Ile- R 212 313 412 511 609 723 835	y_{12} y_{11} y_{10} y_9 y_8 y_7 y_6 5MeHex- Val- Thr- Val- Val- Pro- Orn- Ile- R 1266 1165 1066 967 868 756 643
Kahalalide R 1464 $C_{74}H_{122}N_{14}O_{16}$	5MeHex-Val- Thr- Val- Val- Pro- Orn- Val- R 212 313 412 511 609 723 821	5MeHex- Val- Thr- Val- Val- Pro- Orn- Val- R 1252 1151 1052 953 $-NH_3$ 742 643 839
Kahalalide S 1492 $C_{76}H_{126}N_{14}O_{16}$	5MeHex-Val- Thr- Val- Val- Pro- Lys- Ile- R 212 313 412 511 609 764 849	5MeHex- Val- Thr- Val- Val- Pro- Lys- Ile- R 1280 1179 1080 981 882 756 643 $-NH_3$ 864

R-Cyclo (Thr-Ile-Val-Phe-Dhb-Val)

[37]. The tandem MS technique used in this study is much faster compared to NMR. It is highly sensitive and does not require a highly purified material, and therefore detection of minor peptides was possible.

Several other studies have examined the power of tandem MS. Williams [38] used MS^n experiments involving low-energy collisionally activated dissociation (CAD) in a quadrupole ion trap to characterize the fragmentation of alkali, alkaline-earth, and transition metal complexes of five cyclic peptides (cyclosporin A, valinomycin, enniantin A, iturin A, and surfactin). It showed that the diagnostic utility of the fragmentation information of the metal complexes is metal-specific. Of all the metal ions tested, those that provide the most useful complementary structural information to that of the protonated peptides are NaC and Pb_2C ions for the depsipeptides, and Ni_2C, Sr_2C, AgC, and LiC for the other cyclic peptides. A detailed description of the fragmentation behavior of beauverolides using post source decay (PSD), MS^n in the quadrupole ion trap, and in-source collision-induced dissociation (CID) in atmospheric pressure chemical ionization (APCI) on a sector instrument has been reported. Contrary to laborious PSD and tricky online LC/MS experiments, the in-source CID in APCI represents a powerful and robust tool for providing unambiguous sequence results [39]. The peptide amino acid sequence for aspergillicins A–E was unambiguously assigned by MS^n ion-trap ESI-MS [40]. Jegorov [41] identified natural roseotoxins A–C and minor destruxins A–B using an ion-trap ESI-MS. He proposed the dissociation pathway that involved opening of the cyclopeptide ring via formation of oxazolone derivatives, and the fragments of the resulting linear species have a free *N*-terminus and an oxazolone ring at the *C*-terminus.

Eight new peptaibols, namely trichobrachin A I–IV (Aib9-Pro10 sequence) and as trichobrachin B I–IV (Val9-Pro10 sequence), were identified from the marine-related *Trichoderma longibrachiatum* strain [42]. The use of ESI-MS^n-IT technology seems to represent a simpler method for peptaibol sequencing and leads to the establishment of peptaibol sequences from a mixture of short peptaibols. The methodology is used to establish the sequences of short peptaibols in a mixture without previous individual separation. ESI-MS^n-IT does not allow the differentiation of Val/Iva, Valol/Ivaol, Leu/Ile, or Leuol/Ileol. Gas chromatography/EI-MS analysis of amino acids such as *N*-trifluoroacetylisopropyl esters allowed this differentiation. In this case of the mixture, only Val, Valol, Leu, and Leuol were detected, allowing the establishment of final sequences. Two closely related proteins, β-thymosin^{His} and β-thymosin^{Gln}, have been *de novo* sequenced by top-down MS in the common neurobiology model, *Aplysia californica* [43]. As determined by nanoelectrospray quadrupole enhanced Fourier-Transform MS with collisionally activated and ECD, both of these *Aplysia* β-thymosins are acetylated and differ by a single residue in the central actin-binding domain. Profiling of individual cells and tissue by MALDI reveals that these proteins are widely expressed in the *Aplysia* central nervous system, including in individual identified neurons, neuronal clusters, nerves, and connective tissues.

The peptide library present in the venom of the piscivorous marine snail *Conus achatinus* has been probed using a combination of mass spectrometry and cDNA sequencing methods. MALDI-MS analysis, before and following global reduction/alkylation of peptide mixtures, permits the rapid classification of individual components on the basis of the number of disulfide bonds. Mass fingerprinting and the reverse phase HPLC retention times permit a further deconvolution of the library in terms of peptide size and hydrophobicity. Sequencing of cDNA derived using O-superfamily specific primers yielded five complete conotoxin precursor sequences, ranging in polypeptide length from 75 to 87 residues containing six Cys residues at the *C*-terminus [44]. Sequence analysis permits classification of the five putative mature peptides as δ, ω, and ω-like conotoxins. The presence of these predicted peptides in crude venom was established by direct MALDI-MS/MS sequencing following trypsin digestion of the peptide mixture after global reduction/alkylation. The determination of partial peptide sequences and comparison with the predicted sequences resulted in the identification of four of the five predicted conotoxins. The characterization of posttranslationally modified analogs, which are hydroxylated at proline or amidated at the *C*-terminus, is also demonstrated. Crude venom analysis should prove powerful in studying both the inter- and intraspecies variation in peptide libraries.

Diketopiperazines (DKPs) corresponding to cyclic dipeptides have been reported to exhibit antimicrobial, antitumor, antimutagenic, and antiviral properties. These compounds are commonly isolated from microorganisms, sponges, and from a variety of tissues and body fluids. In this work, the use of ESI-MS/MS to investigate the fragmentation of a series of DKPs previously isolated from *Aspergillus fumigatus*, which exhibit the same structural core, has been used [45]. Loss of CO directly from the protonated molecule was found to be a fragmentation process common to all the compounds analyzed. However, the results revealed a series of ions that are diagnostic for the substituents at C(4) and C(9).

8.7 ALKALOIDS

Alkaloids are naturally occurring, nitrogen-containing, biologically active heterocyclic compounds. Over the last few years, a large number of biologically important alkaloids with antiviral, antibacterial, anti-inflammatory, antimalarial, antioxidant, and anticancer activities have been isolated from various marine sources.

8.7.1 Case Study: Bromotyrosine Alkaloids

The research group at CSIR-NIO, Goa, [46] have investigated extracts of marine sponge *Psammaplysilla purpurea* during three collections from Mandapam (Tamil Nadu, India) and Okha (Gujarat, India), and have indentified two new bromotyrosine alkaloids, purpurealidin I and J using ESI-MS/MS. This sponge has tremendous chemical diversity of bromotyrosine alkaloids. Here, the proteomics approach has been used in identifying related bromotyrosine alkaloids based on the predicated mass fragmentation pattern. The focus is on the examination of detailed product ion spectra of six known compounds that allowed identification of new compounds based on its mass fragmentation pattern. The isotopic pattern of the peaks for protonated molecules indicated the number of bromine atoms present in the molecule. During MS/MS studies, the most prominent product ion peak is for the presence of side chain propane with either free NH_2, NHMe, or NMe_2. The cleavage at C–C bond between oxime-amide carbonyl and amide-phenoxy moiety also gave characteristic product ions. The ESI-MS spectra for all three collections show that the bromotyrosine metabolites vary during different seasons and also by geographical location, although some common metabolites were observed during the three collections. Thus, ESI-MS/MS is a method of choice in identifying the related compounds.

The positive ion ESI mass spectra of three methanolic extracts of marine sponge *Psammaplysilla purpurea* (14S007, OK22, and NIO759) obtained by direct infusion showed several clusters of peaks for protonated molecules with diagnostic characteristic patterns for the presence of bromotyrosine alkaloids. For illustration, Figure 8.5 shows expansion of ESI-MS spectrum of only NIO759 sample. The different ratios of the cluster indicate presence of two, three, or four bromine atoms in the molecule. By comparing Figure 8.5 with ESI-MS spectra of the previous study [47] and literature survey, a tentative assignment of the protonated molecule peaks to some of the known bromotyrosine alkaloids, purpuramine H-I, aplysamine 2-3, purealidin Q and 14-debromoprearaplysillin has been made. Further the structures of the known compounds were confirmed by studying fragmentation pattern with the aid of MS/MS. By using the MS/MS information, structure assignment of the two new minor bromotyrosine alkaloids, purpurealidin I-J having protonated molecule peaks at *m/z* 712; 714; 716;718;720 (1:4:6:4:1) and *m/z* 728; 730; 732; 734; 736 were possible. Purpurealidin I was present in the methanolic extract of 14S007 sample (ESI-MS spectrum not shown). The diagnostic fragment ions are labeled a, x, y, z, and b for two aromatic rings and the side chain, respectively. Intense fragment ion b at *m/z* 58 indicates a propyl amine side chain. The cleavage at the *p*-methoxy benzyl group gave peaks at *m/z* 199, 201 (a) indicating presence of one bromine atom on the ring. The fragment ion a contains only one bromine atom, which can be present either as Br^{79} or Br^{81} in the molecule.

To illustrate the fragmentation pattern herein, the CID of set of cluster ions at *m/z* 728, 730, 732, 734, and 736 (1:4:6:4:1) for the new compound, purpurealidin J, has been explained. The molecule is 14 amu lighter than purealidin Q. The CID spectrum shows that y and z fragment ions identical with purealidin Q evidencing intact two bromine atoms on second aromatic ring. Figure 8.5 illustrates the characteristic fragment ions formed during MS/MS study. The other fragment ions observed were less than 14 amu. Fragment ion x gave

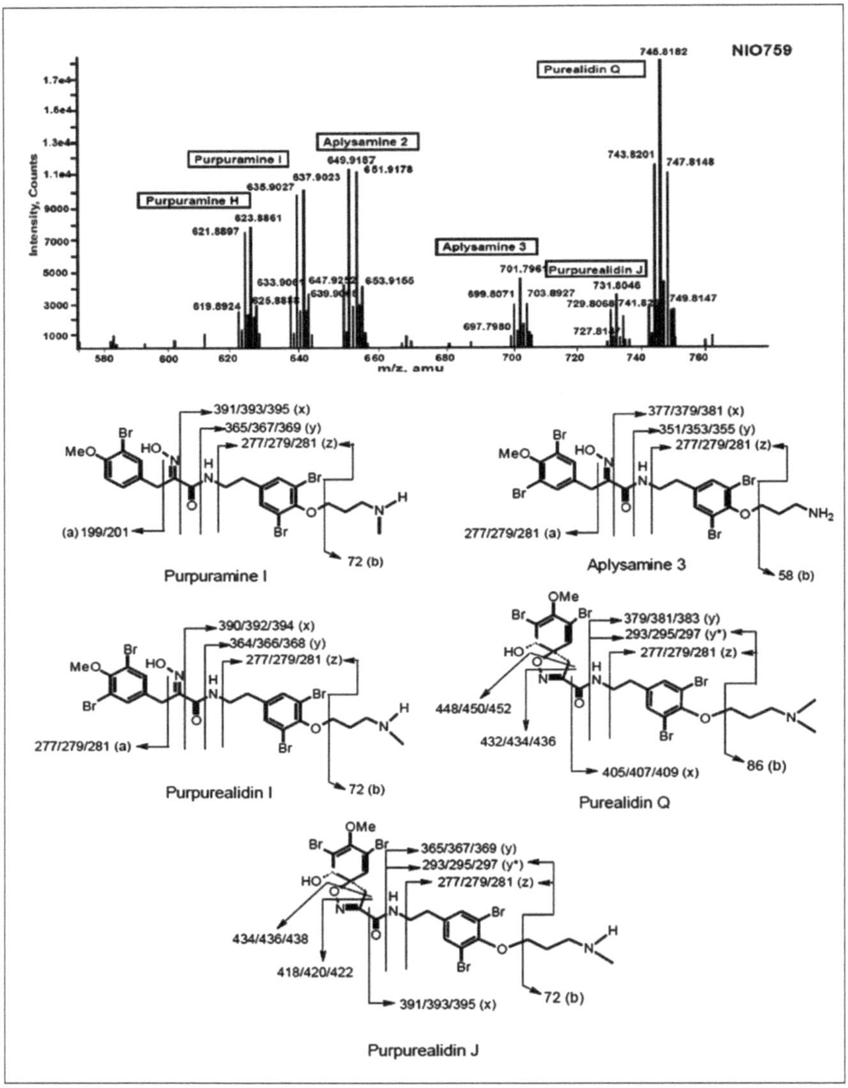

FIGURE 8.5 ESI-MS spectrum of methanolic extract of marine sponge *Psammaplysilla purpurea* (NIO759) and schematic representation of fragmentation pattern in bromotyrosine alkaloids, purpuramine I, aplysamine 3, purpurealidin I, purealidin Q, and purpurealidin J.

peaks at 391, 393, 395, and y at *m/z* 365, 367, 369. This revealed change must be on the side chain. The removal of one methyl group from the *N,N*-dimethylpropylamine chain fits well with fragment ions observed. Absence of peak at *m/z* 86 and intense peak at *m/z* 72 also confirms the presence of N-methylpropylamine. Cleavage of C–O and O–N in the dihydroisoxazole ring gives ions at *m/z* 434, 436, 438 (P), and 418, 420, 422, respectively. Peaks at *m/z* 306, 308, and 310 were due to the ion $[b\text{-}58]^+$. This represents a new bromotyrosine alkaloid and is designated as purpurealidin J. The fragmentation pattern of the known alkaloids helped in assigning the structure of the new compounds. The detailed fragmentation pattern is given in the publication by Tilvi *et. al* [46].

Various tandem MS studies have been carried out by different research groups for the identification of bioactive compounds. Kerzaon *et al.* [48] has nicely utilized LC/ESI-MS/MS and high-resolution (LC/HRMS/MS) for the identification and characterization of four other known communesins (A, D, E, and F) in the crude extract of marine-derived *Penicillium expansum*. A fragmentation model for dimethyl epoxide communesins was proposed after detailed interpretation of their MS/MS spectra. Further analyses of the extract using the modeled fragmentations led to the detection of seven new communesins (Com470, Com570, Com524, Com622, Com512, Com644, Com522) found as minor compounds. Chemical structural elucidation of these new derivatives was done based on their fragmentation characteristics.

A detection method based on HPLC-ESI-MS/MS was developed to investigate the total alkaloids prepared from the ascidian *Lissoclinum* cf. *badium*. The aromatic alkaloids possessing polysulfide structures are the major bioactive constituents isolated from ascidians of the genera *Lissoclinum*, *Eudistoma*, and *Polycitor* [49]. These dopamine-derived alkaloids have various important biological activities. The ESI-MS fragmentation behavior of this kind of alkaloids were studied, and the fragmentation was characterized by elimination of the $NH(CH_3)_2$ moiety in both in-source fragmentation (MS) and the MS/MS mode. The deaminated ions can then lose the methyl, methoxyl groups, and sulfur atoms in MS^n ($n = 2, 3$) experiments. The data provides useful information on the structure of these dopamine-derived compounds and can be used in the characterization of such marine natural products and their structurally related metabolites. The use of reversed-phase HPLC/ESI-MS allowed the online separation and detection of 25 aromatic alkaloids.

MALDI-TOF-MS was used for the first time as a direct screening method to analyze sponge tissue (*Stylissa caribica*) [50]. In order to validate this method the results were compared with a HPLC-ESI-MS analysis of the crude extract from the investigated sponge. Both methods showed a good agreement in their results, allowing for the rapid identification of 15 pyrrole-imidazole alkaloids: Debromostevensine, monobromoisophakellin, hymenidin, 4-bromopyrrole-2-carboxy-*N*(ε)-lysine, *N*-methylbromo(iso)phakellin, oxocyclostylidol, 4-bromopyrrole-2-carboxyarginine, stevensine, oroidin, *N*-methyldibromoisophakellin, dibromopalau'amine, ageliferin, hymenamide F, tetrabromostyloguanidine, and hymenamide C. The direct MS analysis (MALDI-TOF-MS) of sponge tissue represents a fast approach for the dereplication of secondary metabolites without a time-consuming extraction of the organism. The comparison of the HR-ESI-MS and MALDI-TOF-MS analyses provides information on the general application of the latter method. The results have shown that the information gained from MALDI-TOF-MS are in good accordance with data obtained by HR-ESI-TOFMS techniques. Only in the region of higher molecular weights (here: 800 to 900 amu) was the MALDI-TOFMS technique unable to detect all masses identified by HR-ESI-TOF-MS. Altogether, MALDI-TOFMS in combination with spectroscopic methods allows for fast identification, isolation, and structure elucidation of new natural products from marine sponges and possibly other organisms.

8.8 LIPIDS

Lipids are grouped under the following eight categories that cover eukaryotic and prokaryotic sources: fatty acyls, glycerolipids, glycerophospholipids, sphingolipids, sterollipids, prenollipids, saccharolipids, and polyketides.

Ceramides comprise a kind of sphingolipid consisting of long-chain aliphatic amino alcohols referred to as a sphingoid base, which are amide-linked to a fatty acid [51–54]. Sphingoid bases can usually be categorized by chain length, position and number of hydroxyl groups, and position and number of double bonds. The N-fatty acyl group of ceramides consists of approximately 16 to 26 aliphatic chains [55]. Ceramides are mainly located in the cellular plasma membrane, where they act as signaling molecules for cellular functions [56,57]. Because the terminal primary hydroxyl group of ceramides can be also attached, for example, to carbohydrates and phosphates, they function as key regulators of a wide range of biological process [58]. Many disease states including diabetes, cancer, Alzheimer's disease, and Parkinson's disease are strongly associated with ceramides, which have an important role in cellular signaling, the regulation of apoptosis, cell differentiation, aging, transformation, and proliferation [59,60]. Several mass spectrometric methods, including electron impact ionization mass spectrometry [61,62], MALDI-TOF MS [63], RP-HPLC/MS/MS [64–66], and ESI-MS/MS [67–69], have been used to identify ceramides extracted from biological samples of some ceramide species. To reduce these problems, the structural identification of ceramide species in the fractions, in which they were separated by the solvent extraction and chromatography techniques, using fast atom bombardment (FAB)-MS/MS, has been conducted. FAB-MS/MS is regarded as a useful tool because it can provide important structural information for compounds that have a repetitive similar moiety [70,71]. Ann and Adams [72,73] contributed significantly to the characterization of ceramides and neutral glycosphingolipid structures by using high-energy CID of $[M+Li]^+$ ions desorbed by FAB. Hsu and Turk [74,75] also established the fragmentation pathways of ceramides using low-energy CID of $[M+Li]^+$ and the $[M-H]^-$ ions generated by ESI. In the study, the structures of 10 ceramides (DNC-A1, A2 with two position isomers, DNC-B1, B2, DNC-C1, C3, C4, and DNC-D1, D2) and four cerebrosides (DNC-B3 and DNC-C2 with three isobaric components) isolated from starfish *Distolasterias nipon* extracts were determined using high-energy CID tandem mass analyses of their monosodiated ions.

8.8.1 Case Study: Molecular Species of Ceramides

Tilvi *et al.* [76] used ESI-MS/MS QTOF MS/MS (QSTAR XL Applied Biosystem instrument, Canada) for the analysis of eight molecular species of ceramide from the lipid fraction of marine sponge *Haliclona cribricutis* without chromatographic purification. The eight molecular species identified are *N*-eicosanoic (20:0), *N*-hencicosanoic (21:0), *N*-docosanoic (22:0), *N*-tricosanoic (23:0), *N*-tetrasanoic (24:0), *N*-pentasanoic (25:0), *N*-hexasanoic (26:0), and *N*-heptasanoic (27:0) sphingosine. The ESI-MS/MS was carried out in positive ionization mode. The sample dissolved in MeOH:$CHCl_3$ (99.5:0.5) containing traces of 0.1% TFA were directly infused at a constant flow rate of 10 μl/min into the ion spray source using integrated syringe pump. The ESI-MS exhibited several molecular $[M+H]^+$ ions at *m/z* 595, 609, 623, 637, 651, 665, 679, and 692 (Table 8.2).

TABLE 8.2 Molecular Ion and Fragment Ion Peaks for Molecular Species of Ceramide in the ESI-MS/MS Spectra of *Haliclona cribricutis*

Long Chain Base	$[M+H]^+$ (*m/z*)	P (*m/z*)	Q (*m/z*)	R (*m/z*)
C20:0	595	577	559	547
C21:0	609	591	573	561
C22:0	623	605	587	575
C23:0	637	619	601	589
C24:0	651	638	615	603
C25:0	665	647	629	611
C26:0	679	661	643	631
C27:0	692	675	657	645

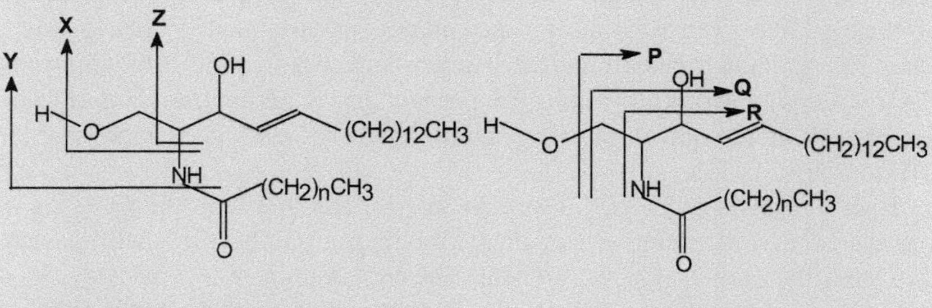

n =18-25 (C20:0 – C27:0)
X = *m/z* 282; X′= X-H_2O = *m/z* 264
Y = *m/z* 296.35; Y′= Y- H_2O = *m/z* 278
Z = *m/z* 252

The MS/MS experiment of each molecular ion $[M+H]^+$ was carried out at a collision energy of 40V. All the molecular ions showed some common fragment ions. The fragment ion corresponding to cleavage between the carbon and nitrogen bond of the sphingosine base are observed at *m/z* 282 (X) and with loss of an additional neutral water molecule resulted in the abundant ion at *m/z* 264 $[X'=X\text{-}H_2O]^+$. The cleavage between the amide bond resulted in the fragment ion at *m/z* 296 (Y) and *m/z* 278 $[Y'=Y\text{-}H_2O]^+$ for the loss of water molecule. It also showed product ions at *m/z* 252 (Z) due to loss of 31 amu from the sphingosine base (Table 8.2) [75,76]. Product ion peaks at *m/z* at 252, 264, 278, 282, and 296 in the spectra of all the compounds confirmed the sphingosine as being d 18:1 and the heterogeneity as being in the acyl group. It demonstrated the identification of ceramides using minimum chromatographic separation and with minimum concentration of sample.

Glycerol-based alkyl ether lipids are ubiquitous components in marine sediments. In order to explore their structural diversity and biological sources, marine sediment samples from diverse environments were analyzed and the mass spectra of widely distributed, novel glycerol di- and tetraethers were examined systematically. Lipid extracts were analyzed by APCI-MS and tandem MS spectra of compounds by a quadrupole time-of-flight (qTOF) mass spectrometer [77]. In addition to the well-established isoprenoidal glycerol dialkyl glycerol tetraether (isoprenoidal GDGT) and branched GDGT, suites of novel lipids were detected in all studied samples. These lipids include the following classes of tentatively identified compounds: isoprenoidal glycerol dialkanol diether (isoprenoidal GDD), hydroxylated isoprenoidal GDGT (OH-GDGT), hybrid isoprenoidal/branched GDGT (IB-GDGT), hydroxylated isoprenoidal GDD (OH-GDD), overly branched GDGT (OB-GDGT), sparsely branched GDGT (SB-GDGT), and an abundant H-shaped GDGT with the $[M+H]^+$ ion of *m/z* 1020 (H-1020). Characteristic MS/MS fragmentation patterns provided mass spectral fingerprints for the recognition of diverse and prominent glycerol ether lipids. The ubiquitous distribution and substantial abundance of these glycerol ethers, as well as their structural variability, suggest a significant ecological role of their source organisms in various marine environments. Their glycerol tetraether skeleton yielded characteristic product ions in MS2 through loss of H_2O, glycerol, and one alkyl unit. The product ions resulting from loss of a single alkyl chain provide clues to the carbon number of each alkyl moiety of the OB- and SB-GDGTs.

P. de la Iglesia *et al.* [78] have detected for the first time the presence of a range of acyl ester derivatives of gymnodimine (GYMs) in shellfish samples from the Gulf of Gabes, Tunisia. The MS fragmentation pathways of 1 and its esters were also elucidated. Evidence of acyl ester metabolites of gymnodimine-B and -C were also obtained. A series of mass spectrometric experiments involving precursor and product ion scans, selected reaction monitoring (SRM), and HR-MS were used to detect and confirm 10-O-acyl esters of gymnodimine-A.

Gabriella Pocsfalvi *et al.* [79] reported structural characterization of a highly heterogeneous mixture of glucosylceramides (GlcCers) isolated from a deep-water Mediterranean dendrophylliid coral, *Dendrophyllia cornigera*. The neutral glycosphingolipid (GSL) components of the coral were separated into three HPLC fractions that were structurally characterized by NMR and MS. NMR analysis revealed a β-glucosylpyranose, a methyl branched conjugated sphingadienine, and α-hydroxy fatty acid moieties characteristic for the species. Molecular mass distributions of the HPLC fractions were monitored using single-stage MS. At least 17 different GlcCer constituents with variable long-chain base and fatty acid residues were observed based on the molecular ion peaks in the liquid secondary ion (LSI) survey spectra. Structures of the individual components were revealed by product ion spectra of the alkali cationized molecules $[M+Cat]^+$, which resulted in two characteristic fragment ions, FF and FS. Tandem MS of the same fragment ions formed in the ion source showed that FF carries the hydoxy fatty acid, while FS carries the long-chain sphingoid base, thus providing complementary structural information for the characterization of ceramide composition. Based on the tandem mass spectra of the molecular ions $[M+Na]^+$, 26 different GlcCers of the coral were identified. The ceramide moiety showed heterogeneity in both the sphingoid portion (d18:2, d19:2, d20:2, and d20:3) and the α-hydroxy fatty acid chain (h19-h24, either saturated or unsaturated), forming an extremely heterogeneous mixture. The method is generally applicable to the characterization of structurally heterogeneous GlcCer mixtures.

Total lipid was extracted from *Skeletonema* microalgae. Separation of lipids was achieved via RP-HPLC and MS analysis of the lithium adduct of glycosphingolipids (GSLs) was conducted to determine their structures [80]. Two types of novel GSLs were identified from three strains of *Skeletonema* microalgae. The N-acyl groups were primarily long-chain saturated or monoenoic fatty acids (monounsaturated fatty acids) with 16C and 20 to 24C. The sphingoid long-chain bases (LCB) were sphingosine, sphingadienine, and sphingatrienine. The saccharide polar head groups of glycosphingolipids were disaccharide (heptose-hexose) or trisaccharide (heptose-hexose-hexose). Semiquantitative analysis of GLSs in the three strains of microalgae showed that GSL contents ranged from 0.09 to 8.79 nmol/mg dry microalgae powder.

ESI and liquid secondary ionization (LSI) mass spectrometry were applied for characterization of glycosphingolipids (GSLs) isolated in their peracetylated form from four *Agelas* marine sponge species [81]. Since peracetylated glycosphingolipids are not soluble in solvents traditionally used for ESI, lithium chloride was added to the samples in order to obtain lithium cationized molecules. Although the preferred fragmentation seems to be the sequential loss of acetic acid molecules, it was found that tandem mass spectra obtained from peracetylated diglycosyl ceramides might provide direct information about the structure of the long-chain base (formation of $W0^0$/Z0 fragments). The utility of ESI and LSI in the analysis of these compounds has also been compared.

It was found that the tandem mass spectra obtained by LSI-MS/MS experiments could provide information about the chain-length (carbon atom number) variations within a certain ceramide mass. Thus, from one of the samples, 25 different ceramide compositions have been identified from eight precursor (Z0) ions. Comparison of the two ionization modes (LSI and ESI) highlights the fact that molecular mass distributions obtained by LSI-MS, especially the presence of unsaturated species, have to be interpreted carefully. For the first time a direct HPLC/ESI-MS method was used for characterization of complex mixtures of peracetylated GSLs. The results demonstrated that HPLC/ESI-MS is able to analyze mono- and diglycosylated GSLs, and other kinds of glycolipids that are actually present in the sample.

Six cerebrosides were isolated from the eggs of the starfish *Asterias amurensis* using solvent extraction, silica gel column chromatography, and RP-HPLC [82]. The study demonstrated that the structures of cerebrosides could be completely characterized, based on their sodium-adducted molecules, using FAB tandem MS. The high-energy CID of the sodium-adducted molecule, $[M+Na]^+$, of each cerebroside molecular species generated abundant ions, providing information on the compositions of the 2-hydroxy fatty acids and long-chain sphingoid bases, as well as the sugar moiety polar head group. Each homologous ion series along the fatty acid and aliphatic chain of the sphingoid base was useful for locating the double-bond positions of both chains and the methyl branching position of the long-chain base. The *N*-fatty acyl portions were primarily long-chain saturated or monoenoic acids (C16 to C24) with an α-hydroxy group. The sphingoid long-chain base portions were aliphatic chains (C18 or C22) with two or three degrees of unsaturation and with or without methyl branching.

Three cyclitol derivatives were isolated from the marine sponge *Sarcotragus sp*. by RP-HPLC and analyzed by FAB-MS [83]. FAB-MS spectra produced a significant abundance of the sodium adducts $[M+Na]^+$ and $[M+2Na-H]^+$ from a mixture of m-NBA and NaI. In addition, trifluoroacetylation of the cyclitol derivatives was used for confirmation of the presence of the cyclitol ring. High abundance $[M-5H+5CF_3CO+Na]^+$ ions were observed in the FAB-MS spectra of the trifluoroacetyl-cyclitol derivatives. CID of the $[M+Na]^+$ ions produced diverse product ions via a series of dissociative processes. Charge-remote fragmentation (CRF) patterns of $[M+Na]^+$ ions were very useful for the identification of product ions, which are characteristic for the cyclitol ring and long hydrocarbon chains substituted at the glycerol backbone. Moreover, the CID-MS/MS spectra of the $[M+Na]^+$ ions yielded characteristic product ions at *m/z* 53, 83, 113, 155, and 171 for the cyclitol moiety, and at *m/z* 213, 229, and 245 for the glycerol backbone attached to the cyclitol ring.

Five new monoacylglycerols (MAGs) were isolated from the marine sponge *Stelletta sp*. by reversed phase HPLC and analyzed by positive ion FAB-MS. FAB mass spectra of these compounds produced abundant sodium-adducted molecules $[M+Na]^+$ from a mixture of 3-nitrobenzyl alcohol and sodium iodide [84]. The structural elucidation of these sponge MAGs was carried out by

FAB tandem MS. To find diagnostic ions for the characterization of the MAGs, authentic MAGs were initially analyzed by CID-MS/MS. The CID MS/MS of $[M+Na]^+$ precursor ions resulted in the formation of numerous characteristic product ions via a series of dissociative processes. The product ions formed by charge-remote fragmentation (CRF) provided important information for the characterization of acyl chains substituted at the glycerol backbone, and product ions at *m/z* 84, 97, 113, and 139 were diagnostic for the sodiated glycerol backbone. On the basis of these fragmentation patterns, the structures of five MAGs extracted from marine sponge were elucidated. In addition, high-resolution mass measurement was performed to obtain the elemental compositions of the MAGs.

The pharmaceutical industry is interested in identifying novel target compounds. Due to their versatile pharmacological activities (e.g., antiviral, anticarcinogen, and immunosuppressive) sulfoquinovosyldiacylglycerides (SQDGs) are potential drug candidates [85]. The present publication deals with the purification and structural characterization of SQDGs from three different strains of *Phaeodactylum tricornutum*. Besides detection of SQDGs (sn-1: C16:1/sn-2: C16:0 and sn-1: C20:5/sn-2: C16:0), two novel 2'-O-acylsulfoquinovosyldiacylglyerides (Ac-SQDGs, sn-1:C16:0/sn-2: C16:0/2' C20:5 and sn-1: C20:5/sn-2: C16:0/2' C20:5) were identified by using MALDI QTrap ToF hybrid mass spectrometry (MS) with multistage MSn. The analytical method enables the sn-position verification of fatty acids (MS^2) as well as the confirmation of the regioposition of eicospentanoic acid at the sulfoquinovose (MS^3).

Bacterial fatty acid profiling is a well-established technique for bacterial identification. Ten bacteria were analyzed using both positive- and negative-ion modes with a modified matrix-assisted laser desorption ionization mass spectrometry (MALDI MS) approach using CaO as a matrix replacement (metal oxide laser ionization MS (MOLI MS) [86]. The results show that reproducible lipid cleavage similar to thermal *in situ* tetramethyl ammonium hydroxide saponification/derivatization had occurred. Principal component analysis showed that replicates from each organism grouped in a unique space. Cross-validation (CV) of spectra from both ionization modes resulted in greater than 94% validation of the data. When CV results were compared for the two ionization modes, negative-ion data produced a superior outcome. MOLI MS provides clinicians a rapid, reproducible, and cost-effective bacterial diagnostic tool.

8.9 CONCLUSION

Currently, many natural products chemists still ignore MS in the isolation process and instead employ NMR data to facilitate isolation of as many unique (and structurally interesting) compounds as possible. It is clear from the discussion that tandem mass spectroscopy has extraordinary capabilities for the structural analysis of individual components of complex mixtures with the first MS working as the separation device and the second one used for chemical structure determination, and showed to be very effective for structural analysis of peptides, bromotyrosine alkaloids, and molecular species of ceramides. Indeed, in the future there could be mass spectrometers that can simultaneously solve natural product structures and identify those that are biologically active. Such accomplishments will require sustained collaborative efforts involving (but not limited to) analytical chemists, natural products chemists, instrument and software developers, and biologists. Nonetheless, as we see how far MS technique has progressed in recent decades, these advances are possible, and the future of MS as a tool for natural products research is unquestionably bright.

ACKNOWLEDGEMENTS

The authors thank the Director, CSIR-National Institute of Oceanography for constant encouragement. Financial assistance provided by the OCEAN FINDER and EU-FP7-KBBE-2009-3-245137 MAREX is highly acknowledged. Author MSM is grateful to CSIR-NIO for the award of Scientist Fellow-QHS.

REFERENCES

[1] J.B. Fenn, Electrospray wings for molecular elephants (Nobel lecture), Angew. Chem. Int. 42 (2003) 3871.

[2] K. Tanaka, The origin of macromolecule ionization by laser irradiation (Nobel lecture), Angew, Chem. Int. 42 (2003) 3861.

[3] I.X. Peng, R.R. Ogorzalek Loo, J. Shiea, J.A. Loo, Reactive-electrospray-assisted laser desorption/ionization for characterization of peptides and proteins, Anal. Chem. 80 (2008) 6995.

[4] R.B. Cody, J.A. Larame, H.D. Durst, Versatile new ion source for the analysis of materials in open air under ambient conditions, Anal. Chem. 77 (2005) 2297.

[5] Z. Taka'ts, J.M. Wiseman, B. Gologan, R.G. Cooks, Mass spectrometry sampling under ambient conditions with desorption electrospray ionization, Science 306 (2004) 471.

[6] D.R. Ifa, A.U. Jackson, G. Paglia, R.G. Cooks, Forensic applications of ambient ionization mass spectrometry, Anal. Bioanal. Chem. 394 (2009) 1995.

[7] R. Aebersold, M. Mann, Mass spectrometry-based proteomics, Nature 422 (2003) 198.

[8] K.L. Busch, G.L. Glish, S.A. McLuckey (Eds.), Mass Spectrometry/Mass Spectrometry: Techniques and Applications of Tandem Mass Spectrometry, VCH, New York, 1988.

[9] S.A. McLuckey, Principles of collisional activation in analytical mass spectrometry, J. Am. Soc. Mass Spectrom 3 (1992) 599.

[10] A.K. Shukla, J.H. Futrell, Tandem mass spectrometry: dissociation of ions by collisional activation, J. Mass Spectrom 35 (2000) 1069.

[11] P.D. Schnier, W.D. Price, R.A. Jockush, E.R. Williams, Blackbody Infrared Radiative Dissociation of Bradykinin and Its Analogues: Energetics, Dynamics, and Evidence for Salt-Bridge Structures in the Gas Phase, J, Am. Chem. Soc. 118 (1996) 7178.

[12] R.A. Zubarev, N.L. Kelleher, F.W. McLafferty, Electron Capture Dissociation of Multiply Charged Protein Cations—a Nonergodic Process, J. Am. Chem. Soc. 120 (1998) 3265.

[13] C.M. Whitehouse, R.N. Dreyer, M. Yamashita, J.B. Fenn, Electrospray interface for liquid chromatographs and mass spectrometers, Anal. Chem. 57 (1985) 675.

[14] R.B. Cole (Ed.), Electrospray ionization mass spectrometry. Fundamental, Instrumentation, and Applications, Wiley, New York, 1997.

[15] B.N. Pramanik, A.K. Ganguly, M.L. Gross (Eds.), Applied Electrospray Mass Spectrometry, Marcel Dekker, New York, 2002.
[16] M. Karas, F. Hillenkamp, Laser desorption ionization of proteins with molecular masses exceeding 10,000 daltons, Anal. Chem. 60 (1988) 2299.
[17] R.C. Beavis, B.T. Chait, Matrix-assisted laser desorption ionization mass-spectrometry of proteins, Methods Enzymol. 270 (1996) 519.
[18] K.L. Busch, G.L. Glish, S.A. McLuckey (Eds.), Mass spectrometry/Mass spectrometry: Techniques and Applications of Tandem Mass Spectrometry, VCH, New York, 1988.
[19] Tandem Mass Spectrometry; F.W. McLafferty (Ed.), Wiley: New York, 1983.
[20] B. Paizs, S. Suhai, Fragmentation pathways of protonated peptides, Mass Spectrom. Rev. 24 (2005) 508.
[21] P. Roepstorff, J. Fohlman, Proposal for a common nomenclature for sequence ions in mass spectra of peptides, Biomed. Mass Spectrom 11 (1984) 601.
[22] K. Biemann, Nomenclature for peptide fragment ions (positive ions), Methods Enzymol. 193 (1990) 886.
[23] I.A. Papayannopoulos, The interpretation of collision-induced dissociation tandem mass spectra of peptides, Mass Spectrom. Rev. 14 (1995) 49.
[24] D.R. Muller, M. Eckersley, W. Richter, Hydrogen transfer reactions in the formation of "Y + 2" sequence ions from protonated peptides, Org. Mass Spectrom 23 (1988) 217.
[25] M.M. Cordero, J.J. Houser, The neutral products formed during backbone fragmentations of protonated peptides in tandem mass spectrometry, Anal. Chem. 65 (1993) 1594.
[26] L.C. Ngoka, M.L. Gross, A nomenclature system for labeling cyclic peptide fragments, J. Am. Soc. Mass Spectrom 10 (1999) 360.
[27] M. Mann, M.S. Wilm, Errora-tolerant identification of peptides in sequence databases by peptide sequence tags, Anal. Chem. 66 (1994) 4390.
[28] J.K. Eng, A.L. McCormack, I.J.R. Yates, An approach to correlate tandem mass spectral data of peptides with amino acid sequences in a protein database, J. Am. Soc. Mass Spectrom 5 (1994) 976.
[29] M.S. Boguski, The turning point in genome research, Trends Biochem. Sci. 20 (1995) 295.
[30] M.T. Hamann, P.J. Scheuer, Kahalalide F.: a bioactive depsipeptide from the sacoglossan mollusk *Elysia rufescens* and the alga *Bryopsis sp*., J. Am. Chem. Soc. 115 (1993) 5825.
[31] L. L. Lopez, J. M. Trigo, J. Jimeno, L. Pazares, C. A. Ruiz, M. A. Izquierdo, J. Schellens, Kahalalide compounds for use in cancer therapy. Patent Number EP 1435990, 2004.
[32] B. Nuyen, S. Weitman, G. T. Faircloth, Kahalalide F formulation. Patent Number US 2004067895, 2004.
[33] M. Garcia-Rocha, P. Bonay, J. Avila, The antitumor compound kahalalide F acts on cell lysosomes, Cancer Lett. 99 (1996) 43.
[34] E. Ciruelos, J.M. Trigo, B. Pardo, L. Paz-Ares, N. Estaun, C. Cuadra, M.J. Domωnguez, A. Marin, A. Ruiz, J. Jimeno, M.A. Izquierdo, A phase I clinical and pharmacokinetic (PK) study with Kahalalide F (KF) in patients (pts) with advanced solid tumours (AST) with a continuous weekly (W) 1-hour iv infusion schedule, Eur. J. Cancer 38 (2002) S33.
[35] S.G. Gomez, J.A. Bueren, G.T. Faireloth, J. Jimeno, B. Albella, In vitro toxicity of three new antitumoral drugs (Trabectedin, Aplidin and Kahalalide F) on hemotopoietic progenitors and stem cells, Exp. Hematol. 31 (2003) 1104.
[36] J. C. Jimenez, I. Rodrigues, E. Giralt, A. Lopes, M. Royo, F. Albericio, I. Manzanares. Kahalalide compounds, Patent Number PL 356800, 2004.
[37] S. Tilvi, C.G. Naik, Tandem mass spectrometry of kahalalides: identification of two new cyclic depsipeptides, kahalalide R and S from *Elysia grandifolia*, J. Mass Spectrom 42 (2007) 70.

[38] S.M. Williams, J.S. Brodbelt, MSn characterization of protonated cyclic peptides and metal complexes, J. Am. Soc. Mass Spectrom 15 (2004) 1039.
[39] M. Kuzma, A. Jegorov, P. Kacer, V. Havlicek, Sequencing of new beauverolides by high performance liquid chromatography and mass spectrometry, J. Mass Spectrom 36 (2001) 1108.
[40] R.J. Capon, C. Skene, M. Steward, J. Ford, R.A.J. O'hair, L. Williams, E. Lacey, J.H. Gill, K. Heiland, T. Friedel, Aspergillicins A-E: five novel depsipeptides from the marine-derived fungus Aspergillus carneus, Org. Biomol. Chem. 1 (2003) 1856.
[41] A. Jegorov, B. Paizs, M. Zabka, M. Kuzma, V. Havlicek, A.E. Giannakopulos, P.J. Derrick, Profiling of cyclic hexadepsipeptides roseotoxins synthesized *in vitro* and *in vivo*: a combined tandem mass spectrometry and quantum chemical study, Eur. J. Mass Spectrom 9 (2003) 105.
[42] M. Mohamed-Benkada, M. Montagu, J.-F. Biard, F. Mondeguer, P. Verite, M. Dalgalarrondo, J. Bissett, Y.–F. Pouchus, New short peptaibols from a marine Trichoderma strain, Rapid Commun. Mass Spectrom 20 (2006) 1176.
[43] E.V. Romanova, M.J. Roth, S.S. Rubakhin, J.A. Jakubowski, W.P. Kelley, M.D.K. Neil, L. Kelleher, J.V. Sweedler, Identification and characterization of homologues of vertebrate b-thymosin in the marine mollusk Aplysia californica, J. Mass Spectrom 41 (2006) 1030.
[44] K.H. Gowd, K.K. Dewan, P. Iengar, K.S. Krishnan, P. Balaram, Probing peptide libraries from *Conus achatinus* using mass spectrometry and cDNA sequencing: identification of d and ω-conotoxins, J. Mass Spectrom 43 (2008) 791.
[45] N. Araçari, J.C. Furtado, R. Vessecchi, J.C. Tomaz, S.E. Galembeck, J.K. Bastos, N.P. Lopes, A.E.M. Crotti, Fragmentation of diketopiperazines from Aspergillus fumigatus by electrospray ionization tandem mass spectrometry (ESI-MS/MS), J. Mass Spectrom 42 (2007) 1279.
[46] S. Tilvi, L. D'Souza, Identifying the related compounds using ESI-MS/MS: Bromotyrosine alkaloids from marine sponge *Psammaplysilla purpurea*, Eur. J. Mass Spectrum 18 (2012) 333.
[47] S. Tilvi, C. Rodrigues, C.G. Naik, P.S. Parameswaran, S. Wahidhulla, New bromotyrosine alkaloids from the marine sponge Psammaplysilla purpurea, Tetrahedron 60 (2004) 10207.
[48] I. Kerzaon, Y.F. Pouchus, F. Monteau, B. Le Bizec, M.-R. Nourrisson, J.-F. Biard, O. Grovel, Structural investigation and elucidation of new communesins from a marine-derived Penicillium expansum Link by liquid chromatography/electrospray ionization mass spectrometry, Rapid Commun. Mass Spectrom 23 (2009) 3928.
[49] H. Liu, H. Gao, M. Namikoshi, H. Kobayashi, R.E.P. Mangindaan, N. Wang, X. Yao, Characterization and online detection of aromatic alkaloids in the ascidian *Lissoclinum cf. badium* by liquid chromatography/UV detection mass spectrometry, Rapid Commun. Mass Spectrom 21 (2007) 199.
[50] A. Grube, T. Maier, M. Kostrzewa, M. Köck, MS-guided Fractionation as a Fast Way to the Identification of New Natural Products – MALDI-TOF-MS Screening of the Marine Sponge *Stylissa caribica*, Z. Naturforsch. 62b (2007) 600.
[51] S. Wartewig, R.H. Neubert, Properties of ceramides and their impact on the stratum corneum structure: a review, Skin Pharmacol. Physiol. 20 (2007) 220.
[52] M.E. Stewart, D.T. Downing, A new 6-hydroxy-4-sphingenine containing ceramide in human skin, J. Lipid Res. 40 (1999) (1434).
[53] Y. Masukawa, H. Narita, E. Shimizu, N. Kondo, Y. Sugai, T. Oba, R. Homma, J. Ishikawa, Y. Takagi, T. Kitahara, Y. Takema, K. Kita, Characterization of overall ceramide species in human stratum corneum, J. Lipid Res. 49 (2008) 1466.
[54] Y. Mizutani, S. Mitsutake, K. Tsuji, A. Kihara, Y. Igarashi, Ceramide biosynthesis in keratinocyte and its role in skin function, Biochimie 91 (2009) 784.

[55] W. Zheng, J. Kollmeyer, H. Symolon, A. Momin, E. Munter, E. Wang, S. Kelly, J.C. Allegood, Y. Liu, Q. Peng, H. Ramaraju, M.C. Sullards, M. Cabot, A.H. Merrill, Ceramides and other bioactive sphingolipid backbones in health and disease: lipidomic analysis, metabolism and roles in membrane structure, dynamics, signaling and autophagy, Biochim. Biophys. Acta 1758 (2006) 1864.

[56] S. Pandey, R.F. Murphy, D.K. Agrawal, Recent advances in the immunobiology of ceramide, Exp. Mol. Pathol. 82 (2007) 298.

[57] Y.A. Kim, D.M.C. Ramirez, W.J. Costain, L.J. Johnston, R. Bittman, A new tool to assess ceramide bioactivity: 6-bromo-7-hydroxycoumarinylcaged ceramide, Chem. Commun. 47 (2011) 9236.

[58] J. Spot, F.M. Goni, A. Alonso, Molecular associations and surface active properties of short- and long-N-acyl chain ceramides, Biochim. Biophys. Acta 1711 (2005) 12.

[59] Y.A. Hannun, L.M. Obeid, Many ceramides, J. Biol. Chem. 286 (2011) 27855.

[60] Z.F. Yu, M. Nikolova-Karakashian, D. Zhou, G. Cheng, E.H. Schuchman, M.P. Mattson, Pivotal role for acidic sphingomyelinase in cerebral ischemia-induced ceramide and cytokine production, and neuronal apoptosis, J. Mol. Neurosci. 15 (2000) 85.

[61] B. Samuelsson, K. Samuelsson, Gas-liquid chromatography-mass spectrometry of synthetic ceramides, J. Lipid Res. 10 (1969) 41.

[62] B. Samuelsson, K. Samuelsson, Gas–liquid chromatographic separation of ceramides as di-O-trimethylsilyl ether derivatives, Biochim. Biophys. Acta 164 (1968) 421.

[63] O. Shadyro, I. Yurkova, M. Kisel, O. Brede, J. Arnhold, Formation of phosphatidic acid, ceramide, and diglyceride on radiolysis of lipid: identification by MALDI-TOF mass spectrometry, Free Radical Biol. Med. 36 (2004) 1612.

[64] T.S. Haynes, P.J. Duerksen-Hughes, M. Filippova, V. Filippov, K. Zhang, C18 ceramide analysis in mammalian cells employing reversed-phase high-performance liquid chromatography tandem mass spectrometry, Anal. Biochem. 378 (2010) 80.

[65] R. Wieboldt, D.A. Campbell, J. Henion, Quantitative liquid chromatographic–tandem mass spectrometric determination of orlistat in plasma with a quadrupole ion trap, J. Chromatogr. B Biomed. Sci. Appl. 708 (1998) 121.

[66] G. Liebisch, W. Drobnik, M. Reil, B. Trumbach, R. Arnecke, B. Olgemoller, A. Roscher, G.J. Schmitz, Quantitative measurement of different ceramide species from crude cellular extracts by electrospray ionization tandem mass spectrometry (ESI-MS/MS), J. Lipid Res. 40 (1999) 1539.

[67] X. Han, Characterization and direct quantitation of ceramide molecular species from lipid extracts of biological samples by electrospray ionization tandem mass spectrometry, Anal. Biochem. 302 (2002) 199.

[68] A.P. Bruins, Mass spectrometry with ion sources operating at atmospheric pressure, Spec. Rev. 10 (1991) 53.

[69] J.C. Schwartz, I. Jardine, Quadrupole ion trap mass spectrometry, Methods Enzymol. 270 (1996) 552.

[70] T. Park, Y.S. Park, J.-R. Rho, Y.H. Kim, Structural determination of cerebrosides isolated from *Asterias amurensis* starfish eggs using high-energy collision-induced dissociation of sodium-adducted molecules, Rapid Commun. Mass Spectrom 25 (2011) 572.

[71] N.J. Jensen, M.L. Gross, A comparison of mass spectrometry methods for structural determination and analysis of phospholipids, Mass Spectrom. Rev. 7 (1988) 41.

[72] Q. Ann, J. Adams, Structure-specific collision-induced fragmentations of ceramides cationized with alkali-metal ions, Anal. Chem. 65 (1993) 7.

[73] Q. Ann, J. Adams, Structure determination of ceramides and neutral glycosphingolipids by collisional activation of $[M+Li]^+$ ions, J. Am. Soc. Mass Spectrom 3 (1992) 260.

[74] F.F. Hsu, J. Turk, Structural studies on ceramides as lithiated adducts by low energy collisional-activated dissociation tandem mass spectrometry with electrospray ionization, J. Am. Soc. Mass Spectrom 13 (2002) 680.

[75] F.F. Hsu, J. Turk, Characterization of ceramides by low energy collisional-activated dissociation tandem mass spectrometry with negative-ion electrospray ionization, J. Am. Soc. Mass Spectrom 13 (2002) 558.

[76] S. Tilvi, M. Majik, C.G. Naik, Tandem mass spectrometric approach for determining structure of molecular species of ceramide in the Marine Sponge *Haliclona cribricutis*, Eur. J. Mass Spectrom 11 (2005) 346.

[77] X.–L. Liu, R.E. Summons, K.-U. Hinrichs, Extending the known range of glycerol ether lipids in the environment: structural assignments based on tandem mass spectral fragmentation patterns, Rapid Commun. Mass Spectrom 26 (2012) 2295.

[78] P. de la Iglesia, P. McCarron, J. Diogène, M.A. Quilliam, Discovery of gymnodimine fatty acid ester metabolites in shellfish using liquid chromatography/mass spectrometry, Rapid Commun. Mass Spectrom 27 (2013) 643.

[79] G. Pocsfalvi, A. Malorni, I. Mancini, G. Guella, F. Pietra, Molecular characterization of a highly heterogeneous mixture of glucosylceramides from a deep-water Mediterranean scleractinian coral Dendrophyllia cornigera, Rapid Commun. Mass Spectrom 14 (2000) 2247.

[80] F. Zhao, J. Xu, J. Chen, X. Yan, C. Zhou, S. Li, X. Xu, F. Ye, Structural elucidation of two types of novel glycosphingolipids in three strains of Skeletonema by liquid chromatography coupled with mass spectrometry, Rapid Commun. Mass Spectrom 27 (2013) 1535.

[81] Z. Szilágyi1, G. Pócsfalvi, V. Costantino, A. Mangoni, A. Malorni, E. Fattorusso, Partial characterization of glycosphingolipids of Agelas sponges in their peracetylated form by liquid secondary ionization mass spectrometry and high-performance liquid chromatography combined with direct electrospray ionization mass spectrometric detection, Rapid Commun. Mass Spectrom 18 (2004) 2989.

[82] T. Park, Y.S. Park, J.–R. Rho, Y.H. Kim, Structural determination of cerebrosides isolated from *Asterias amurensis* starfish eggs using high-energy collision-induced dissociation of sodium-adducted molecules, Rapid Commun. Mass Spectrom 25 (2011) 572.

[83] J.H. Gil, J. Seo, K.–J. Kim, J.H. Jung, O.–S. Jung, M.-S. Kim, J. Hong, Structural determination of cyclitol derivatives by fast-atom bombardment tandem mass spectrometry, Rapid Commun. Mass Spectrom 20 (2006) 1253.

[84] J.-H. Gil, J.–Y. Hong, J.H. Jung, K.–J. Kim, J. Hong, Structural determination of monoacylglycerols extracted from marine sponge by fast atom bombardment tandem mass spectrometry, Rapid Commun. Mass Spectrom 21 (2007) 1264.

[85] I. Naumann, B.C. Klein, S.J. Bartel, K.H. Darsow, R. Buchholz, H.A. Lange, Identification of sulfoquinovosyldiacyglycerides from Phaeodactylum tricornutum by matrix-assisted laser desorption/ionization QTrap time-of-flight hybrid mass spectrometry, Rapid Commun. Mass Spectrom 25 (2011) 2517.

[86] K.J. Voorhees, K.R. Jensen, C.R. McAlpin, J.C. Rees, R. Cody, M. Ubukata, C.R. Cox, Modified MALDI MS fatty acid profiling for bacterial identification, J. Mass Spectrom 48 (2013) 850.

Chapter 9

Chromatography Coupled to Various Detectors as a Tool for Separation and Determination of Bioactive Compounds

Regina M.B.O. Duarte and Armando C. Duarte
Department of Chemistry & CESAM, University of Aveiro, Portugal

Contents

9.1 INTRODUCTION

The successful reports of using marine samples as potential sources of bioactive compounds encourages the continuous investigation of screening potential candidates from such natural sources. Profiling active components from a given marine sample is not, however, an easy task, particularly if we consider the remarkable diversity of chemical structures included within each sample. Moreover, taking into account that the complexity and heterogeneity of these matrices is likely to be influenced by the habitat, maturity, and environmental conditions, the prospect of mining the diversity of their chemical structures

Analysis of Marine Samples in Search of Bioactive Compounds, Vol. 65. DOI: 10.1016/B978-0-444-63359-0.00009-4

is even more challenging. Nuclear magnetic resonance (NMR) spectroscopy and mass spectrometry (MS; addressed in Chapters 7 and 8 of this book, respectively) are among the most promising techniques for deciphering in detail the structure and composition of marine bioactive compounds (e.g., fatty acids, sugars, sterols, amino acids, or phenols) that occur at vastly different concentrations. However, the inherent complexity of these matrices hampers an extensive structural assignment by means of those high-resolution techniques. Therefore, any separation prior to a structural inquiry is extremely beneficial in reducing the complexity of marine matrices and in turn providing additional structural information on the existing active components.

Chromatography, either liquid or gas, has developed into the most widely applied analytical method for the separation of real-world samples. Nowadays, one-dimensional (1D) chromatography plays an increasingly important role not only in target analysis of well-known compounds (e.g., pharmaceutical drugs; food components such as carbohydrates, proteins, lipids, and artificial additives; polymers; organic pollutants; and natural bioactive compounds) [1,2], but also in more explorative applications (i.e., nontarget analysis) such as life sciences (e.g., proteomics and metabolomics) [3] and in the analysis of complex environmental samples [4,5].

Regardless of the field of application, the samples under study are very complex, often exceeding the separation capacity of a single chromatographic system. The most obvious consequence is that many of the peaks appearing in the chromatogram may contain several coeluting target and nontarget analytes. This challenge motivates researchers to develop more sophisticated chromatographic methods that enable a greater resolution and, therefore, an improved peak capacity. In the quest for better separation efficiency, multidimensional separation schemes have become clearly attractive for the analysis of multicomponent mixtures of real-life samples [6,7]. Multidimensional chromatography combines two or more separation mechanisms that fractionate samples based on widely different and independent (i.e., orthogonal) properties. Two-dimensional (2D) chromatography is the simplest example of a multidimensional separation scheme. The main advantage of combining two separation mechanisms with completely different selectivity is the dramatic increase of the peak capacity and, consequently, the reduction of analytes overlap. When coupled to multichannel detectors (e.g., photodiode array (PDA) or MS), these 2D separation schemes definitely represent a dramatic improvement in chromatographic sciences by successfully allowing the qualitative and quantitative assessment of targeted and nontargeted analytes in various types of samples.

Several reviews and books on different aspects of 1D and multidimensional chromatography have been published in the last five years [1,3,6–14]. This chapter aims at building upon these earlier review works and it includes a critical assessment of the most recent improvements of the various branches of chromatography as analytical platforms able to reduce the complexity of natural marine extracts, thus maximizing the subsequent structural elucidation of their molecular components. This chapter focuses on the specific contributions and potentials of

1D liquid and gas chromatography (1D LC and 1D GC, respectively) for searching bioactive compounds in complex marine extracts. This chapter also addresses the potential of using multidimensional chromatography, specifically 2D liquid and gas chromatography (2D LC and 2D GC, respectively), as powerful separation techniques capable of bringing a major fraction of the complex marine extracts within an adequate analytical window. The principal aim of this chapter is to provide the marine biochemistry research community with insights into the capabilities of the recent developments in chromatography, so that they can be used advantageously in the discovery of bioactive compounds in marine samples.

9.2 ONE-DIMENSIONAL CHROMATOGRAPHY FOR BIOACTIVE COMPOUND ANALYSIS

9.2.1 Setting the Scene and General Trends on the Use of One-Dimensional Liquid Chromatography

Nowadays, the dominant position of 1D LC in analytical strategies aiming at developing a characteristic profile of complex natural samples hardly needs stating. When developing a chromatographic fingerprint procedure for screening bioactive compounds from any natural source, including marine products, researchers typically adopt two strategies: (1) they focus on a particular class of compounds (i.e., target analysis), or (2) they focus on the analysis of the whole sample (i.e., nontarget analysis). Figure 9.1 attempts to set the scene on

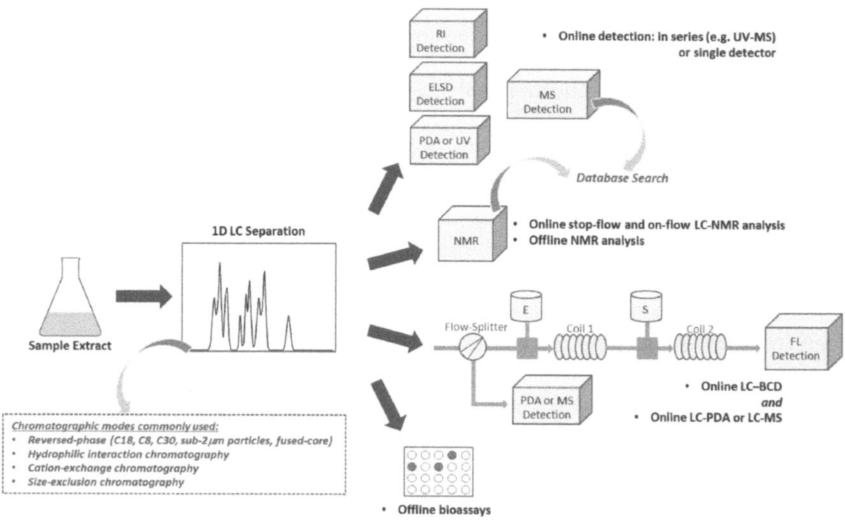

FIGURE 9.1 Schematic illustration of the experimental platform for one-dimensional liquid chromatography (1D LC)-based methods for profiling bioactive extracts. RI = refractive index; ELSD = evaporative light scattering detector; PDA = photodiode array; UV = ultraviolet; FL = fluorescence; MS = mass spectrometry; NMR = nuclear magnetic resonance; BCD = biochemical detection; E = enzyme source/receptor; S = substrate.

the different online and offline 1D LC-based approaches developed thus far for screening natural active components in diverse natural extracts, including marine samples. Table 9.1 is focused on marine samples and it summarizes a few examples of the chromatographic conditions employed in the analysis of different marine bioactive compounds. The use of LC for bioactive compounds analysis has been performed at two different levels: (1) to isolate and purify sufficient amounts the desired analyte(s) or fractions of unknown analyte(s) from complex samples extract (i.e., preparative LC) before offline bioactive and/or chemical screening [15,17,20,21,26,34]; and (2) to separate the analyte(s) according to a predefined separation mechanism followed by chemical screening (i.e., analytical LC) in order to identify and quantify specific classes of compounds or to detect as many different analytes as possible according to their bioactivity [16,18,19,22–25,27–35]. The following subsections will address the stationary phases, the mobile phases, and detection systems currently in use for marine bioactive compounds analysis by means of 1D LC. Where new stationary or mobile phases and detection systems are yet to be applied to marine samples, their potential applicability is presented.

9.2.1.1 Stationary Phases

Regardless of the purpose of the chromatographic step, one of the most important parameters is the choice of the most appropriate stationary phase. In preparative LC, the most widely used stationary phase is the reversed-phase (RP) octadecyl-silica (C18) [15,17,20,21,26], with which the analytes will undergo hydrophobic interactions. Nevertheless, Treschow *et al.* [20] presented a more elaborate and time-consuming purification scheme for achieving the identification of a homologous series of novel omega 3 polyunsaturated fatty acids, with significant anti-inflammatory activity, from marine mollusk *Perna canaliculus* (Table 9.1). At a first stage, the authors used a preparative normal-phase (NP) silica-based stationary phase for isolating the free fatty acids (FFAs) from the mussel extract. The FFAs were subsequently separated on the basis of the length of their carbon chain by a preparative C18 stationary phase. The obtained fractions were then screened by *in vitro* assay and the bioactive fractions, which were converted into their methyl esters, were separated on the basis of their degree of unsaturation (with the most saturated analytes being eluted first) by means of a preparative silver impregnated C18 column, in a technique known as preparative argentation-RP-LC (Ag-RP-LC). This preparative chromatographic technique is a powerful method for the isolation and purification of unsaturated isomers [36], with its analytical counterpart finding now widespread diffusion for the separation of molecular species of simple lipids, such as triacylglycerols [1]. In the work of Treschow *et al.* [20], the fractions obtained from the Ag-RP-LC procedure were further screened for bioactivity by *in vitro* assays, and those shown to be bioactive were analyzed by GC-MS to ascertain the structures of the novel compounds.

TABLE 9.1 Chromatographic Conditions for the Analysis of Bioactive Compounds in Diverse Marine Samples Using LC-Based Methodologies

Sample	Analyte(s)	Stationary Phase	Mobile Phase	Detector(s)[a]	Objective	Reference
Microalgae *Scrippsiella* sp.	Polyunsaturated fatty acids (methyl esters, C16–C28)	Alltima C18, Alltech (Analytical: 250 mm × 4.6 mm, 5 μm) Alltima C18, Alltech (Preparative: 300 mm × 22 mm, 10 μm)	Isocratic: Acetonitrile:Water (97.5:2.5, v/v) Gradient: A. Acetonitrile, B. Trichloromethane	ELSD (at 40 °C)	Purification	[15]
Mussels *Mytilus galloprovincialis*	Domoic acid	TSK-GEL Amide-80 HILIC column, TosoHaas (Analytical: 250 mm × 2 mm, 5 μm)	Isocratic (binary mixture, 75% of B): A. Water (2 mM Ammonium formate + 3.6 mM Formic acid) B. Acetonitrile:Water (95:5, v/v), 2 mM Ammonium formate + 3.6 mM Formic acid	MS	Identification Quantification	[16]
Marine actinomycete strain	Streptokordin (Nonactic acid, Dilactone, Trilactone, Nonactin)	YMC ODS-A C18, YMC (Preparative: 250 mm × 10 mm, 10 μm)	Gradient: A. Water B. Methanol	UV at 210 nm	Purification	[17]
Ascidian *Lissoclinum* cf. *badium*	Aromatic alkaloids	Mightysil C18, Kanto Chemical Co. (Analytical: 250 mm × 4.6 mm, 5 μm)	Gradient: A. 0.1% Trifluoroacetic acid B. Methanol	PDA-ESI-ITMS	Identification	[18]

(Continued)

TABLE 9.1 Chromatographic Conditions for the Analysis of Bioactive Compounds in Diverse Marine Samples Using LC-Based Methodologies *(cont.)*

Sample	Analyte(s)	Stationary Phase	Mobile Phase	Detector(s)[a]	Objective	Reference
Microalgae *Dinophysis acuta*	Okadaic acid Dinophysistoxin-2	Symmetry C18 column, Waters (Analytical: 50 mm × 2.1 mm, 3 μm)	Gradient: A. Acetonitrile + 2 mM Ammonium formate + 0.01% Formic acid B. Water + 2 mM Ammonium formate + 0.01% Formic acid	Online LTQ-ITMS Online Quattro Ultima Pt triple quadrupole MS Offline NMR	Identification Quantification Identification	[19]
Marine mollusk *Perna canaliculus*	Omega 3 polyunsaturated fatty acids	Normal-Phase Nova-Pak® HR, Waters (Preparative: 100 mm × 40 mm, 6 μm) Ultrasphere™ C18, Beckmann (Semipreparative: 250 mm × 10 mm, 5 μm) Chromsphere™ C18 silver impregnated, Chrompack International (Semipreparative: 250 mm × 10 mm, 5 μm)	Gradient: A. Hexane B. Tetrahydrofuran Gradient: A. Water B. Acetonitrile C. Methanol Gradient: A. Dichloromethane: Dichloroethane, 1:1 (v/v) B. Dichloromethane: Dichloroethane: Acetonitrile:Methanol, 40:40:10:10 (v/v/v/v)	ELSD (at 80 °C) *(with a stream splitter)* ELSD (at 45 °C) ELSD (at 70 °C)	Purification Separation Separation	[20]
Marine algae *Dictyota acutiloba*	Possible terpenoid secondary metabolites	Shim-pack CLC ODS, C18, Shimadzu (Preparative: 15 mm × 4.6 mm, particle size unspecified)	Isocratic: Acetonitrile:Water (8.5:1.5)	UV at 355 nm	Purification	[21]

Seagrasses *Zostera noltii* *Zostera marina*	Zosteric acid	Hypersil GOLD C8, Thermo Finnigan (Analytical: 250 mm × 4.6 mm, 5 μm)	Gradient: A. 0.1% Trifluoroacetic acid in water B. 0.1% Trifluoroacetic acid in Methanol	PDA at 280 nm	Quantification	[22]
Microalgae *Synechocystis* sp. Marine algae *Himanthalia elongata*	Carotenoids (β-carotene, zeaxanthin, myxoxanthophyll, echinenone, fucoxanthin, zeaxanthin) NI Chlorophyll	Novapack C18 column, Waters (Analytical: 150 mm × 3.9 mm, 4 μm)	Gradient: A. Methanol:Ammonium acetate 0.1 N (7:3) B. Methanol	PDA	Identification Quantification	[23]
Marine algae *Ulva lactuca*	Essential amino acids (e.g., valine, leucine, lysine, threonine) Nonessential amino acids (aspartic and glutamic acid)	Cation –exchange column (specifications not mentioned) (Analytical: 200 mm × 4.6 mm, particle size unspecified)	Gradient: A. Sodium citrate 200 mM (pH 3.2) B. Sodium citrate 200 mM (pH 4.25) C. Sodium citrate 1.2 M (pH 6.45) D. NaOH 400 mM	UV-Vis at 440 and 570 nm	Identification Quantification	[24]
Mussels	Lipophilic toxins (dinophysistoxins, pectenotoxins, and azaspiracids) Domoic acid Saxitoxins (Paralytic shellfish poisoning toxins)	Hypersil Gold C18 column, Thermo Fisher Scientific (Analytical: 100 mm × 2.1 mm, 1.9 μm) Acquity BEH Amide HILIC column, Waters (Analytical: 100 mm × 2.1 mm, 1.7 μm)	Gradient: A. 0.1% Formic acid in Water B. 0.1% Formic acid in 90% (v/v) Acetonitrile Gradient: A. 2 mM Ammonium formate pH 3.5 B. 0.1% Formic acid in Acetonitrile	MS	Identification	[25]

(Continued)

TABLE 9.1 Chromatographic Conditions for the Analysis of Bioactive Compounds in Diverse Marine Samples Using LC-Based Methodologies *(cont.)*

Sample	Analyte(s)	Stationary Phase	Mobile Phase	Detector(s)[a]	Objective	Reference
Marine sponge *Spongosorites halichondriodes*	Sterol esters	KROMASIL C18 Column (Analytical: 250 mm × 4.6 mm, 5 μm)	Isocratic: Methanol:Water (10:90, v/v)	UV at 254 nm	Separation	[26]
Marine algae *Saccharina latissima* *Mastocarpus stellatus*	Polysaccharides	TSK-Gel G 5000 PW, SEC, TosoHaas (Analytical: 300 mm × 7.5 mm, 17 μm)	Isocratic: Sodium nitrate 0.1 M	RI	Molecular weight distribution	[27]
Mussels	Paralytic shellfish poisoning toxins (saxitoxin-like congeners) (with postcolumn derivatization before detection)	Gemini C18 column, Phenomenex (Analytical: 150 mm × 4.6 mm, 5 μm) Kinetex™ C18 column, Fused-core, Phenomenex (Analytical: 100 mm × 4.6 mm, 2.6 μm) Poroshell™ C18 column, Fused-core, Agilent (Analytical: 100 mm × 4.6 mm, 2.7 μm)	Gradient: A. 0.1 M Ammonium formate (pH 6) B. 0.1 M ammonium formate and 5% (v/v) Acetonitrile	Fluorescence ($\lambda_{Excitation}$ = 340 nm /$\lambda_{Emission}$ = 395 nm)	Identification Quantification	[28]

Marine snail *Conus consors*	Peptides	XBridge C18 column (Analytical: 150 mm × 3.0 mm, 3.5 μm) Acquity UPLC BEH C18 columns (Analytical: 150 mm × 2.1 mm, 1.7 μm; Analytical: 150 mm × 3.0 mm, 1.7 μm)	Gradient: A. 0.1% Formic acid in water B. 0.1% Formic acid in Acetonitrile	TOF-MS	Identification	[29]
Marine sponges *Callyspongia* sp. *Lamellodysidea* sp	Tetra- to hexa-brominated diphenyl ethers metabolites	Shim-pack FC-ODS, Shimadzu (Analytical: 150 mm × 4.6 mm, 3 μm)	Isocratic: Acetonitrile:Water (9:1, v/v)	Qtrap®-MS/MS	Identification Quantification	[30]
Marine sponge *Agelas mauritiana*	Diterpenoid derivatives (agelasine B and (−)-agelasine D)	Zorbax SB-C18 column, Agilent (Analytical: 250 mm × 4.6 mm, 5 μm)	Gradient: A. 0.05% Formic acid in water B. Acetonitrile	UV-ITMS Offline bioassay	Identification Antitumor activity	[31]
Marine red algae *Gracilaria caudata*	Polysaccharides	Ultrahydrogel™ column, Linear, Waters (Analytical: 300 mm × 7.8 mm, 10 μm)	Isocratic: Sodium nitrate 0.1 M	RI	Molecular weight distribution	[32]
Japanese Spanish mackerel Oysters Blood clam	Paralytic shellfish poisoning toxins (saxitoxin-like congeners)	TSK-GEL Amide-80 HILIC column, Tosoh Bioscience (Analytical: 150 mm × 2.0 mm, 3 μm)	Gradient: A. 0.1% Formic acid in 2 mM Ammonium formate B. 0.1% Formic acid in Acetonitrile	MS/MS	Identification Quantification	[33]

(Continued)

TABLE 9.1 Chromatographic Conditions for the Analysis of Bioactive Compounds in Diverse Marine Samples Using LC-Based Methodologies *(cont.)*

Sample	Analyte(s)	Stationary Phase	Mobile Phase	Detector(s)[a]	Objective	Reference
Sea cucumber *Acaudina molpadioides*	Oligosaccharides (Fucoidan, consisting of fucose and sulphate ester groups	One Superdex Peptide column, GE Healthcare (Technical details of the column not specified) Two Superdex Peptide columns connected in series, GE Healthcare (Technical details of the column not specified)	Isocratic: 25% (v/v) Acetonitrile containing 50 mM ammonium formate	MS RI Offline ESI-CID-MS/MS Offline NMR	Profiling Separation and purification Structural characterization of purified fractions	[34]
Marine Algae *2 green algae* (Chloropyceae), *2 red algae* (Rhodophyceae), *4 brown algae* (Pheophyceae)	Triacylglycerols Phospholipids Carotenoids	Ascentis Express C18 Fused-core column, Supelco (Analytical: 150 mm × 4.6 mm, 2.7 μm) Ascentis Express HILIC Fused-core column, Supelco (Analytical: 150 mm × 4.6 mm, 2.7 μm) YMC30 column, YMC Europe (Analytical: 250 mm × 4.6 mm, 5 μm)	Gradient: A. Acetonitrile B. 2-Propanol Gradient: A. Acetonitrile:Ammonium formate (90:10, v/v) B. Acetonitrile:Metahnol: Ammonium formate (55:35:10, v/v/v) Isocratic (binary mixture, 10% of B): A. Methanol:Acetone (60:40, v/v) B. Acetonitrile:Water (60:40, v/v)	APCI-MS ESI/IT-TOF-MS PDA-MS	Identification Quantification Identification Identification	[35]

[a]*ELSD = evaporative light scattering detector; UV = ultraviolet; UV-Vis = ultraviolet-visible; PDA = photodiode array; ESI-ITMS = electrospray ionization-ion trap mass spectrometry; LTQ-ITMS = Linear trap quadrupole-ion trap mass spectrometry; TOF-MS = time-of-flight mass spectrometry; QTrap®-MS/MS = QTrap® tandem mass spectrometry; APCI-MS = atmospheric-pressure chemical ionization mass spectrometry; NMR = nuclear magnetic resonance; RI = refractive index*

Besides preparative RP, NP, and Ag-RP stationary phases, size-exclusion chromatography (SEC) can also be used to fractionate complex organic mixtures on the basis of the molecular weight of their components. In this regard, Yu *et al.* [34] used two size-exclusion columns connected in series to separate and purify a mixture of oligosaccharides extracted from sea cucumber *Acaudina molpadioides* (Table 9.1). With this procedure, four major oligosaccharide fractions were collected and lyophilized for subsequent offline structural studies by NMR and tandem MS.

In analytical LC for bioactive compounds analysis, the most commonly used stationary phase is also the RP C18 [19,23,25,28–31,35]. Other stationary phases employed include the RP octyl-silica (C8) [22] and triacontyl-silica (C30) [35], cation-exchange [24], size-exclusion [27,32,34], and hydrophilic interaction chromatography (HILIC) [16,25,33,35]. Compared to the C18 stationary phase, the retention of highly hydrophobic solutes is reduced by using the C8 stationary phase, whereas the highly hydrophobic stationary phase of the C30 provides enhanced interaction with long chained molecules. The C30 is the preferred stationary phase for resolving complex mixtures of carotenoid isomers, aiming at their subsequent identification and quantification [1]. Very recently, Ragonese *et al.* [35] employed a YMC30 column for the separation of carotenoids and chlorophylls present in different species of lyophilised seaweeds (green, red, and brown algae), in combination with online PDA-MS detection (Table 9.1).

Over the last few years, the developments in stationary phases led to introduction in the market of different types of highly efficient LC columns. For example, columns packed with sub-2 μm porous silica particles are now being preferred over the conventional columns packed with 3 or 5 μm RP silica particles for herbal fingerprinting [11] and for the analysis of phenolic compounds in food samples [12,13]. These types of columns are employed for ultrahigh performance LC (UHPLC); they generate a maximum plate number of 300,000 and can withstand a maximum pressure drop of 150 MPa [1]. The columns packed with 3 to 5 μm particles, employed in the conventional high-performance LC (HPLC) analyses, generate approximately 10,000 to 20,000 theoretical plates and the maximum column pressure drop is 40 MPa [1]. It has been stated that the use of solid state particles of sub-2 μm diameter allows achieving superior sensitivity and higher separation efficiency than that of the columns packed with 3 to 5 μm particles [11].

Aiming at maximizing the peak capacity of peptide separations by LC from the venom of the predatory marine snail *Conus consors* (Table 9.1), Eugster *et al.* [29] showed that the strategy of using columns packed with 1.7 μm particles (namely, the Acquity UPLC BEH C18 column) systematically surpassed the conventional HPLC method using 3.5 μm particles (XBridge C18 column). The authors also studied the effect of mobile phase temperature, column internal diameter (I.D.), mobile phase flow rate, and gradient time in achieving a high resolution profiling of the complex mixture of peptides from the *C. consors* venom. The use of high mobile phase temperatures (up to 90 °C) enhanced the chromatographic performance under UHPLC conditions. An increase in the calculated peak capacity was also reported when using a column with an I.D. of 3.0 mm compared to that with an I.D. of 2.1 mm. It is also noteworthy that Eugster *et al.* [29] verified a slight relative decrease in peak capacity at 90 °C for the larger I.D. column, which was attributed to the frictional heating that occurred at high flow rates (1450 μL min^{-1}) with sub-2 μm diameter particles. The authors also reported that the best performance in UHPLC gradient mode was obtained with the highest mobile phase flow rate (1460 μL min^{-1}). For safety reasons, Eugster *et al.* [29] recommended a flow rate that generates 90% of the maximal backpressure tolerated by the UHPLC system in order to extend the column and instrument lifetimes while obtaining maximal peak capacity. In terms of the effect of the gradient time, and in order to find a good compromise between peak capacity and analysis time, Eugster *et al.* [29] also suggested the selection of a gradient time providing between 50% and 80% of the maximum peak capacity of an infinite time. By employing these so-called "optimized" chromatographic conditions to the UHPLC-QTOF-MS profiling of *C. consors* venom, Eugster *et al.* [29] reported a peak capacity enhancement of 2- to 3-fold when compared to conventional LC–MS profiling studies usually performed in the field of venomics.

Before the work of Eugster *et al.* [29], a LC-MS platform using a C18 column packed with sub-2 μm porous silica particles was already employed by Blay *et al.* [25] for screening marine lipophilic toxins, such as dinophysistoxins, pectenotoxins, and azaspiracids, which are produced by naturally occurring microalgae and pose a significant food safety risk when bioaccumulated in shellfish. Using gradient elution (see Table 9.1) and a flow rate of 400 μL min^{-1}, the authors reported the successful separation at the baseline of 10 lipophilic toxin standards in just 6 min. These results clearly demonstrate that the use of LC columns with sub-2 μm particles could be attractive to laboratories that deal with routine analyses of target bioactive analytes or for more explorative nontarget analysis [29].

Another development in stationary phases is the use of superficially coated porous particles in fused-core silica columns. Fused-core particles are described as having a 1.7 μm solid core with a surrounding "fused" 0.5 μm porous silica shell (total diameter of 2.7 μm) [1]. This type of column provides similar high speed and high efficiency as sub-2 μm particles but at half the back pressure, thus circumventing the requirements of using dedicated UHPLC instrumentation [1]. C18 columns with fused-core particle technology were employed by Robert and Turner [28] for the analysis of paralytic shellfish poisoning (PSP) toxins using the AOAC *Official Method*^{SM} 2005.06 (accepted by the European Union legislation) with postcolumn derivatization before fluorescence detection (FLD), and more recently by Ragonese *et al.* [35] for the separation of triacylglycerols occurring in various algal species in combination with APCI-MS detection (Table 9.1). Robert and Turner [28] reported that the use of fused-core particle technology offered the potential to reduce the time of analysis of PSP toxins to less than half of that of the conventional method using the standard porous C18 bead with 5.0 μm particles. The use of fused-core particle columns also allowed reducing mobile phase consumption in approximately 65% when compared to that required by the standard 5.0 μm C18 column [28], which is a significant relative reduction in chemical waste disposal. Besides allowing a shorter analysis time, the use of fused-core particle columns also resulted in an improved analytical sensitivity (lower limits of detection and quantification) for almost all toxins [28]. Nevertheless, Robert and Turner [28] concluded that the most significant negative issue affecting fused-core particle columns is their reduced lifetime when compared to the standard 5.0 μm C18 column due to the increase in back pressure. The authors reported that the fused-core particle columns were able to perform 150 to 500 sample analysis cycles before reaching the predetermined 300 bar threshold against more than 3000 sample analysis cycles reached in the standard 5.0 μm C18 column.

These findings suggest that more work needs to be performed to improve fused-core column lifetimes in order to include them in cost-effective chromatographic-based platforms, particularly within the framework of the routine analysis of marine bioactive compounds. Indeed, the application of fused-core columns on marine bioactive compounds analysis is still very incipient, particularly when compared to the standard porous C18 bead. However, the future impact of this new column technology on this research field seems to be promising, particularly judging from the several successful applications of fused-core columns described in the literature, including herbal fingerprinting (analysis of flavonoids) [37], analysis of emerging pollutants belonging to personal care products (antimicrobials, preservatives), benzotriazole ultraviolet (UV) stabilizers and organophosphorus compounds in fish tissue [38], analysis of sulphonamide residues in grass carp tissues [39], analysis of bisphenol A and its chlorinated metabolites in wastewaters, river water and drinking water plants [40], and analysis of pharmaceuticals and their metabolites in wastewater [41].

Although RP is overall the most commonly used separation technique for marine bioactive compounds, the highly hydrophilic, ionic, and polar compounds are usually too weakly retained in such stationary phases to allow their separation. This specific problem can be partly solved with the use of a stationary phase based on the conventional NP separation mechanism. However, these hydrophilic analytes are retained too strongly in nonaqueous mobile phases used in NP and/or may not be sufficiently soluble in NP solvent systems [42]. In order to help solve these problems, hydrophilic interaction chromatography (HILIC), introduced by Alpert in 1990 [43], has emerged as one of the preferred analytical techniques for the separation of very polar compounds from diverse application areas, including the analysis of biomarkers [44], analysis of nutritional compounds (proteins, carbohydrates, lipids, or water soluble vitamins), pesticides and toxins in food matrices [16,25,33,45], in metabonomic/metabolomic studies of samples of varying nature (from polar biological fluids to organic extracts of plants or biological tissue) [46], glycomics [47], and separation of natural organic matter from different environments prior to a structural inquiry [4,48].

The reason for the success of HILIC is that it offers a polar stationary phase (comparable to NP) for the separation of hydrophilic compounds using aqueous-organic mobile phases (comparable to RP), typically acetonitrile-rich eluents [14]. Another reason for the success of this chromatographic mode is that the use of a mobile phase containing low aqueous (5–40%, v/v)-to-high polar organic solvent content is almost ideal for MS detection, resulting in increased sensitivity [49]. This obviously rules out the use of mobile phases containing high concentrations of salts such as sodium chloride or potassium hydrogen phosphate. In this case, the use of MS-compatible aqueous buffers (e.g., ammonium hydrogencarbonate) should be taken into consideration as a good alternative for HILIC separations of hydrophilic analytes.

Although HILIC has been widely applied, retention mechanisms of this chromatographic mode are still debated. As recently reviewed by Cavazzini and Felinger in a chapter dedicated to HILIC in the book *Liquid Chromatography: Fundamentals and Instrumentation* [14] and by Nguyen and Schug [49] in their review work, the mechanism of HILIC involves the partitioning of an analyte between the bulk eluent (predominantly polar organic mobile phase) and a thin water-enriched layer partially immobilized on the polar HILIC stationary phase. However, a combination of mechanisms also involving charge interaction, hydrogen-bonding, and dipole–dipole interactions can further affect analytes selectivity [14,42,49]. In general, the packing materials for HILIC can be grouped into the following categories: straight unbonded silica with decreased surface concentration of silanol groups, neutral silica-based packings bearing amino-, amide-, cyano-, carbamate-, diol-, or polyol- functionality, and ionizable silica-based packings bearing anion-exchange, cation-exchange, or zwitterionic functionality [14,42]. Modern mixed-mode HILIC stationary phases, featuring hydrophilic interaction and RP characteristics or HILIC and ion-exchange (IEC) mechanism, were also introduced into the market to fulfill the requirements of separating samples containing a complex mixture of hydrophilic and hydrophobic species or to distinguish analytes by lipophilicity and charge differences, respectively [42,50]. A complete survey of more than 50 stationary phases currently available for HILIC mode separations is well beyond the scope of this chapter. For an in-depth description of the characteristics of the polar stationary phases commonly used in HILIC, including the type, name, and functional group structure of each phase together with the commercial names of the corresponding columns, readers are referred to the review works of Cavazzini and Felinger in the book *Liquid Chromatography: Fundamentals and Instrumentation* [14], and Jandera [42] as well as the literature cited therein.

Due to the huge variety of HILIC stationary phases commercially available, the selection of the most appropriate phase system is far from being an easy task. As recently demonstrated by Fountain *et al.* [51], screening multiple columns and mobile phases (including the concentration of water, pH, and ionic additives) is mandatory for a successful HILIC separation method. Indeed, all these parameters should be flexibly adjusted, preferably with the support of chemometric tools (i.e., design of experiments), to address specific separation problems.

Similar to RP modes, HILIC columns are also available with sub-2 μm particles and fused-core particle technology [1]. A HILIC column packed with sub-2 μm particles (namely, the Acquity BEH Amide HILIC column) was applied by Blay *et al.* [25] for the separation of domoic acid and PSP toxins from extracts of lyophilized mussel tissues. The authors verified that the employed HILIC-MS platform was suitable for the combined analysis of domoic acid and PSP toxins in a single 6 min chromatographic run, using gradient elution. On the other hand, a few years earlier, Ciminiello *et al.* [16] applied a HILIC-MS method, using a standard 5 μm Amide HILIC column (Table 9.1) and isocratic elution, for the analysis of domoic acid in *Mytilus galloprovincialis* mussels. The adopted methodology allowed the analysis of domoic acid in 10 min. With gradient elution, the same authors were able to perform the combined analysis of domoic acid and PSP toxins in 35 min. Once again, these results clearly show the benefits of using HILIC columns with sub-2 μm particles for accomplishing a fast and cost-effective separation of target highly polar bioactive compounds. A HILIC column with fused-core particle technology was also very recently employed by Ragonese *et al.* [35] for the separation of phospholipids occurring in various algae species, in combination with ESI/IT-TOF-MS detection (Table 9.1).

The separation of 10 different polar lipid classes, eluting according to their decreasing polarity, was achieved under gradient conditions within a run time of 20 min. The lipid classes were defined by the polar head-group nature, and different molecular species are likely to occur within any single class, according to the nature (length and degree of saturation) of the fatty acids [35]. From a more general perspective, it seems that the use of sub-2 μm particles and fused-core HILIC columns has found its way only marginally into marine polar bioactive compounds analysis. The reason for this may lie in the difficulty in choosing both a fit-for-purpose stationary phase and an optimal mobile phase composition. However, these difficulties should be viewed as a strong driving force for the development of new chromatographic methods offering high-efficiency separations in less time, which combined with high-resolution detection systems (e.g., MS, discussed in Chapter 8) will constitute good resources for the identification and quantification of marine bioactive compounds in both target and nontarget approaches.

Additional stationary phases already employed in chromatographic-based methods for target and nontarget analysis of marine bioactive compounds include: (1) the cation-exchange, combined with ultraviolet-visible (UV-Vis) detection for a qualitative and quantitative assessment of the amino acid composition in *U. lactuca* seaweeds [24]; and (2) the size-exclusion, which allowed the assessment of the molecular weight distribution of polysaccharides from marine algae using refractive index (RI) detection [27,32], as well as a preliminary profiling of the oligosaccharides present in an enzymatic hydrolysate from sea cucumber, using MS detection [34]. In the particular case of SEC, the analyst should be particularly careful in the interpretation of the obtained results because too often the obtained separation is not strictly based on the molecular size and shape of the analyte with respect to the average pore size of the stationary phase. In fact, the application of the SEC technique still suffers from a number of significant drawbacks. First, the occurrence of nonsize-exclusion effects (e.g., hydrophobic interactions, ion-exchange and ion-exclusion effects) due to the interaction between column's stationary phase and the analytes; second, the lack of adequate standards of the same molecular type as the samples under study (e.g., similar molecular weight and charge distribution) to calibrate the column; and third, the different sensitivity exhibited by multiple online detectors (e.g., UV absorbance, dissolved organic carbon, RI, and fluorescence detectors) [5,47]. In the case of marine-derived polysaccharides, pullulan standards (straight chain of maltotriose units) usually are used to determine the separation range of polysaccharides, whereas an aqueous solution of 0.1 M sodium nitrate has been used as a mobile phase [27,32]. Nonetheless, depending of the pursued aims and analytes under study, there are a set of parameters (sometimes with opposite effects) that should be appropriately tuned to get the maximal capability of the SEC system and ensure the success of the separation, namely the column features (e.g., type of stationary phase), mobile phase composition, working conditions (e.g., flow-rate and column temperature), and detector. Statistical experimental designs are efficient tools to accomplish this objective and they should be included in the development of optimum SEC separation strategies for the analysis of marine-derived bioactive compounds.

9.2.1.2 Mobile Phases

For the separation of marine bioactive compounds by analytical RP, the mobile phase is usually a binary or ternary phase consisting of an organic modifier solvent in combination with acidified water, usually by adding a small amount of trifluoroacetic acid [18,22] or formic acid [19,25,29,31] and/or by using MS-compatible ammonium formate [19,25,28] or ammonium acetate [23] buffers. The separation of analytes with similar polarity is accomplished by gradient elution using methanol [18,22,23], acetonitrile [19,25,28,29,31,35], and 2-propanol [35]. Although gradient is the mode of elution preferred, isocratic elution can also be applied by using acetonitrile-rich [30] or aqueous-rich [26] mobile phases or by using mixtures comprising 90% of methanol/acetone (60:40, v/v) and 10% of acetonitrile/water (60:40, v/v) [35]. The chromatographic HILIC mode is usually performed in gradient mode with organic-rich (75-55% acetonitrile) mobile phases [25,33,35]. In HILIC, the water is introduced as the stronger eluting solvent, thus replacing the polar organic phase traditionally used in the RP mode [13].

As an answer to the increasing demand for developing environmentally friendly analytical methodologies, ionic liquids (ILs) have emerged as green "designer solvents" with a wide range of applications in many areas of chemistry, including organic chemistry, inorganic chemistry, biochemistry, physical chemistry, and analytical chemistry [52]. As recently reviewed by Tang *et al.* [52], ILs have also attracted considerable attention as solvents in the extraction and separation of bioactive compounds from plants. As a mobile phase additive in LC, ILs can decrease band tailing by removing the effects of free silanols, existing at the surface of the stationary phase, on the strong retention of basic analytes (e.g., amines), thus improving their resolution and quantitative analysis [52]. As a modifier of the stationary phase, ILs-based silica and polymers can enhance the separation of target bioactive analytes from plants due to their polarity and low volatility associated with nonpolar and ionic interactions [53]. Furthermore, when compared to ILs-based silica material, ILs-based polymers exhibit a higher selectivity and a wider operating pH range [52]. A complete survey of the unique properties of ILs, as well as the preparation and application of ILs or ILs-based materials in bioactive compounds analysis, is well beyond the scope of this chapter. For an in-depth description of the available data and references to examine the advantages of ILs and IL-based materials for the chromatographic separation of bioactive compounds, refer to the review work of Tang *et al.* [52]. ILs and IL-based materials are gaining acceptance in LC and the continuous improvements in this area open new possibilities for their application in the development of more environmentally friendly chromatographic methodologies for the analysis of marine bioactive compounds.

9.2.1.3 Detection Systems

In order to detect and quantify active components, online photodiode array (PAD) is the easier and quicker choice because it gives detailed chromatographic information on the basis of the full spectra collected in a wide wavelength range (≈ 200–400 nm) [18,22,23,35]. On the other hand, when performing target analysis of well-known compounds, detection at a specific UV wavelength can be also employed [17,21,22,24,26]. The active components must contain chromophoric groups in their structure, otherwise a postcolumn derivatization procedure must be applied before detection [28] or another type of detection system must be applied. For example, an evaporative light scattering detector (ELSD) has been employed to track the purification process of polyunsaturated fatty acids [15,20], whereas an RI detector was chosen for SEC studies of polysaccharides [27,32,34].

As shown in Table 9.1, online and offline MS-based platforms have been the detection systems of choice as they simultaneously offer the possibility of analyzing compounds belonging to structurally different groups. MS is also particularly useful for nontarget analysis of bioactive compounds. For more information on the use of MS-based platforms for bioactive compounds analysis, refer to Chapter 8. Also, offline NMR studies of the obtained chromatographic fractions could lead to the complete structure assignment of bioactive compounds [19,34]. Depending on the sample amount, the coupling of NMR to LC is possible using either stop-flow or on-flow NMR analysis. For a detailed description of the use of NMR spectroscopy within the framework of bioactive compound analysis, refer to Chapter 7.

The classical approach for bioactive screening of a given prefractionated compound or group of compounds involves offline biological screening in pharmacologically relevant assays [31]. Alternatively, online biochemical detection (BCD) assays used in tandem or in parallel with LC-PDA and/or –MS platforms (see Figure 9.1 for a schematic representation) can be used for fast-track identification of bioactives in extracts or mixtures without the need for lengthy separation and isolation procedures [53]. The power of this type of screening procedure and its applicability for detection of bioactive compounds will not be addressed here since Chapter 10 is devoted to this topic.

9.2.2 Application of One-Dimensional Gas Chromatography-Based Methods for Bioactive Compound Analysis

The 1D GC is a popular preference for achieving the separation of volatile and nonvolatile (but capable of being volatilized by a derivatizing agent) compounds present in a sample. When coupled to MS or a flame ionization detector (FID), it becomes a useful tool to obtain as much information as possible about metabolites from different matrices [11,20,23,24,35,54–62]. GC analysis of marine-derived bioactive compounds has been usually limited to fatty acids [20,23,24,35,60–63], but the target and nontarget analysis of other bioactive compounds have also been reported, such us that of amino-acids, sterols, lupanes, monosaccharide, alkanes, phenols, phytol, neophytadiene, methanephrine, cis-1-chloro-9-octadecene, 13-bromotetradecanoic acid, 6-amino-2-phenazinol, and morphinan-2,4-diol-6-one [23,24,55,59,62].

The analysis of most of the aforementioned compounds usually requires a previous derivatization step in order to render them suitable for GC analysis by making them sufficiently volatile (at least, the analysis of those compounds where the information about a derivatization step is provided). Derivatization reactions used for GC fall into three general reaction types: (1) alkylation of which the general process is esterification, and it reduces the polarity of the compounds by substituting labile hydrogens for an aliphatic or aliphatic-aromatic (e.g., benzyl) group; (2) acylation, where compounds containing a labile hydrogen (–OH, –SH and –NH) are transformed into esters, thioesters, and amides through the introduction of an acyl group by the loss of the labile hydrogen; and (3) silylation, where the polarity of the compounds is reduced by substituting labile hydrogens for a trimethylsilyl group [63]. Alkylation using boron trifluoride (BF3) in methanol has been most commonly used to convert fatty acids into their methyl ester derivatives before identification by means of GC-FID [20,24,35,60,61] and/or individual structural assignment by GC-MS

[20,23,35,59–61]. However, Pereira *et al.* [59] successfully used a silylation agent, the N-methyl-N-(tri-methylsilyl)-trifluoroacetamide (MSTFA), at a mild temperature (40 °C), for the detection and absolute quantification of not only fatty acids, but also amino acids, sterols, and lupanes in ethanolic extracts from echinoderm *Marthasterias glacialis* Linnaeus spiny sea star. The authors concluded that derivatization with MSTFA is preferable to standard BF3, which required the use of a temperature around 90 °C. Moreover, the derivatives obtained through the standard BF3 method must be purified before injection into the GC in order to assure the total absence of water.

In opposition to this acylation method, silylation with MSTFA did not require a purification step, and the derivatives were injected directly into the GC. The same research group has also applied the silylation with MSTFA for the nontarget GC-MS analysis of ethanolic extracts from 18 macroalgae of the Portuguese coast, which led to the characterization of 14 compounds: proline, phloroglucinol, mannitol, eight fatty acids, and three sterols [62]. Finally, acylation has also been used by Yaich *et al.* [24] for studying the monosaccharide composition (glucose, rhamnose, xylose, galactose, mannose, and arabinose) of the U. lactuca seaweeds. After chemical hydrolysis of the algae powder, the authors adopted a time-consuming and laborious procedure [64] for converting the neutral sugars into their alditols acetate derivatives before GC-FID analysis.

The recent availability of capillary columns coated with ILs has paved the way for achieving improved selectivities for the analysis of methyl ester derivatives of fatty acids. Fardin-Kia *et al.* [61] investigated the separation of fatty acid methyl esters prepared from menhaden fish oil using the more polar SLB-IL111 ionic liquid capillary column. Identifications of fatty acids were achieved by applying Ag-RP-LC fractionation and GC-time-of-flight (TOF)-MS analysis. This methodology allowed the identification of 125 fatty acids in menhaden oil, including isoprenoid and furanoid fatty acids, and the novel 7-methyl-6-hexadecenoic and 7-methyl-6-octadecenoic fatty acids. Overall, the SLB-IL111 ionic liquid capillary column provided the resolution of 107 fatty acids and the coelution of 18 peaks in a single 90 min analysis. These results were compared to those obtained by means of a conventional capillary column coated with 100% cyanopropyl siloxane, which was able to separate these fatty acids into only 70 peaks, as well as the coelution of 13 peaks (9 small peaks could not be identified) [61].

Another development in GC columns that is likely to benefit the target analysis of marine bioactive compounds by reducing the time of analysis (as short as 40–100 seconds) without impairing the resolution is the use of micro-bore capillary columns with reduced stationary phase film thickness (I.D. $\leq$ 100 μm) [11]. To the best of the author's knowledge, fast GC analysis has never been applied for screening marine bioactive compounds. However, the work of Mondello *et al.* [65], who successfully applied a 5 m x 50 μm capillary column with 0.05 μm stationary phase film thickness for very fast high resolution GC analysis of lime essential oil, should be taken as an example for a more widespread use of narrow bore GC capillary columns in the routine analysis of marine bioactive compounds.

9.3 MULTIDIMENSIONAL CHROMATOGRAPHY FOR BIOACTIVE COMPOUNDS DISCOVERY

9.3.1 Concepts and Considerations

The challenge of analyzing multicomponent mixtures with extreme complexity from diverse fields has promoted an increase in the application of multidimensional separation systems (either LC or GC) as an effort to substantially improve the resolution of such complex samples [6–10,47,66–69]. These multidimensional separation systems are capable of discriminating the components from a mixture, using different separation mechanisms that are connected (by means of a transfer system) but do not interact among themselves. In practice, multidimensional LC and GC chromatography involves transferring a fraction or fractions from one chromatographic medium (usually a column) to a secondary (or third) fast chromatographic medium (also a column or columns) with different selectivity for further separation. Although there is no inherent limitation to the number of independent separation methods that can be coupled, the 2D version of multidimensional separation systems has gained popularity in recent years by offering higher resolution, selectivity, and peak capacity than single-column separation methods.

In both 2D LC and 2D GC, there are two modes of operation: heart-cutting and comprehensive. In the heart-cutting mode, only some selected fractions are transferred from the first into the second separation system, and the results become two separate 1D data sets. On the other hand, a separation is comprehensive when the whole sample is subjected to two different separation mechanisms, the separation (resolution) obtained in the first dimension is essentially maintained, and the chromatogram obtained is representative of the entire sample (after pretreatment), which requires that either no sample goes to waste (everything passes through the detector) or a sufficient number of second dimension chromatograms are recorded very frequently across the width of a first-dimension peak [70,71].

Heart-cutting 2D LC and 2D GC methods (depicted by LC-LC and GC-GC, respectively) are basically target analysis methods, which either alone or in conjunction with a multichannel detector (e.g., MS or PDA), can be powerful tools when the analyte(s) of interest is present in a complex sample [8,72]. The heart-cutting methods are useful when it is known *a priori* where the chromatographic zone containing the analyte(s) of interest is overlapped [8], thus requiring an additional separation step. Obviously, this is not the case for samples whose composition is highly variable and unknown. Furthermore, these heart-cutting approaches can be time-consuming [73], and where heart-cutting 2D LC is concerned, changes in analyte retention time and peak width with column age and fouling are a serious problem that could move out the analyte from the preselected time window in the first dimension for automated valve switching, thus compromising quantitation in the second dimension [72]. Comprehensive 2D LC and 2D GC methods (depicted by LC×LC and GC×GC, respectively) allow gathering far more information by totally deconvoluting a complex sample via repetitive injection into the second dimension column of consecutive cuts of the first dimension column effluent [8]. An increased resolution can be further achieved if combining either LC×LC or GC×GC with a third selective detection dimension (e.g., MS), thus making these chromatographic techniques powerful analytical devices for the analysis of complex organic samples. In comparison to GC×GC, the applicability of LC×LC for bioactive compounds discovery is by far more appealing since it offers a wide variety of options in terms of efficient separation mechanisms and stationary phases with truly different selectivities, thus allowing the separation of a wide range of compounds with different chemical characteristics and without the need of a previous derivatization procedure. As such, this third section is concerned mostly with LC×LC and its potential for bioactive compounds discovery.

9.3.1.1 Fundamentals and Instrumental Aspects of Comprehensive Two-dimensional Liquid Chromatography

Details regarding the fundamentals, instrumentation, operation, and data handling of LC×LC and its applicability for various complex matrices have been described in depth in several review articles, of which references [6,8–10,71,74,75], and references cited therein, are valuable examples. In this section, the main aspects of interest concerning LC×LC will be briefly outlined, including peak capacity, orthogonality, sampling frequency, and basic LC×LC setup. Aspects concerning data processing will be outlined in Section 3.1.2, whereas the most important features in the selection of LC×LC experimental conditions, including column dimensions, mobile phase compatibility, and mode of elution will be addressed in Section 3.2.

As already stated in the introductory section of this chapter, the combination of two separation mechanisms with completely different selectivities offers one great advantage: a dramatic increase of the peak capacity. The peak capacity, n_c, is simply stated as the maximum number of peaks that can theoretically be resolved in a given timeframe between the first and last eluting peak [10]. When a sample is separated using two dissimilar (i.e., orthogonal) separation mechanisms, the theoretical peak capacity ($n_{c,2D}$) is the product of the peak capacity of each dimension as shown in Eq. (9.1) [10].

$$n_{c,2D} = {}^{1}n_c \times {}^{2}n_c \qquad (9.1)$$

where ${}^{1}n_c$ and ${}^{2}n_c$ are the peak capacities for the theoretical first and second dimension separations, respectively. The real peak capacity is, however, usually lower than predicted from Eq. (9.1) and it depends mostly on the degree of similarity of the coupled dimensions (i.e., orthogonality) and the rate of sampling time per fraction in the first dimension [10]. Unfortunately, most of the applications of LC×LC encountered in the literature do not report the value of the real peak capacity achieved by the developed method, which means that the reader can only draw an idea about the potential of the employed instrumental setup.

The orthogonality between the first and second dimensions in LC×LC is of utmost importance. For orthogonal separations to be obtained, it is necessary to use two separation mechanisms in the first and second dimensions that are independent of each other and, therefore, provide different selectivities [10,73]. Figure 9.2 shows a schematic representation of the concept and meaning of orthogonality. A high correlation (i.e., low orthogonality) between the black dots is observed in Figure 9.2A, thus indicating that the use of a second separation dimension is totally unnecessary. In Figure 9.2B, the dots are randomly distributed over the entire separation space, thus producing a low correlation. This feature indicates that a high orthogonality has been achieved as a consequence of the different selectivity provided by the second separation mechanism. As stated by François *et al.* [10], in practice, a true orthogonality is technically difficult to achieve, as this parameter not only depends on the separation mechanisms, but also on the properties of the solutes (which differs with sample origin) and the separation conditions (e.g., mobile phase composition). Therefore, when facing a separation problem in LC×LC, the analyst should carefully choose the appropriate stationary and mobile phases with respect to the physicochemical properties (e.g., size, specific charge, polarity, and/or hydrophobicity) of the sample constituents in order to achieve as much information as possible [10]. Various combinations of RP, NP, IEC, HILIC, or SEC can be used in LC×LC. Section 3.2 provides additional details on the practicability of these possible combinations for LC×LC. As also reviewed by François *et al.* [10], very few reports discuss the importance of the concept of orthogonality in their own LC×LC experiments from a mathematical point of view. Despite the existence of diverse approaches for orthogonality determination [75], in practical terms, their application is not straightforward, and the most important result is typically related to whether the 2D plane of separation is or is not appropriately filled.

As reviewed by François *et al.* [10], in order to maintain the resolution obtained in the first dimension, a large number of fractions should be transferred along a peak eluting from the first dimension. The first dimension peaks are

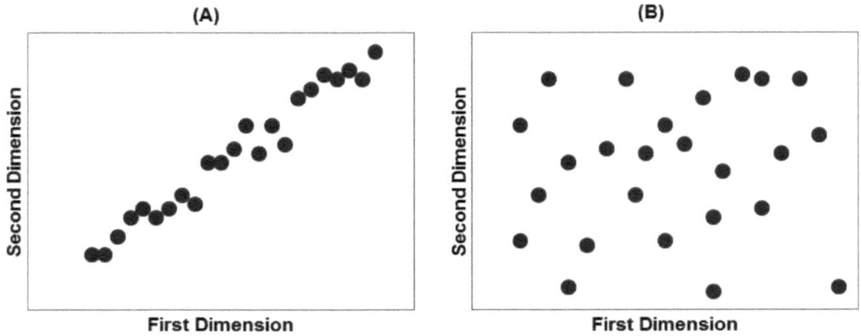

FIGURE 9.2 Schematic representation of orthogonality in comprehensive two-dimensional chromatography: (A) high correlation, low orthogonality; (B) low correlation, high orthogonality.

usually broadened by operating the first dimension column at a low-speed mobile phase, thus achieving a sufficient amount of fractions across the eluting peak. As a rule, three to four samplings across the first dimension peak are considered to be sufficient. Obviously, fast second-dimension analyses are a prerequisite in LC×LC in order to ensure that prior to the introduction of a successive fraction onto the second dimension column, the analysis of the previous fraction is already finished. The effect of proper sampling (or modulation) time is easily understood. If using longer modulation periods, substantially larger than the peak widths emerging from the first dimension column, the resolution in the first dimension will be partially lost, since the compounds that were (partially) separated are being collected in the same fraction [10]. On the other hand, if the modulation period is too short, significant wrap-around may take place as a consequence of the compounds only being eluted during the next modulation period alongside the compounds in that fraction, thus yielding peak tailing and a loss of resolution [76].

As previously mentioned, the instrumentation and operation of LC×LC has been described in the literature [6,8,10,71,74,75]. In-house built LC×LC systems are typically used in the majority of the research work already published, although very recently Agilent has launched a user-friendly LC×LC system whose application is yet to be reported in the literature. An LC×LC system can be easily assembled using conventional commercial LC setups equipped with more than one pump, columns, and an automated switching valve [47]. Basically, an LC×LC system consists of two pumps, two columns, injector, interfacing switching valve, and one (or more) detector(s) [47]. A typical example of an LC×LC setup is shown in Figure 9.3. The heart of the system is the interfacing valve that connects the two columns. The interface is usually an 8-port or 10-port switching valve equipped with two equal sampling loops, which are alternately used for the collection of the first dimension effluent in aliquots of predefined volumes, and allows the reinjection of these fractions into the second dimension column in an automated way [10,74]. The time available for analysis in the second dimension, therefore, is equal to the sampling period. There are other sampling loop interfaces, such as the packed loop interface, stop–flow interface, interface with parallel second dimension, and vacuum evaporation interface. A complete survey of the different available interface configurations is well beyond the scope of this chapter. For an in-depth description of their features, refer to the review works of Malerod *et al.* [6] and François *et al.* [10] as well as the literature cited therein.

9.3.1.2 Graphical Representation and Data Processing in Comprehensive Two-dimensional Liquid Chromatography

Data obtained from a LC×LC analysis include a three-dimensional dataset, representing the first dimension retention time, the second dimension retention time, and the detector response. Figure 9.4A shows a simulated output from a detector positioned at the end of a LC×LC separation system, with a modulation

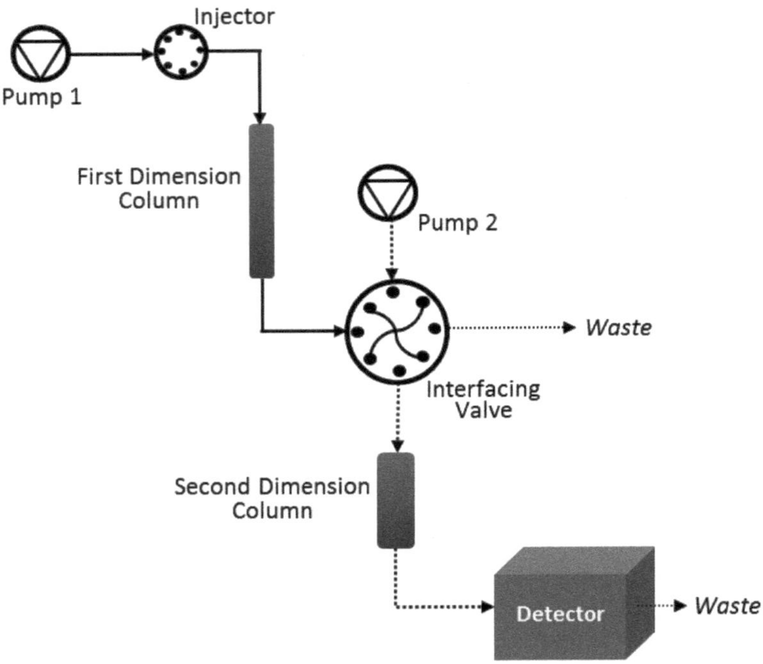

FIGURE 9.3 Basic LC×LC setup with a two-position/8-port switching valve.

time of 2 min. Since the modulation time is known, then the application of the slicing methodology is straightforward, and the resulting Figure 9.4B shows the same data set but now under in the form of a 2D chromatogram. After data modulation, it may be necessary to smooth the chromatogram by interpolation of data points between 1D chromatograms, mainly along the first dimension [71], in order to obtain a 2D chromatogram (Figure 9.4C). For representation of a 2D data matrix, a contour plot shown in Figure 9.4C (or even a 3D plot) can easily allow visualization of the separation and obtaining an initial indication of how well the LC×LC separation has performed. Such a graph does not provide quantitative information but it is an excellent visualization tool, which becomes the first essential step for further processing of the data obtained [71].

The huge amount of data generated in LC×LC applications creates the corresponding difficulties in data handling in order to become useful for extracting the required chemical information on a given sample. These data sets require the use of computational systems to process and extract the maximum chemical information possible. However, this processing requires different approaches and methodologies depending on the different ends or goals set by the analyst. Recently, Matos *et al.* [71] presented an in-depth review how state-of-the-art mathematical tools for data processing of multidimensional data sets obtained from comprehensive 2D chromatography (including both LC×LC and GC×GC)

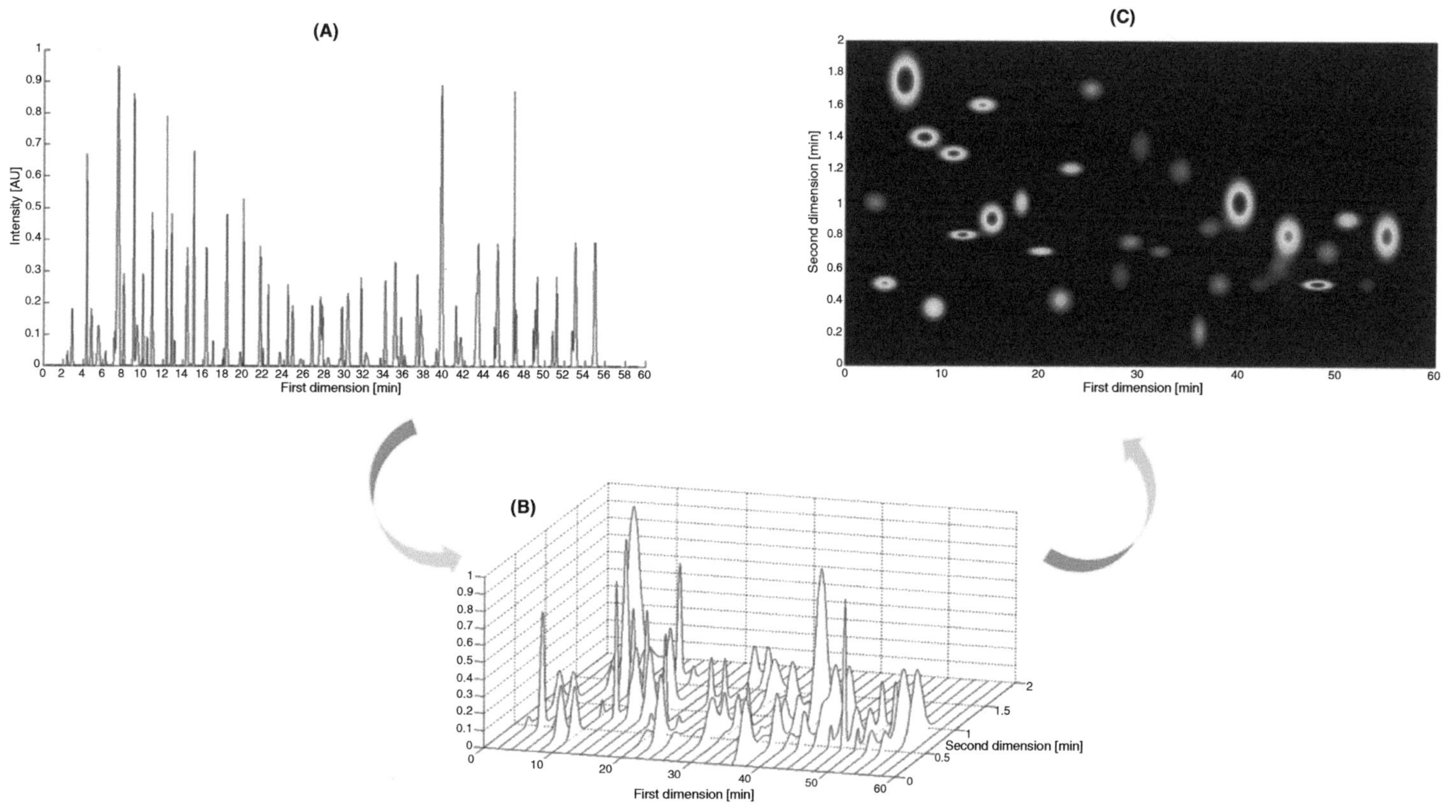

FIGURE 9.4 Representation of a simulated data set for comprehensive two-dimensional liquid chromatography: (A) raw data from the detector positioned at the end of the second dimension column; (B) layout of the sliced 1D chromatograms defined by the modulation time (2 min); (C) contour plot of a 2D chromatogram after smoothing.

aid analysts in data interpretation and analysis, from the pretreatment until the quantification of the identified chromatographic peaks (this includes also peak detection and quantification, and measurement of overlapping of 2D peaks—resolution in 2D chromatography). This review work should be consulted for further details on mathematic and chemometric tools since they will not be discussed in detail in this chapter.

The huge data sets produced by 2D separations make their analysis almost impossible without using some type of computer software in order to transform the 2D chromatographic data into usable information. Quite often, many researchers have developed their own algorithms, usually in MatLab (Mathworks, Natick, MA, USA), well-known commercial software for numerical computation [71,77]. As reviewed by Matos *et al.* [71], one of the few examples of commercial software developed and available for dealing with 2D chromatographic data, particularly for GC×GC, is the software system GC Image, developed as a spin-off at the University of Nebraska-Lincoln, which incorporates the inverted watershed algorithm and allows a digital image processing for visualization, processing, analysis, and reporting the GC×GC chromatographic data. Matos *et al.* [71] also point out that a version of this software has been recently developed and released for dealing with LC×LC data.

9.3.2 Exploring the Use of Comprehensive Two-Dimensional Liquid Chromatography

To the best of our knowledge, LC×LC has never been applied to the analysis of marine bioactive compounds. Therefore, to develop an LC×LC method that fits the purpose of bioactive compounds discovery, several works published earlier on other complex matrices (food, proteins, plant extracts, and polymers) can be used as guidelines (e.g., references [6,8,11,47,74] and literature cited therein).

The development of an LC×LC method should start with the selection of the separation mechanism in both dimensions, which must have different selectivities for the sample components. Ideally, the best separation is achieved when there is no relationship between the retention features of the analytes in the two dimensions. Several approaches to LC×LC have been described in the literature:

- IEC×RP for the analysis of terpene derived acids and mono- and dicarboxylic acids in atmospheric aerosols [78]
- Ag-coated-IEC×RP for the analysis of triacylglycerols in corn oil [79]
- Ion chromatography (IC)×RP for the separation of low-molar-mass organic acids in food and beverages [80]
- IC×IC for the analysis of inorganic anions [81]
- RP×SEC for the characterization of synthetic polymers [82,83]
- SEC×RP for the screening of green and black teas with the identification of novel flavonol glycosides and several saponin isomers [84]

- HILIC×SEC for the separation of natural organic matter from aquatic environments [47]
- HILIC×RP for the quantification of polyphenols in red wines [85] as well as proanthocyanidins, phenolic acids, flavonols, and flavonol derivatives in green tea [86] and phospholipids in cow's milk [87]
- RP×RP for the separation of phenolic and flavone natural antioxidants [88] as well as proteome samples [89]
- NP×RP for the analysis of the carotenoid composition of orange essential oil and juice [87] as well as the analysis of a lemon oil extract (furocoumarines and coumarines) [90]

Overall, in these combinations, the separation is based on the differences in the molecular structure of the separated compounds such as the size and shape, polarity, hydrophobicity, and ionic charge. In practice, however, the selectivities of the two LC modes can be correlated to some degree since secondary interaction mechanisms occurring between the analyte(s) and the stationary and mobile phases may also govern the analytes' retention pattern, and these additional interaction mechanisms can be correlated in the two LC modes. Such a case is likely to occur in combinations involving IEC and RP, RP and SEC, HILIC and RP, or HILIC and SEC modes. For example, in IEC×RP, useful for the separation of ionic compounds, acids, or bases, the analytes are mainly separated by electrostatic interactions in the first IEC dimension and hydrophobic interactions in the second RP dimension. Nevertheless, in practice, the IEC phase often has additional characteristics, such as hydrophobicity, thus reducing the orthogonality [91]. To minimize the role of hydrophobicity in IEC, the addition of an organic modifier to the mobile phase is often a good choice [91]. Nevertheless, care should be taken regarding the amount of organic modifier used in the first dimension in order to maintain weak elution strength in the second RP dimension. In peptides analysis, 10% to 15% of organic modifier (either acetonitrile or methanol) can normally be included in the elution buffer at low pH (<4.0) [91].

Another example is the combination SEC×RP, which is useful for mapping the chemical characteristics versus molecular weight distribution (and separation as well) of macromolecules such as biopolymers. Although the compatibility of mobile phases in a SEC×RP system is high (the mobile phase used in SEC can be aqueous, thus having low elution strength in the second RP dimension), it could be necessary to add an organic modifier to the mobile phase of the first SEC dimension in order to minimize the occurrence of nonsize-exclusion interactions [6]. By increasing the elution strength in the second dimension, the combination SEC×RP could become less compatible with the subsequent reduction of both peak capacity and orthogonality. On the other hand, with RP in the first dimension in a combination RP×SEC, the analytes will be focused in the first dimension and not in the second SEC dimension, and hence the total peak capacity can increase [6]. For separating complex mixtures containing both hydrophilic and hydrophobic compounds, the combination of HILIC and RP stationary phases in an HILIC×RP system may allow achieving a comprehensive separation since the two phases are claimed to be opposite separation mechanisms regarding polarity. In theory, the use of aqueous-organic mobile phases in both dimensions makes the HILIC and RP modes compatible. In practice, however, this could be an example of nonmatching solvent strengths, where the high elution strength of the mobile phase of the first HILIC dimension may impair the resolution in the second RP dimension. To operate in the HILIC mode it is necessary to use organic-rich mobile phases, which are very strong eluents in RP thus reducing the resolution of a HILIC×RP separation system. A likely possible solution to overcome this problem is to use a low volume for the transferred fraction alongside weak elution strength in the second RP dimension in order to focus the fraction in a narrow zone on the top of the secondary column before the elution step.

The combination NP×RP is more demanding, although NP is highly orthogonal with RP and the combined peak capacity is expected to be high [6]. Indeed, the nonaqueous organic mobile phases in NP are often immiscible with the aqueous mobile phases in RP, being not only very strong eluents in RP but also creating large viscosity differences between the two dimensions, thus deteriorating the separation and producing signal interferences [10]. A possible solution to deal with this compatibility problem is to employ a vacuum evaporation interface, which not only enables the collection of the fractions eluting from the first dimension, but also allows the evaporation of the mobile phase solvents by applying high temperatures under vacuum conditions [10]. Nevertheless, this solution comes with the cost that not all sample components are able to be fully recovered since they can be lost due to poor solubility in the second dimension mobile phase [6]. An alternative to a vacuum evaporation interface was employed by Dugo *et al.* [87] for the elucidation of carotenoid patterns in citrus products by means of NP×RP. The authors applied a low flow rate (0.01 mL min^{-1}) in the first NP dimension and a high flow rate (4.7 mL min^{-1}) in the second RP dimension, whereas the volume of the transferred fractions was only 3 μL. Under these experimental conditions, only a small aliquot of the nonpolar NP mobile phase (consisting of *n*-hexane (A) and ethyl alcohol (B) in gradient mode) was introduced on the second RP dimension where it suffered an extensive dilution effect (mobile phase consisting of 2-propanol (A) and 20% (v/v) of water in acetonitrile (B) in gradient mode), thus promoting the analytes to be focused on the RP and minimizing the band broadening [87].

Apparently, the combination of RP in both dimensions seems to be marginally beneficial in terms of orthogonality, but the use of different RP columns and/or mobile phases with different organic modifier content [88] or pH values [89] allows different selectivities between the two dimensions, while keeping the hydrophobic interactions as their main separation mechanism. Once again, the selection of the mobile phase plays a key role in RP×RP systems, and it is mandatory that the first dimension mobile phase has weak elution strength in the second dimension. For the analysis of complex samples, Malerod *et al.* [6] suggested the use of gradient elution with a rather low percentage of organic modifier in the first dimension (e.g., up to 14% (v/v) of acetonitrile [88]), which typically provides a higher peak capacity than the isocratic mode, and a gradient increasing to a higher percentage of organic modifier in the second dimension (e.g., up to 38% (v/v) of acetonitrile [88]).

Undoubtedly, the selection of the mobile phases in the first and second dimensions is an important issue in any LC×LC method development. The mobile phase used in both first and second dimensions alongside the selected separation mechanisms influence the selectivity, resolution, and the peak capacity obtained in the LC×LC method. Furthermore, in order to avoid excessive peak broadening with the subsequent wraparound and poor resolution in the second dimension, it is advisable that the mobile phase used in the first dimension has low elution strength in the second dimension. The compatibility requirement for the mobile phases of the first and second dimensions should also be followed. This is important to avoid the occurrence of the viscous fingering phenomenon when using two solvents with different viscosities in the two dimensions (e.g., when the less viscous acetonitrile penetrates into methanol), which together with problems of solvent strength and/or immiscibility are very likely to negatively influence the retention and selectivity of the LC×LC method [10,74]. Moreover, the mobile phase of the second dimension should be compatible with the detection system, most often the MS detector [74]. Wherever possible, programmed elution techniques should be used in LC×LC separations since they provide significantly improved peak capacity with respect to isocratic elution [6]. Gradient elution is widely used in the first dimension in comprehensive LC×LC, but is equally useful in the second dimension since it improves the orthogonality of the 2D system and the regularity of coverage of the 2D retention plane [88]. It should be emphasized, however, that using fast and repeatable gradients in a very short time in the second dimension can be difficult, but it can be solved with the aid of UHPLC instrumentation. The use of supra-ambient temperatures could also allow the use of high-speed mobile phases by reducing the viscosity of the mobile phase [10]. Nevertheless, the choice of such temperatures should take into account the stability of the stationary phases and analytes.

Long micro- or narrow-bore columns (I.D. = 0.5–2 mm) with a low mobile phase rate (between 0.05 and 0.1 mL min^{-1}) are typically used in the first dimension in order to significantly reduce band broadening and diminish signal interferences caused by potential solvent incompatibilities in the second dimension [6,10,74]. In the second dimension, a short column with an I.D. larger than that used in the first dimension should be used at a high-speed mobile phase (typically between 0.5 and 5 mL min^{-1}) for achieving rapid separation and to minimize extra column peak broadening [10,74]. As a rule, the I.D. of the second dimension column is always larger than the one of the column used for the first dimension [10]. As the analysis speed of the second dimension is also of fundamental importance for the success of the separation, columns packed with sub-2 μm or fused-core particles are most suitable to achieve this purpose [74]. The analyst should be aware, however, that the use of columns packed with sub-2 μm particles requires the use of specific instrumentation capable of withstanding high pressures. In this sense, columns packed with fused-core particles technology could be a better choice, since it allows operating at high-speed mobile phases with lower pressure drops using conventional LC instrumentation [10].

9.4 CONCLUDING REMARKS

Nowadays, the interest in isolating marine bioactive compounds is increasing and, therefore, there is a growing need for developing analytical methods useful for chemical characterization and quantification. Within this context, chromatography coupled to a sensitive detection system (e.g., MS) is a powerful tool for separating and discovering novel bioactive compounds. Over the past decades a number of innovative 1D LC approaches have been conceived and implemented for tracking bioactivity in complex marine matrices. The most recent developments in this area have been, in particular, in the field of new LC stationary phases using sub-2 μm and fused-core particles technology. The latter is undoubtedly an added value for the routine analysis of complex samples in today's laboratories. This type of column provides similar high speed and high efficiency as sub-2 μm particles but at half the back pressure, thus circumventing the requirements of using the more expensive and sophisticated UHPLC instrumentation. Several recent applications to various targets (e.g., triacylglycerols phospholipids, and peptides) have confirmed the potential of this new technology for the separation and further detection and identification of bioactive compounds in marine sample extracts. Additionally, the ILs as mobile phase additives in LC and IL-based materials in both LC and GC are gaining acceptance for the separation of complex organic matrices. The continuous improvements in this area opens new possibilities for their application in the development of more environment-friendly chromatographic methodologies with higher selectivity for the analysis of marine bioactive compounds.

Notwithstanding all the innovative 1D LC and GC approaches that have been implemented for screening marine bioactive compounds, the main advances in their characterization are likely to come with the application of multidimensional chromatographic techniques, namely LC×LC. Today, the principles and LC×LC instrumental setup are well established. As in 1D LC, hyphenation of LC×LC with spectroscopic or spectrometric techniques enables obtaining detailed information on the eluting fractions. With the advent of the first commercially available LC×LC system from Agilent, this chromatographic technique is likely to become more suitable to routine analysis and screening of complex samples. Therefore, as new tools for data handling develop, the coming decades will undoubtedly see major advances in the characterization and quantification of new marine bioactive compounds.

ACKNOWLEDGMENTS

This work was supported by European Funds through COMPETE and by National Funds through the Portuguese Science Foundation (FCT) within project PEst-C/MAR/LA0017/2013 (Centre for Environmental and Marine Studies—CESAM). The FCT, through the European Social Fund (ESF) and Programa Operacional Potencial Humano—POPH, is further acknowledged for financial support to Regina Duarte. This work was also funded by FEDER under the Programa de Cooperação Territorial Europeia INTERREG IV B SUDOE within the framework of the research project ORQUE SUDOE, SOE3/P2/F591.

REFERENCES

[1] S. Fanali, P.R. Haddad, C.F. Poole, P. Schoenmakers and D. Lloyd (Eds.), Liquid Chromatography: Applications, Waltham, Massachusetts, 2013.
[2] O. Potterat, M. Hamburger, Nat. Prod. Rep. 30 (2013) 546–564.
[3] K.K. Unger, R. Ditz, E. Machtejevas, R. Skudas, Angew. Chem. Int. Ed. 49 (2010) 2300–2312.
[4] G.C. Woods, M.J. Simpson, P.J. Koerner, A. Napoli, A.J. Simpson, Environ. Sci. Technol. 45 (2011) 3880–3886.
[5] R.M.B.O. Duarte, A.C. Duarte, Anal. Chim. Acta 688 (2011) 90–98.
[6] H. Malerod, E. Lundanes, T. Greibrokk, Anal. Methods 2 (2010) 110–122.
[7] C. Meinert, U.J. Meierhenrich, Angew. Chem. Int. 51 (2012) 10460–10470.
[8] L. Mondello, A.C. Lewis and K.D. Bartle (Eds.), Multidimensional Chromatography, Chichester, West Sussex, 2002.
[9] G. Guiochon, N. Marchetti, K. Mriziq, R.A. Shalliker, J. Chromatogr. A 1189 (2008) 109–168.
[10] I. François, K. Sandra, P. Sandra, Anal. Chim. Acta 641 (2009) 14–31.
[11] C. Tistaert, B. Dejaegher, Y.V. Heyden, Anal. Chim. Acta 690 (2011) 148–161.
[12] K.M. Kalili, A. Villiers, J. Sep. Sci. 34 (2011) 854–876.
[13] M.-J. Motilva, A. Serra, A. Macià, J. Chromatogr. A 1292 (2013) 66–82.
[14] S. Fanali, P.R. Haddad, C.F. Poole, P. Schoenmakers and D. Lloyd (Eds.), Liquid Chromatography: Fundamentals and Instrumentation, Waltham, Massachusetts, 2013.
[15] M.P. Mansour, J. Chromatogr. A 1097 (2005) 54–58.
[16] P. Ciminiello, C. Dell'Aversano, E. Fattorusso, M. Forino, G.S. Magno, L. Tartaglione, M.A. Quilliam, A. Tubaro, R. Poletti, Rapid Commun. Mass Spectrom. 19 (2005) 2030–2038.
[17] S.-Y. Jeong, H.J. Shin, T.S. Kim, H.-S. Lee, S.-k. Park, H.M. Kim, J. Antibiot. 59 (2006) 234–240.
[18] H. Liu, H. Gao, M. Namikoshi, H. Kobayashi, R.E.P. Mangindaan, N. Wang, X. Yao, Rapid Commun. Mass Spectrom. 21 (2007) 199–206.
[19] T. Rundberget, M. Sandvik, K. Larsena, G.M. Pizarro, B. Reguera, T. Castberg, E. Gustad, J.I. Loader, F. Rise, A.L. Wilkins, C.O. Miles, Toxicon 50 (2007) 960–970.
[20] A.P. Treschow, L.D. Hodges, P.F.A. Wright, P.M. Wynne, N. Kalafatis, T.A. Macrides, Comp. Biochem. Physiol., Part B: Biochem. Mol. Biol. 147 (2007) 645–656.
[21] R.D.J. Solomon, V.S. Santhi, World J Microbiol Biotechnol 24 (2008) 1747–1752.
[22] S. Achamlale, B. Rezzonico, M. Grignon-Dubois, J. Appl. Phycol. 21 (2009) 347–352.
[23] M. Plaza, S. Santoyo, L. Jaime, G.G.-B. Reina, M. Herrero, F.J. Señoráns, E. Ibáñez, J. Pharm. Biomed. Anal. 51 (2010) 450–455.

[24] H. Yaich, H. Garna, S. Besbes, M. Paquot, C. Blecker, H. Attia, Food Chem. 128 (2011) 895–901.
[25] P. Blay, J.P.M. Hui, J. Chang, J.E. Melanson, Anal. Bioanal. Chem. 400 (2011) 577–585.
[26] M.S. Kumar, N.S. Pandita, A.K. Pal, Toxicon 60 (2012) 1135–1147.
[27] E. Gómez-Ordóñez, A. Jiménez-Escrig, P. Rupérez, Talanta 93 (2012) 153–159.
[28] R.G. Hatfield, A.D. Turner, J. AOAC Int. 95 (2012) 1089–1096.
[29] P.J. Eugster, D. Biass, D. Guillarme, P. Favreau, R. Stöcklin, J.-L. Wolfender, J. Chromatogr. A 1259 (2012) 187–199.
[30] Y. Kato, S. Okada, K. Atobe, T. Endo, K. Haraguchi, Anal. Bioanal. Chem. 404 (2012) 197–206.
[31] Y. Zhang, S. Xiao, L. Sun, Z. Ge, F. Fang, W. Zhang, Y. Wang, Y. Cheng, Anal. Chim. Acta 777 (2013) 49–56.
[32] F.C.N. Barros, D.C. Silva, V.G. Sombra, J.S. Maciel, J.P.A. Feitosa, A.L.P. Freitas, R.C.M. Paula, Carbohydr. Polym. 92 (2013) 598–603.
[33] L. Zhuo, Y. Yin, W. Fu, B. Qiu, Z. Lin, Y. Yang, L. Zheng, J. Li, G. Chen, Food Chem. 137 (2013) 115–121.
[34] L. Yu, X. Xu, C. Xue, Y. Chang, L. Ge, Y. Wang, C. Zhang, G. Liu, Chen He, Food Chem. 139 (2013) 702–709.
[35] C. Ragonese, L. Tedone, M. Beccaria, G. Torre, F. Cichello, F. Cacciola, P. Dugo, L. Mondello, Food Chem. 145 (2014) 932–940.
[36] F. Zhang, X. Chu, X. Wang, M. Yang, Y. Ling, L. Sun, X. Li, W. Yong, J. Chromatogr. A 1213 (2008) 176–180.
[37] M.C. Pietrogrande, F. Dondi, A. Ciogli, F. Gasparrini, A. Piccin, M. Serafini, J. Chromatogr. A 1217 (2010) 4355–4364.
[38] J.-W. Kim, B.R. Ramaswamy, K.-H. Chang, T. Isobe, S. Tanabe, J. Chromatogr. A 1218 (2011) 3511–3520.
[39] Y. Lu, Q. Shen, Z. Dai, H. Zhang, H. Wang, J. Chromatogr. A 1218 (2011) 929–937.
[40] H. Gallart-Ayala, E. Moyano, M.T. Galceran, J. Chromatogr. A 1217 (2010) 3511–3518.
[41] I. Tarcomnicu, A.L.N. van Nuijs, W. Simons, L. Bervoets, R. Blust, P.G. Jorens, H. Neels, A. Covaci, Talanta 83 (2011) 795–803.
[42] P. Jandera, J. Sep. Sci. 31 (2008) 1421–1437.
[43] P. Hemström, K. Irgum, J. Sep. Sci. 29 (2006) 1784–1821.
[44] J.W. Treadway, G.S. Philibert, S.V. Olesik, J. Chromatogr. A 1218 (2011) 5897–5902.
[45] J. Bernala, A.M. Ares, J. Pól, S.K. Wiedmer, J. Chromatogr. A 1218 (2011) 7438–7452.
[46] K. Spagou, H. Tsoukali, N. Raikos, H. Gika, I.D. Wilson, G. Theodoridis, J. Sep. Sci. 33 (2010) 716–727.
[47] R.M.B.O. Duarte, A.C. Barros, A.C. Duarte, J. Chromatogr. A 1249 (2012) 138–146.
[48] M. Wuhrer, A.R. de Boer, A.M. Deelder, Mass Spectrom Rev. 28 (2009) 192–206.
[49] H.P. Nguyen, K.A. Schug, J. Sep. Sci. 31 (2008) 1465–1480.
[50] M. Lämmerhofer, M. Richter, J. Wu, R. Nogueira, W. Bicker, W. Lindner, J. Sep. Sci. 31 (2008) 2572–2588.
[51] K.J. Fountain, J. Xu, D.M. Diehl, D. Morrison, J. Sep. Sci. 33 (2010) 740–751.
[52] B. Tang, W. Bi, M. Tian, K.H. Row, J. Chromatogr. B 904 (2012) 1–21.
[53] C.J. Malherbe, D. de Beer, E. Joubert, Int. J. Mol. Sci. 13 (2012) 3101–3133.
[54] L. Miao, T.F.N. Kwong, P.-Y. Qian, Appl. Microbiol. Biotechnol. 72 (2006) 1063–1073.
[55] H. Xiong, S. Qi, Y. Xu, L. Miao, P.-Y. Qian, J. Hydro-environ. Res. 2 (2009) 264–270.
[56] J. Selvin, S. Shanmughapriya, R. Gandhimathi, G.S. Kiran, T.R. Ravji, K. Natarajaseenivasan, T.A. Hema, Appl. Microbiol. Biotechnol. 83 (2009) 435–445.

[57] M. Gomez-Romero, A. Segura-Carretero, A. Fernandez-Gutierrez, Phytochem. 71 (2010) 1848–1864.
[58] E. Humston, K. Dombek, B. Tu, E. Young, R. Synovec, Anal. Bioanal. Chem. 401 (2011) 2387–2402.
[59] D.M. Pereira, J. Vinholes, P.G. Pinho, P. Valentão, T. Mouga, N. Teixeira, P.B. Andrade, Talanta 100 (2012) 391–400.
[60] M. Tabarsa, M. Rezaei, Z. Ramezanpour, J.R. Waaland, J. Sci. Food Agric. 92 (2012) 2500–2506.
[61] A.R. Fardin-Kia, P. Delmonte, J.K.G. Kramer, G. Jahreis, K. Kuhnt, V. Santercole, J.I. Rader, Lipids 48 (2013) 1279–1295.
[62] P.B. Andrade, M. Barbosa, R.P. Matos, G. Lopes, J. Vinholes, T. Mouga, P. Valentão, Food Chem. 138 (2013) 1819–1828.
[63] C. Schummera, O. Delhomme, B.M.R. Appenzeller, R. Wennig, M. Millet, Talanta 77 (2009) 1473–1482.
[64] A.B. Blakeney, P.J. Harris, R.J. Henry, B.A. Stone, Carbohydr. Res. 113 (1983) 291–299.
[65] L. Mondello, R. Shellie, A. Casilli, P.Q. Tranchida, P. Marriott, G. Dugo, J. Sep. Sci. 27 (2004) 699–702.
[66] M. Herrero, E. Ibañez, A. Cifuentes, J. Bernal, J. Chromatogr. A 1216 (2009) 7110–7129.
[67] X. Zhang, A. Fang, C.P. Riley, M. Wang, F.E. Regnier, C. Buck, Anal. Chim. Acta 664 (2010) 101–113.
[68] D.R. Stoll, Anal. Bioanal. Chem. 397 (2010) 979–986.
[69] M. Farré, L. Kantiania, M. Petrovic, S. Pérez, D. Barceló, J. Chromatogr. A 1259 (2012) 86–99.
[70] P. Schoenmakers, P. Marriott, J. Beens, LC-GC Europe 16 (2003) 335–339.
[71] J.T.V. Matos, R.M.B.O. Duarte, A.C. Duarte, J. Chromatogr. B 910 (2012) 31–45.
[72] F. Regnier, G. Huang, J. Chromatogr. 750 (1996) 3–10.
[73] J. Dallüge, J. Beens, U.A.Th. Brinkman, J. Chromatogr. A 1000 (2003) 69–108.
[74] J. Pól, T. Hyötyläinen, Anal. Bioanal. Chem. 391 (2008) 21–31.
[75] D.R. Stoll, X. Li, X. Wang, P.W. Carr, S.E.G. Porter, S.C. Rutan, J. Chromatogr. A 1168 (2007) 3–43.
[76] M. Kivilompolo, T. Hyötyläinen, J. Chromatogr. A 1145 (2007) 155–164.
[77] J.T.V. Matos, R.M.B.O. Duarte, A.C. Duarte, Anal. Chim. Acta 804 (2013) 296–303.
[78] J. Pól, B. Hohnová, M. Jussila, T. Hyötyläinen, J. Chromatogr. A 1130 (2006) 64–71.
[79] E.J.C. van der Klift, G. Vivó-Truyols, F.W. Claassen, F.L. van Holthoon, T.A. van Beeka, J. Chromatogr. A 1178 (2008) 43–55.
[80] S.S. Brudina, R.A. Shellie, P.R. Haddad, P.J. Schoenmakers, J. Chromatogr. A 1217 (2010) 6742–6746.
[81] R.A. Shellie, E. Tyrrell, C.A. Pohl, P.R. Haddad, J. Sep. Sci. 31 (2008) 3287–3296.
[82] A. van der Horst, P.J. Schoenmakers, J. Chromatogr. A 1000 (2003) 693–709.
[83] A. Greiderer, L. Steeneken, T. Aalbers, G. Vivó-Truyols, P. Schoenmakers, J. Chromatogr. A 1218 (2011) 5787–5793.
[84] C.T. Scoparo, L.M. de Souza, N. Dartora, G.L. Sassaki, P.A.J. Gorin, M. Iacomin, J. Chromatogr. A 1222 (2012) 29–37.
[85] P. Dugo, F. Cacciola, P. Donato, D. Airado-Rodriguez, M. Herrero, L. Mondello, J. Chromatogr. A 1216 (2009) 7483–7487.
[86] K.M. Kalili, A. Villiers, J. Sep. Sci. 33 (2010) 853–863.
[87] P.Q. Tranchida, P. Donato, F. Cacciola, M. Beccaria, P. Dugo, L. Mondello, TrAC - Trends Anal. Chem. 52 (2013) 186–205.

[88] P. Cesla, T. Hajek, P. Jandera, J. Chromatogr. A 1216 (2009) 3443–3457.
[89] P. Donato, F. Cacciola, E. Sommella, C. Fanali, L. Dugo, M. Dach, P. Campiglia, E. Novellino, P. Dugo, L. Mondello, Anal. Chem. 83 (2011) 2485–2491.
[90] I. François, A. de Villiers, B. Tienpont, F. David, P. Sandra, J. Chromatogr. A 1178 (2008) 33–42.
[91] A. Motoyama, J.R. Yates, Anal. Chem. 80 (2008) 7187–7193.

Chapter 10

Online Combination of Bioassays with Chemical and Structural Characterization for Detection of Bioactive Compounds

Ana C. Freitas[1,2], Sofia Isabel G.H.M. Montalvão[3], Armando C. Duarte[1] and Teresa Rocha-Santos[1,2]

[1]*CESAM - Centre for Environmental and Marine Studies & Department of Chemistry, University of Aveiro, Campus Universitrio de Santiago, Aveiro, Portugal*

[2]*ISEIT/Viseu, Instituto Piaget, Estrada do Alto do Gaio, Galifonge, Lordosa, Viseu, Portugal*

[3]*Centre for Drug Research, Faculty of Pharmacy, Viikki Biocenter, P.O. Box 56 (Viikinkaari 5 E), FIN, University of Helsinki, Finland*

Contents

10.1 INTRODUCTION

The screening and identification of bioactive compounds from complex mixtures without the need for unwieldy purification is difficult and challenging [1]. Around 60% of all available small molecules with anticancer activity from the

Analysis of Marine Samples in Search of Bioactive Compounds, Vol. 65. DOI: 10.1016/B978-0-444-63359-0.00010-0

1940s to today are of natural product origin [2]. The classic analytical approach, a subject explored in previously chapters, is to search with time-consuming bioassay guided fractionation until a single bioactive compound can be isolated. Alternatively, the online combination or simultaneous screening of bioassays with chemical and structural characterization allows for rapid screening and identification of individual bioactive compounds with several biological activities requiring no prior purification steps [3].

Several strategies have been attempted in screening bioactive compounds in complex mixtures, such as marine extracts, involving total or partial online screening. These systems integrate separation methods, chemical detection methods (e.g., diode-array detection, mass spectrometry (MS), nuclear magnetic resonance (NMR)), and biochemical assays [4]. Two main strategies have been used by researchers, each with its own advantages and drawbacks: (1) high-throughput screening (HTS) and (2) high-resolution screening (HRS). The approaches that have been used for HTS can generally be categorized into precolumn and postcolumn methodologies. Precolumn methodologies (precolumn affinity-recognition mode) have been attempted based on the detection of the ligand-protein interaction principle; that is, based on the fact that a bioactive compound in a complex mixture must interact first with a target (protein) before separation, followed by chemical detection and identification [5,6]. HTS postcolumn methodologies basically consist of fractionating complex mixtures, collecting the resulting fractions and their evaporation, followed by microplate-based bioassays to detect the bioactive fractions with parallel chemical detection and identification. This approach is not totally online but it will be considered in this chapter due to its importance and use for screening bioactive compounds in complex mixtures. HRS in general involves the online coupling of a bioassay of the chromatographic separation [4]. A schematic representation of both HTS and HRS strategies is depicted in Figures 10.1 and 10.2.

In HTS postcolumn screening, the high resolution obtained from the chromatographic separation steps is often lost in lower resolution fraction collection for the bioactivity screen [3]. In HRS, rapid and straightforward online biochemical detection (BCD) assays (e.g., antioxidant screening assays, enzymes, and receptor-based assays) must be included to reach high resolution and sensitivity; the detection of bioactive compounds in simulated and nonsimulated biochemical reactions is the basis of BCD assays [7,8]. This screening strategy is a way to surpass limitations associated with preisolation since it directly measures the effects of bioactive compounds after separation (postcolumn) and cuts down on *in vitro* tests because only fractions or new compounds demonstrating specific activity on a biological marker need to be isolated and tested further [8].

10.2 HIGH-THROUGHPUT SCREENING METHODS

HTS-based methods appeared in the 1990s when technological advances such as HPLC-coupled spectroscopy had tremendous impact, making abundant information available on the structure of the compounds present in natural

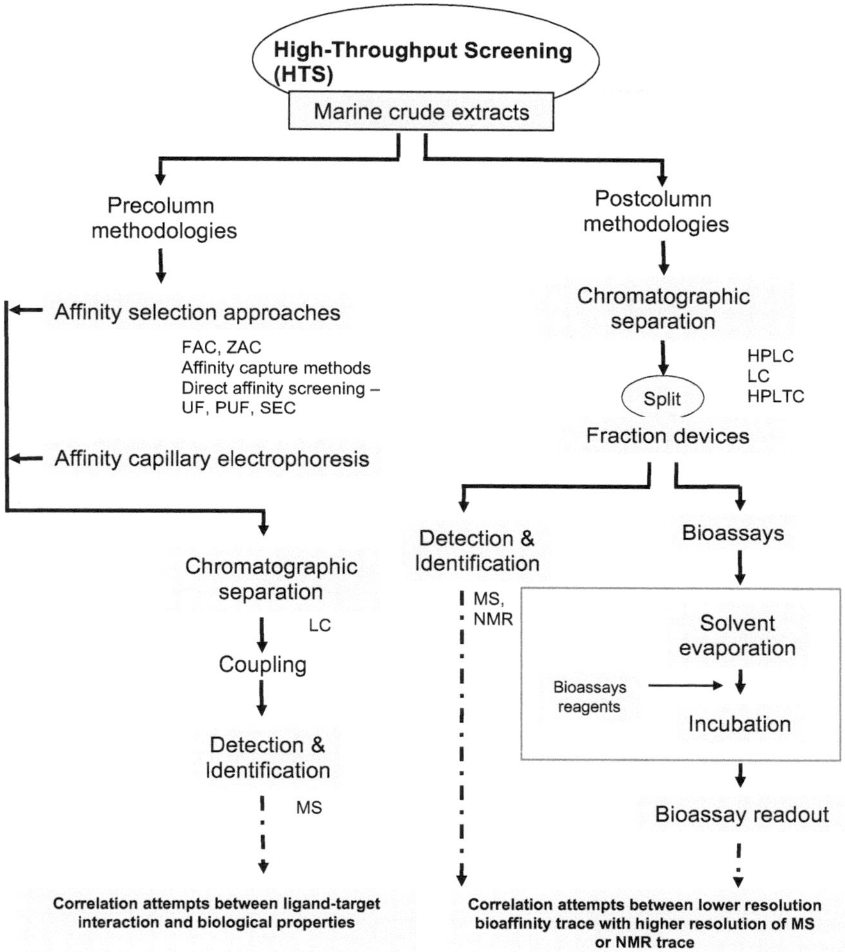

FIGURE 10.1 Schematic representation of high-throughput screening strategies for bioactive compounds from complex mixtures such as marine natural extracts.

extracts or their fractions, which in turn, allows efficient dereplication of known molecules [6].

The inherent complexity of extracts and their dereplication time as well as the lag in prioritization of lead compounds in comparison to synthetic libraries are some of the difficulties faced by HTS applied to crude natural extracts [9]. These difficulties led to the search for new methodologies to produce high-purity natural product libraries for HTS in order to facilitate and accelerate the possibility of drug discoveries in natural products. HTS and the miniaturization of assays implied the need to optimize natural product samples to better suit these technologies to shorten the time period from hit detection to lead

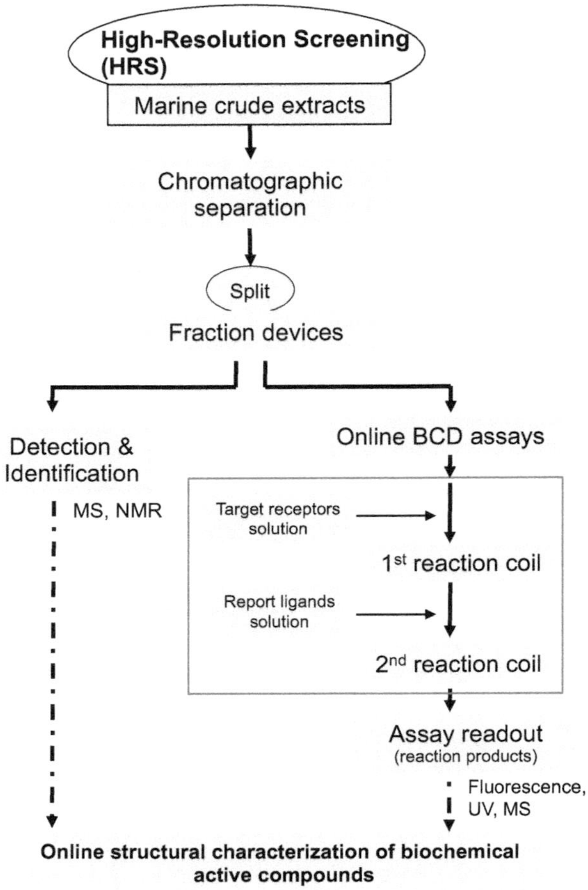

FIGURE 10.2 Schematic representation of high-resolution screening strategies for bioactive compounds from complex mixtures such as marine natural extracts.

characterization [2]. Laboratory automation facilitated the generation of large libraries of prepurified fractions or extracts that are more suitable for HTS but the efficient interfacing of biological data with chemo-analytical information still remains a challenge [6]. The interrelationship between the target and the screening process as well as the the quality of the compound library is the basis of an output from HTS [10]. In addition, the higher number of fractions leads to cleaner samples and increases the chance for discovery of minor bioactive compounds although it increases the costs for the production and screening of these libraries [6]. Croft et al. [11] found that HTS did not achieve a desired success rate due to the large scale involved and the resulting high costs.

Several protocols based on solid phase extraction (SPE), high-performance liquid chromatography (HPLC), or a combination of both are able to generate less than 10 to over 100 fractions per extract [6]. Wyeth Research from the United States has produced a prefractionated natural product library using reverse-phase (RP)-HPLC to complement their library of crude extracts. Although crude libraries have a number of advantages, including a great diversity of chemical compounds, the inherent limitations of their use (e.g., the unsuitability for liquid handling automated systems, minor metabolites may not be detected or may be masked by other components in the complex mixture, among others-these issues are discussed in detail by Koehn [12] and Wagenaar [2]) has boosted a new generation of natural product libraries based on prefractionated extracts. Prefractionation is basically the fractionation of a crude extract prior to biological testing, and prefractionated samples are in general less complex than the crude extract for screening bioactive compounds from natural products. When a bioactive extract is identified, strategies leading to less complex library fractions before bioassay evaluation can pinpoint lead compounds in one step, reducing time and multiple steps such as those involved in bioassay-guided fractionation [13]. Cost, time, resources, number of compounds per fraction, and degree of diversity for libraries of a fixed size are some of the factors that must be considered when planning a prefractionated library [2].

A number of affinity-based approaches have been developed to detect ligand-protein interactions, with most using MS [6]. Frontal affinity chromatography MS and (pulsed) ultrafiltration MS are some affinity selection/MS methods that have been applied to the screening of compound libraries in HTS discovery programs in the field of natural products [6], which will be discussed in the following subsections.

The HTS strategies based on postcolumn analysis applied by several authors on marine natural products (MNP) to search for novel bioactive compounds will also be explored in the following subsection.

10.2.1 High-Throughput Screening Based on Precolumn Methodologies

The precolumn affinity principle is based on the noncovalent or covalent binding of a ligand to a target (protein, antibody, small molecule, or other bioactive agent) immobilized or not on a solid support, followed by the analysis of bound target-ligands [5,14].

Frontal affinity chromatography (FAC) is one of the most used affinity selection methods and it enables affinity calculation on the basis of the saturation time and shape of the breakthrough curve because a known concentration of analyte is infused onto an immobilized protein column [14]. The order of compound elution is a consequence of its affinity to the immobilized target with weak and strong binders having the shortest and longest elution times, respectively. The identification of eluting compounds occurs by coupling MS to FAC, which allows for the determination of its bioactivity or determination of dissociation constants [5]. Besides analysis of target-ligand interactions, FAC-MS is also able to screen complex mixtures for bioactive compounds. The review of FAC-MS on target-based drug discovery by Calleri *et al.* [15] shows that this technique can be used in HTS screening to determine relative binding affinities of a number of ligands in complex mixtures and in a single experiment. This complementary approach has been successfully applied to a wide variety of pharmaceutical targets against new chemical compounds. No studies were found on marine natural extracts, however. Luo *et al.* [16] applied FAC-MS to screen compounds from traditional Chinese herb extracts (*Phyllanthus urinaria* L.). Plant extracts were loaded on immobilized antibody columns (polyclonal antibodies mimicking hepatitis C Virus (HCV) NS3 protease) and analyzed by MS; five compounds with inhibitory activities to HCV NS3 (HCV is known to cause chronic infection) were detected: brevifolin, brevifolin carboxylic acid, corilagin, ellagic acid, and phyllanthusiin U [16].

Considered by several authors to be a complementary approach to FAC, zonal affinity chromatography (ZAC) is based on the injection of a known analyte amount mixed with a competitor. This method is more frequently used as a way to acquire more information on the target-ligand interaction by FAC using a measurement of additional binding data such as the location of a binding site [14]. For a detailed overview on FAC and ZAC, the reviews published by Jonker *et al.* [14] and Calleri *et al.* [15] should be revisited.

A second possible affinity-based selection approach, also known as affinity capture or affinity trapping method, is similar to affinity chromatography because its principle depends on analyte binding to immobilized proteins in a solid support but no ligand chromatographic separation is involved [14]. This approach of immobilization of a target on beads (magnetic or gellan beads) or other surfaces can be used in a selective way to extract ligands or bioactives from complex mixtures [6]. Affinity capture has higher selectivity than FAC and may involve incubation between target and ligand under native conditions [5]. This method could better simulate close conditions to those of biological environments. After incubation, unbound compounds have to be removed by washing, whereas bound compounds are characterized by MS [5]. In our literature review no studies were found in marine natural extracts based on affinity capture. Choi and Breevman [17] developed a screening assay for ligands to the estrogen receptor based on magnetic microparticles followed by LC-MS on plant extracts (*Trifolium pretense* L. and *Humulus lupulus* L.). Extracts were first incubated with functionalized (with estrogen receptors) magnetic particles and followed by magnetic separation of particles containing bound from unbound ligands in the extract. Ligands were separated using methanol and in the plant extracts, LC-MS identified genistein, daidzein, and estrogen 8-prenylnaringenin.

A third affinity-based selection approach, also known as direct affinity screening, is based on binding bioactive compounds to targets in solution that, after incubation, could involve ultrafiltration (UF), pulsed utrafiltration (PUF), separation by size-exclusion chromatography (SEC), or even separation by filtration or centrifugation to separate unbound from ligand-target complexes.

After the removal of unbound ligands, in general, a dissociation step of the ligand-target complexes using methanol or pH shift is conducted and ligands are characterized *a posteriori*, often by LC-MS. A detailed description of the affinity selection, based on binding bioactive compounds to targets in solution, can be found in the review by Jonker *et al.* [14] on developments in protein-ligand affinity mass spectrometry.

Choi *et al.* [18] developed a screening strategy using ultrafiltration LC-MS and applied it in the extracts of marine sediment bacteria to search for inhibitors of quinine reductase-2 (QR-2), which can present antimalarial and antitumor activities or even act as chemoprevention compounds interfering in the metabolic activation of toxic quinines. After incubation of marine extracts with human recombinant QR-2, the extracts were submitted to UF through 10,000 Da molecular weight cutoff UF membrane. After the removal of unbound ligands, the QR2-ligand solution was submitted to a new UF through 10,000 Da molecular weight cutoff UF membrane and the ligands were dissociated from QR-2 using methanol; dried QR2-ligands were reconstituted in 50% aqueous methanol and characterized by LC-MS. This UF LC-MS screening strategy led the authors [18] to discover an inhibitor of QR-2, namely tetrangulol methyl ether. Vu *et al.* [19] in turn developed a screening strategy, also based on direct affinity screening, to search for inhibitors of bovine carbonic anhydrase II (bCAII). They first used 10 alkaloid-enriched plant extracts and eight desalted marine extracts spiked with known specific inhibitors of bCAII, which after incubation with bCAII were analyzed using electrospray ionization (ESI) Fourier transform ion cyclotron resonance MS (ESI-FTICR-MS). After method validation with spiked samples, the authors then screened for active compounds in 85 methanolic plant extracts by direct infusion and online SEC with ESI-FTICR-MS and identified one noncovalent complex and an active compound (6-(1*S*-hydroxy-3-methyl-butyl)-7-methoxy-2H-chromen-2-one) with activity against bCAII [19].

Affinity capillary electrophoresis (ACE) has been used to study protein-ligand interactions, which can be applied to screening bioactive compound in natural extracts [6]. Technological advances have contributed to the enhancement of ACE coupling with MS or by the use of microfluidic chip technology [14]. Hyphenation of ACE with ESI/MS for example, allows direct characterization of complex formation as well as online structural determination of the target-ligand complexes without any isolation or purification steps [20]. Mobility shift and immobilized protein methods are two ACE approaches that have been used to study protein-ligand interactions. In mobility shift mode the protein-ligand forms a noncovalent complex in solution and a shift is detected in the capillary electrophoretic mobility of a target in response to the binding of an active compound, whereas the immobilized protein mode involves the immobilization of target protein onto a solid support [6,14]. Enzyme assays based on CE present advantages such as high-efficiency separation, and low sample volumes; and to separate zones of enzymes, substrates, and products in one single experiment [21]. These authors screened marine bacterial phosphatases by CE with laser induced fluorescence detection. In turn, Fermas *et al.* [20] used ACE-MS to study sulfated oligosaccharides heparin and fucoidan (a type of complex sulphated polysaccharide that contains α-1,3-linked sulphated L-fucose found in brown algae), binding and dimerization of stromal cell-derived factor-1. The authors found that chemokine stromal cell derived factor 1 (SDF-1) is a potent chemoattractant that is involved in leukocyte trafficking and metastasis; sulfated oligosaccharides binding of SDF-1 on cell surfaces could modulate its function as coreceptor of this chemokine. Therefore, the effect of sulfated oligosaccharides was studied on the oligomerization of SDF-1 and of its mutated form by ACE-ESI/MS. The SDF-1/oligosaccharide 1/1 complex was only observed with heparin disaccharide and fucoidan pentasaccharide, indicating the role of specific structural determinants in promoting dimer formation, which is important for chemokine functionality [20].

10.2.2 High-Throughput Screening Based on Postcolumn Methodologies

Gassner *et al.* [22] based the HTS strategy on a combination of HPLC methods with a HT yeast halo assay to search for new compounds with antifungal properties. Compounds from the National Cancer Institute (NCI) Developmental Therapeutics Program libraries, including 167 marine sponge crude extracts and 149 crude marine-derived fungal extracts, were screened for the identification of bioactive compounds with antifungal activity. Since a high hit rate was achieved by the HT halo assay for natural compounds of NCI libraries, the authors produced a screening/fractionation strategy (Table 10.1) based on the fractionation of crude extracts, with previous positive results (strong halo) by RP-HPLC. Fractions were screened for antifungal activity in an HT halo assay and, simultaneously, active wells were analyzed by LC-MS. This strategy allowed for the identification of crambescidin 800 as responsible for unreported antifungal bioactivity in the marine sponge *Monanchora unguifera* dichloromethane extract.

To make MNP more amenable to HTS-based methods, Bugni *et al.* [9] produced a MNP library characterized online using MS, from which a subset was screened in a phenotype-selective screen in order to identify compounds able to inhibit growth of BRCA2-deficient cells. The strategy involved an automated HPLC-MS fractionation protocol to generate natural product libraries pure enough for HTS. Fractions were directed into a 96-well plate format and each well was mapped by MS (Table 10.1). A single injection on the HPLC-MS using a C18 monolithic HPLC column yielded sufficient material for the bioactivity screening (in each of the 96 wells) and also for NMR spectroscopy studies. The sample extract needed to be desalted by drying extracts onto HP20SS (porous polystyrene-based absorbent) followed by solid-phase extraction before introduction to HPLC-MS. The effluent from the HPLC was split after a photo-diode array (PAD) detector to a fraction collector and to a Micromass Q-TOF (or ion-trap time of flight) micro MS. The contents of the collection plate were

TABLE 10.1 High-Throughput Screening Strategies Based on Postcolumn Methodologies Applied to Search for Novel Bioactive Compounds on Marine Natural Products

HTS Systems	Marine Natural Samples	Extraction, Fractionation, and Detection	Biological Screening	Ref.
HPLC methods with a HT yeast halo assay	Marine sponge: *Monanchora unguifera*	Extraction and prefractionation: Sponge samples extracted by accelerated solvent extraction (Hexanes, CH_2Cl_2, and MeOH) generating crude extracts. Detection and plus fractionation: Fractionation with HPLC and fraction collector triggered by ELSD in 96-well plates; evaporated fractions were reformatted in 384-well plates and screened for HT yeast halo assay and in parallel by LC-MS and NMR.	Antifungal screening by HT yeast (*Sacharomyces cerevisae*) halo assay; identification of crambescidin 800 as potent antifungal agent.	[22]
2D chromatographic separation, PDA and MS detection coupled with 96 well plates	Sponge specimens from Fiji: *Plakortis quasiamphiaster; Crella spinulata*	Extraction: Vertebrates extracted with MeOH. Fractionation: MeH extracts loaded on HP20SS; after drying five fractions were obtained and dispersed in 96-well plates for HPLC fractionation with C18 monolithic column. Detection: Effluent from HPLC was split after PDA detector to a fractions collector and to a Micromass Q-Tof micro MS; fractions were obtained using one-minute time slices and were mapped on both mass and PDA chromatograms.	Library was screened in a paired cell line phenotype assay to identify leads with selectivity toward BRCA-2 deficient tumor cells for breast cancer.	[9]

(Continued)

TABLE 10.1 High-Throughput Screening Strategies Based on Postcolumn Methodologies Applied to Search for Novel Bioactive Compounds on Marine Natural Products *(cont.)*

HTS Systems	Marine Natural Samples	Extraction, Fractionation, and Detection	Biological Screening	Ref.
LC-MS-UV-ELSD coupled with a panel of high-throughput 96-well plate bioassays	Marine sponge: *Cacospongia mycofijensis*	<u>Extraction and prefractionation</u>: Sponge samples extracted by accelerated solvent extraction (H_2O, Hexanes, CH_2Cl_2, and MeOH) generating semicrude extract assortments. <u>Detection and plus fractionation</u>: Prioritized active extracts submitted to LC-MS-UV-ELSD for library creation in 96-well plates.	Library was screened in a 4 bioassay: cytoskeletal assay with HeLa cells, *Trypanosoma brucei* assay, antiproliferative biosssays with 4 human cancer cell lines, MTT cytotoxicity assay in murine macrophage and prostate cancer cells. Four new compounds were discovered, namely two derivatives of aignopsane, with variable bioactivity.	[13]
Integrated five-channel chromatographic separation, dual UV-MS detection with *in vitro* 96-well plate based bioassays	Marine sponge: *Agelas mauritania*	<u>Extraction</u>: Frozen sponge extracted with solvents (acetone, Et_2O, butanol). <u>Fractionation</u>: Marine extracts loaded on WellChrom pump followed by splitting equally into five flow streams. One stream passed through an auto-sampler and others by four injectors; all connected to SB-C18 columns. <u>Detection</u>: First stream passed through a UV detector and ion trap MS in series for monitoring and identification; other four streams after HPLC column were split by fraction collectors in four 96-well plates with identical fractions (25 s/well).	Library was screened for antitumor activity toward MCF-7 cells (human breast adenocarcinoma cells): agelasine B and (-)-agelasine D, two diterpenoid derivatives, with antitumor activity.	[23]

split to make identical plates for material archive and screening purposes. Three compounds or less per well were detected and identified. With this approach, the authors [9] performed selective screens in a paired cell-line phenotype assay against BRCA2-deficient tumor cells (breast cancer), confirmed structures of the bioactive compounds, and identified mechanisms of action. NMR data from *C. spinulata* active subsamples revealed the presence of a mixture of plakinidines, not previously reported for this sponge. According to the authors [9], the selective activity of the plakinidines could have been masked in the crude extracts, thus becoming an example of the benefit of using prefractionated extracts. The main drawback is that the high resolution obtained from the separation step (chromatographic peaks of seconds to tens of seconds) is often lost in the low-resolution fraction collection (fraction collection in the minute range is common) for the bioactivity screen [3].

In another study, Bugni *et al.* [24] applied a similar strategy to marine invertebrates such as *Leucetta chagosensis*, *Theonella swinhoei*, or *Pseudoceratina purpurea*, and after the desalting step in HP20SS followed by HPLC-MS separation of fractions, a library of prefractionated extracts was generated; cytotoxicity of crude extracts and fractions were evaluated in human colorectal carcinoma cells (HCT-116 cells) using a MTT (3-(4,5-dimethylthiazol-2-yl)-2,5-diphenyltetrazolium bromide) cell proliferation assay. According to Bugni *et al.* [24], the prefractionation approach with HP20SS provided simultaneously a cost-effective method to desalt marine invertebrate extracts and an effective concentration of organic compounds responsible for higher cytotoxicity activity in the fractions in comparison to crude extracts. Fractions screening revealed a four-fold increase in the detected active cultures in comparison to crude extracts, which generates more leads for evaluation [2]. The integration of the HP20SS library with automated LC-MS fractionation protocol facilitates not only rapid replication and purification of bioactive compounds but also enables the assessment of variability in marine specimens due to seasonal or environmental changes resulting in higher chemical diversity of the library [24].

Johnson *et al.* [13,25] have developed high-throughput (HT) based strategies for the discovery of therapeutic leads. In 2010, they reported an assessing pressurized liquid extraction (accelerated solvent extraction) as an effective HT methodology for extracting marine organisms in order to streamline the discovery of bioactive compounds from marine organisms (namely, sponges) [25]. In 2011, a HTS strategy based on LC-MS-UC-ELSD (evaporative light scattering detector) was reported to generate libraries of marine sponge *Cacospongia mycofijensis* assessed by bioassays involving cytoskeletal profiling, tumor cell lines, and parasites; a schematic display of the HTS developed by Johnson *et al.* [13], involving four major steps, can be visualized in Figure 10.3. Basically, the authors performed the extraction and prefractionation of marine sponge samples, which produced semicrude extracts that were submitted to HT bioassays (Table 10.1). Prioritized active extracts were then submitted to LC-MS-UV-ELSD for library creation in 96-well microplates, and fractions were evaluated by the HT

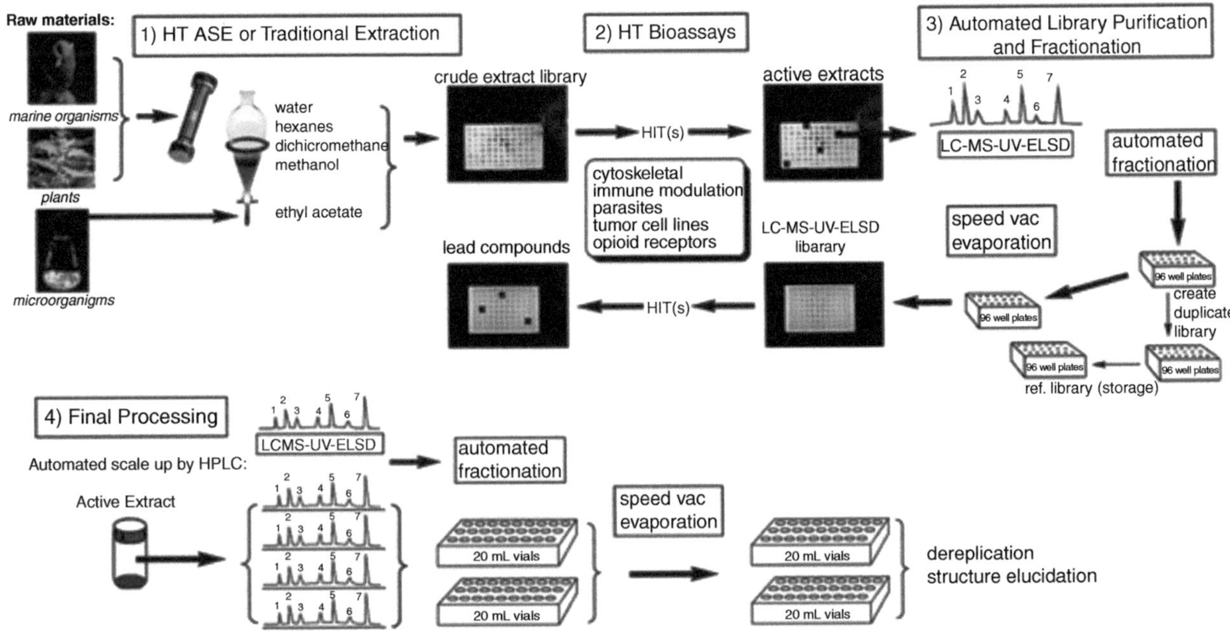

FIGURE 10.3 A schematic representation of a HTS approach involving (1) extraction (ASE), (2) bioassays, (3) automated library purification and fractionation, and (4) final processing using automated scale-up HPLC to initiate dereplication or structure elucidation of lead compounds. Reprinted from Johnson *et al.* [13]. Copyright © 2011, with kind permission of The American Chemical Society.

bioassays. Identified fractions with potential leads were scaled up by HPLC and LC-MS-UV-ELSD followed by evaporation, NMR, dereplication, and structural identification. Twelve known compounds were identified, some of them exhibiting bioactivity not previously reported, and four new compounds were discovered, namely aignopsanoic acid B, apo-latrunculin T, 20-methoxy-fijianolide A, and aignopsane ketal; apo-latrunculin T caused inhibition of parasite *T. brucei* without disrupting microfilament assembly, and 20-methoxy-fijianolide A revealed a modest microtubule-stabilizing effect [13]. This strategy, based on a multiassayed approach, is able to provide rapid identification of lead compounds with biological properties identifying new or known natural compounds with previously unreported biological properties.

A different HTS method was developed recently by Zhang *et al.* [23] in marine sponges integrating five-channel HPLC chromatographic separation, dual UV-MS detection, and microplate based bioassays (Figure 10.4). This method enables four identical sample fractions into 96-well microplates as well as the structural information of each group of fractions by UV-MS detection, simultaneously. This methodology was used to screen bioactive compounds from fractions of a marine sponge *A. mauritania*, and two diterpenoid derivatives (Table 10.1) were identified for the first time as antitumor compounds against MCF-7 cells (human breast adenocarcinoma cells). An example demonstrated that a prefractionation approach could result in a higher quality library with samples more tailored for the high-throughput drug discovery paradigm, as stated by Wagenaar [2]. According to Zhang *et al.* [23] the proposed method based on five-column parallel separations with online UV/MS and microplate assays, implies lower periods of time in comparison to bioactivity-guided isolation approach and, because the bioassays are performed virtually offline in microplates, more versatility in terms of different bioassays may be achieved (different incubation times, different additives, and readout formats). Online bioassays could be more limiting because they rely on simple and straightforward reactions.

According to Potterat and Hamburger [6], thin-layer chromatography (TLC) bioautography involved on the first attempt the correlation of biological information with chromatography separation based on the following principle: a microbial suspension, enzyme solution, or other biological target solution is

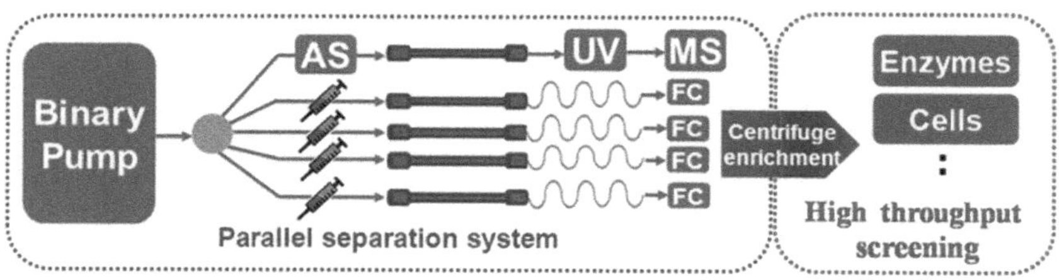

FIGURE 10.4 Schematic representation of the parallel HPLC/UV/MS based screening platform. Reprinted from Zhang *et al.* [23]. Copyright © 2013, with kind permission of Elsevier.

applied onto a developed TLC plate (with a sample that has migrated onto the plate with a suitable solvent), which, after an incubation period, reveals biological activity by appearance of inhibition zones—spot zones [6,26]. Such a technique requires small amounts of sample and it is able to analyze natural complex mixtures and to run several different samples at the same time. In addition, organic solvents can be removed before biological detection, eliminating not only possibilities of enzyme inhibition but also the threat of causing death to living microorganisms. The bioassays that are compatible with TLC such as antifungal and antibacterial assays, enzyme inhibition, antioxidant, and free radical scavenging activity have been reviewed by Marston *et al.* [26] whereas bioautography detection in TLC focused on microbiological screening methods was reviewed by Choma and Grzelak [27].

TLC is one of the most suitable separation methods for HTS analysis [28]. Hyphenation of TLC chromatography with spectroscopy methods such as MS by the development of elution or desorption-based approaches enabled the interfacing of TLC and MS, a subject reviewed by Morlock et al. [29]. Interfaces for the "online" extraction of analytes from the TLC plate prior to ESI-MS have been recently commercialized [6]. Taking into consideration the main differences in TLC-MS operational processes, these have been classified in two categories: (1) indirect MS analyses (or indirect sampling TLC-MS: offline detection), where analytes have to be removed from the TLC plate spot by scraping, extracting, purifying, and concentrating with its posterior analysis by MS-sample treatment processes must be carried out prior to MS analyses; (2) direct MS analyses (or direct sampling TLC-MS: online detection), where the analyte of interest on the TLC plate is characterized directly through MS without scraping, extraction, or concentration steps [28]. Direct TLC-MS analysis can be applied due to the development of MS techniques capable of direct surface sampling and ionization, which could be performed under vacuum but more recently under atmospheric pressure; detailed information on indirect and direct sampling TLC-MS techniques has been reviewed by Cheng *et al.* [28]. The stepwise automated planar chromatographic method enables evaluation first by the detection of spot zones contrasting with online column techniques [30].

Klöppel *et al.* [30] screened marine sponges for bioactivity-based secondary metabolites using high performance TLC (HPTLC) coupled with bioluminescence and MS. Methanolic marine sponges were separated in HPLTC plates and then coated with bioluminescence bacteria (*Vibrio fischeri*) by a dipping procedure. Bioactive compounds were detected by dark zones on the luminescent background; luminescent imaging can be captured rapidly and easily followed by monitoring time-dependent changes. The authors found that this approach was able to reveal new compounds by coupling HPTLC with bioluminescence to a high resolution MS. Mass spectra of bioactive zones were recorded within a small period of time by direct analysis in real time (DART), an ionization interface that works under ambient conditions (each zone was positioned in the excited gas stream of DART ionization interface); mass spectra were acquired in full-scan mode with the settings in positive-ion mode. Twelve different marine sponge species were screened for bioactivity by using *Vibrio fischeri* to detect secondary metabolite markers and avarol, avarone, aeroplysinin-1, and dienon. Identification of avarone was demonstrated by coupling HPTLC with an LTQ Orbitrap XL-MS via DART as ambient ionization interface [30].

10.3 HIGH-RESOLUTION-BASED SCREENING METHODS

Hyphenated techniques coupling online separation and BCD assays to detect and simultaneously identify bioactive compounds from complex mixtures are the basis of HRS methods. Shi *et al.* [1] developed an approach capable of rapid profiling and identification of individual bioactive compounds in complex mixtures, thus providing a powerful method for natural product-based drug discovery. The development of online postcolumn assays delivering bioactivity data for discrete HPLC peaks is of potential interest [6]. Pioneer HRS systems were based on ultraviolet (UV) or fluorescence detection but nowadays several HRS systems include diode array (DA), MS, and NMR detection as an integrated part of the analytical approach to identify and simultaneously quantify the bioactive compounds [1,4]. Generally, complex mixtures or extracts are separated by HPLC followed by a split to the online BCD assay and to parallel detection and identification techniques providing information on the active compounds [1,6]. Postcolumn splitting of the HPLC effluent to a fraction device allows on the one hand for the collection of UV-Vis and MS spectra and on the other hand the screening of the fractions in the BCD assays [8]. A schematic representation of the key elements in a HRS strategy is displayed in Figure 10.2.

Several recent reviews [1,8] focused on the online postcolumn BCD systems for screening and identifying bioactive compounds, and these have been divided in systems coupling antioxidant assays and in those coupling assays based on enzymes and receptors.

10.3.1 Online Postcolumn Antioxidant Assays

The search for novel and natural antioxidant compounds has been increasing in part because of the safety and toxicity problems of some synthetic antioxidants [31]. Antioxidants are compounds known to prevent a number of physiological and pathological damaging processes by free-radical reactions [1]. Antioxidant activity assays online with HPLC have been developed by research teams and have been reviewed specifically by Niederländer *et al.* [32]. The use of online assays for the rapid screening of antioxidant peaks in chromatograms can be divided in those assays involving [1,8,32]:

- Physiologically relevant reactive oxygen species (ROS) such as singlet oxygen, hydrogen peroxide, and superoxide anion, which provide the most realistic measure of antioxidant activity but are difficult to set up and control; ROS is able to oxidize a substrate in concentrations that can be measured.

- Single relatively stable reagents such as 1,1-diphenyl-2-picrylhydrazyl radical (DPPH) or 2,2′-azinobis-(3-ethylbenzothiazoline-6sulfnate) (ABTS), which are more used because they are simpler to set up and control; HPLC-DPPH or ABTS is based on the colorimetric estimation of the radicals DPPH* or ABTS*, which are scavenged by antioxidants resulting in a decrease in absorbance (negative peaks); these are often used for screening radical-scavenging activities of antioxidants of pure substances or complex mixtures.
- Electrochemical detection (coulometric or amperometric) online with a HPLC, which is a versatile detection method for substances with one or more electroactive compounds such as antioxidant polyphenolic compounds in several natural matrices; neither coulometric nor amperometric assays have been applied on a large scale.

In particular, spectrophotometric and chemiluminescense methods have been developed for the online postcolumn detection of antioxidant compounds after chromatographic separation. Chemiluminescense is used particularly to determine initial radical products of lipid oxidation as well as to determine the radical scavenging activity of antioxidants by employing the luminol-oxidation assay [1]; luminol can be oxidized by hydrogen peroxide or superoxide radical anion depending on the ROS generation system used [32]. Regardless of the detection method the basic instrument setup for the online HPLC antioxidant assay can be resumed in a few major steps:

1. Chromatographic separation after sample injection.
2. Flux split by an adjustable splitter.
3. Part of the flux is diverted to detection instrument: UV, MS, or NMR.
4. Part of the flux is mixed with online antioxidant assay reagents (postcolumn reagents) and the reaction occurs in a reaction coil.
5. A proper detector will measure the signal, which will enable the detection of antioxidant activity.
6. A correlation between the antioxidant activity and the presence of specific compounds detected previously by UV, MS, or NMR enables a rapid determination of antioxidant compounds.

HPLC-DPPH or ABTS-based methods have been successfully applied to evaluate radical-scavenging properties of many complex matrices of natural products especially of plant extracts, but not frequently applied to MNP. Chen *et al.* [31] developed a method based on HPLC coupled with ESI-TOF/MS and online ABTS free radical scavenging assay to screen and identify antioxidants in water extracts of *Hippocampus japonicus* Kaup (Table 10.2). The online instrumental system consisted essentially of two hyphenated parts: (1) an HPLC-DAD-MS system composed of one pump system, a C18 column, ESI-TOF-MS, and DAD able to identify chromatographic separated compounds; (2) a second HPLC pump, one mixer, one reaction coil, and a second DAD to detect ABTS free radical antioxidants (Figure 10.5).

TABLE 10.2 High-Resolution Screening Involving the Online Coupling of a Bioassay to Chromatographic Separation Applied to Search for Novel Bioactive Compounds on Marine Natural Products

HRS Systems	Marine Natural Samples	Extraction, Fractionation, and Detection	Biological Screening	Ref.
HPLC-ESI-TOF/MS on-line ABTS assay	Marine organism: *Hippocampus japonicus* Kaup	Extraction: Ultrasonic extraction of 0.5 g powdered sample in 50 mL of water for 2x 30 min; extracts were filtered and concentrated by evaporation. Separation and Detection: Crude extract was separated in HPLC-DAD (254 nm). Then, the LC-separated effluents were split into two parts by a three-way joint. One part was analyzed by ESITOF-MS. The other part was mixed with the ABTS reagent, which was transported by a second pump. The induced bleaching reacted through a coil and was detected as a negative peak photometrically at 734 nm by the second DAD.	Three radical scavenging compounds were identified in *Hippocampus japonicus* Kaup water extracts: hypoxanthine, phenylalanine, and tryptophane.	[31]
HPLC on-line ABTS assay	Green microalga: *Chlorella vulgaris*	Extraction: Maceration, ultrasound-assisted extraction and pressurized liquid extraction of 0.5 g freeze-dried sample in ethanol. Separation and Detection: Extract was separated in HPLC-DAD (254 nm for phenolics, 445 nm for carotenoids, 660 nm for chlorophylls). Then, the HPLC eluent was separated by a "T" piece. Part of the eluent received ABTS reagent and absorbance was measured at 734 nm by UV-Vis detector. Chemical compounds were identified by MS with atmospheric pressure chemical ionization (APCI) in positive mode.	Fifteen major antioxidants were identified: hydrophilic compounds, lutein and its isomers, chlorophylls, and chlorophyll derivatives.	[33]

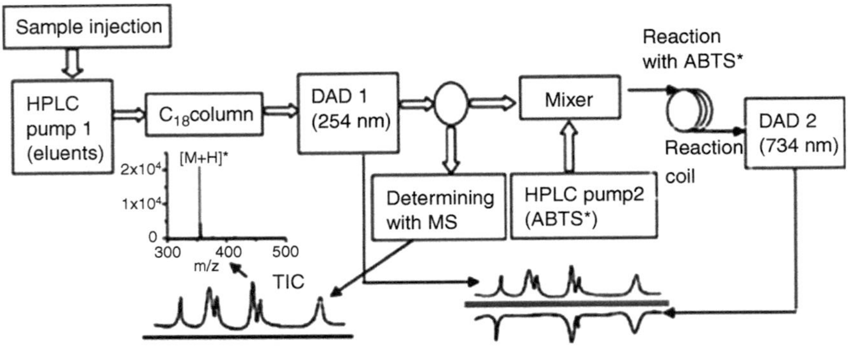

FIGURE 10.5 Schematic representation of the HPLC-DAD coupled online to ESI-TOF/MS and ABTS assay. Reprinted from Chen *et al.* [31]. Copyright © 2010, with kind permission of WILEY-VCH Verlag GmbH & Co. KGaA, Wheinhem.

Chen *et al.* [31] was able to screen and identify antioxidants in *Hippocampus japonicus* Kaup extracts after several steps for selection of suitable HPLC separation conditions in order to obtain symmetrical and well-resolved peaks and optimization, validation of online screening conditions, namely the resolution of negative peaks after reaction coil of several antioxidants such as caffeic acid, catechin hydrate, epigallocatechin, or gallocatechin. UV chromatogram (with positive peaks at 254 nm) and ABTS radical scavenging detection chromatogram (with negative peaks at 734 nm) indicated three possible compounds with inhibitory activity to scavenge ABTS radical (Figure 10.6); these were identified

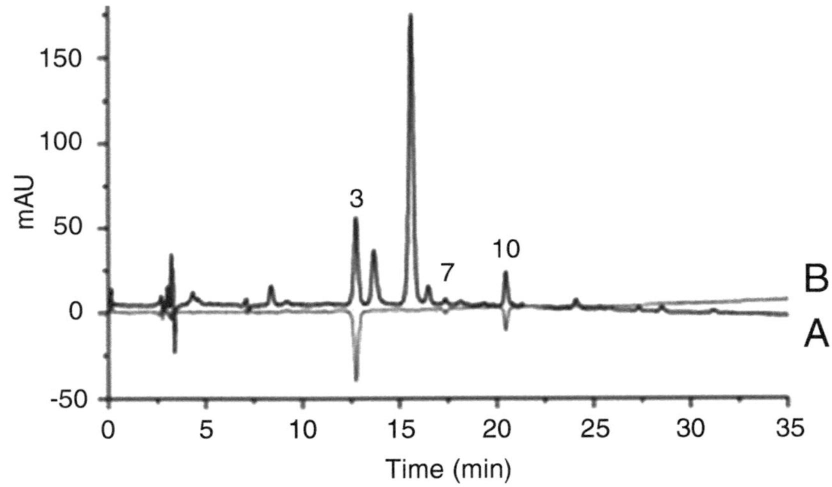

FIGURE 10.6 Chromatograms of water extract from *Hippocampus japonicus* Kaup analyzed by online HPLC-DAD/MS-ABTS; The chromatogram from the antioxidant activity assay was detected by DAD1 at 254 nm (A), DAD2 at 734 nm (B). Reprinted from Chen *et al.* [31]. Copyright © 2010, with kind permission of WILEY-VCH Verlag GmbH & Co. KGaA, Wheinhem.

by using the online ESI-TOF/MS with negative ion mode as hypoxanthine, phenylalanine, and tryptophane (Table 10.2) without any purification step.

Cha *et al.* [33] also designed a HPLC online ABTS assay to screen antioxidant compounds from different extraction methods of green microalgae *Chlorella vulgaris* (Table 10.2). Pressure liquid extraction with ethanol at high temperature (85–160 °C) increased significantly antioxidant extraction, improving the formation of hydrophilic compounds and yielding more antioxidative chlorophyll derivatives. The study found that online HPLC ABTS analysis was a powerful tool for the separation of the main antioxidants from the extracts, and allowed for the identification of hydrophilic compounds, lutein and its isomers, chlorophylls, and chlorophylls derivatives (Table 10.2).

10.3.2 Online Postcolumn Affinity/Activity BCD Methods

The online chromatographic affinity/activity BCD methods include those involving enzyme activity/affinity detection (EAD) or receptor affinity detection (RAD). Enzymes and enzyme inhibitors represent a very important class of drug targets and online EAD assays can be classified according to applied detector mode: colorimeric, fluorometric, and MS [1]. The instrumental configuration is similar to or more complex than those used in the online postcolumn antioxidant assays. In most EAD or RAD assays flow-splitters and more than one reaction coil is mandatory. Part of the flow must be directed to the BCD assay, which could include preincubation phases of the HPLC effluent with receptors or enzymes in an incubated coil before addition of substrates [8]. Furthermore, EAD and RAD can be classified as homogenous (with a distinct signal intensity from bound or unbound ligands) or heterogeneous assays (which need to distinguish bound or unbound ligands by a separation step). Several constraints arise from the combination of biological material such as enzymes, or receptors, and chromatographic conditions, or solvents. Malherbe *et al.* [8] in their review mention some considerations and requirements that must be kept in mind for the successful development of online HPLC-EAD or HPLC-RAD assays:

1. In order to keep the organic modifier content constant or below inhibitory concentrations through the addition of buffer, make-up flow systems are often needed, using, for example, a counteracting gradient.
2. Removal of nonspecific binding biological material from the reaction coil is often necessary.
3. Sensitivity increasing parameters must be optimized, such as reaction times and temperatures with suitable buffers.
4. Biological material has to be incorporated in the HPLC effluent in a nonabrasive way by using, for example, syringe pumps.

Several authors developed EAD assays to search for specific enzyme inhibitors or receptor ligands as possible therapeutics to treat some disease disorders. Acetylcholinesterase inhibitors for Alzheimer's disease treatment, α-glucosidase inhibitors for Diabetes type II treatment, liver cytochrome P540 ligands for cancer prevention, or estrogen receptor ligands from endocrine-disrupting compounds are some examples of application of EAD or RAD assays [34–37]. To the best of our knowledge no studies have yet been performed with MNP. Most of the studies focus on plant extracts. Recently, Falck *et al.* [38] developed an online liquid chromatography-biochemical detection (LC-BCD) method to search for soluble epoxide hydrolase (sHE) inhibitors in mixtures that will be explored as a study example and that could be applied in marine natural examples. Newman *et al.* [39] discuss that sHE action impacts pain and inflammation processes as well as blood pressure regulation, which means that compounds such as epoxides or other substrates could be used as anti-inflammatory, analgesic, and blood pressure regulators. Figure 10.7 schematically represents the developed LC-BCD method, which consists of three main parts: (1) separation, with an auto-injector and a RP-LC column; then, the flow is postcolumn split for (2) the online BCD and for (3) parallel spectrometric detection (by UV at 210 nm and/or by ESI-MS) [38]. The online BCD comprises the mixing of postcolumn eluent with sHE solution, followed by incubation in a reaction coil, and the addition of nonfluorescent substrate (PHOME: (3-phenyl-oxiranyl)-acetic acid cyano-(6-methoxy-naphthalen-2-yl)-methyl ester) followed by another incubation period and fluorescence detection; PHOME is converted to a fluorescent product (6-methoxy-2-naphtaldehyde) by sHE, allowing for the detection of enzyme activity. The enzyme inhibition can be measured by detecting a decrease in the enzymatic formation of the fluorescent product. The splitting of the postcolumn

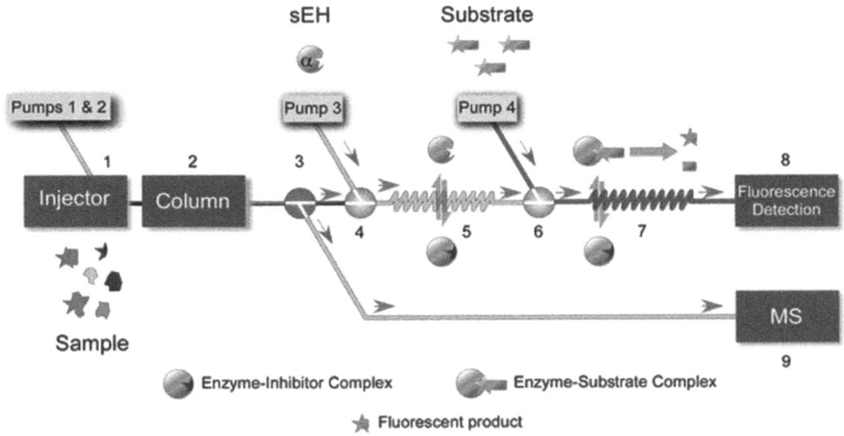

FIGURE 10.7 Setup of the LC–BCD system combining separation, online BCD and additional UV or MS detection in parallel: (1) Auto injector, (2) RP-LC column, (3) Flow-splitting between parallel UV or ESI–MS (9) detection and the online BCD (4–8). The BCD comprises of mixing of LC effluent (4) and an sEH solution, incubation with the enzyme (5), followed by mixing of PHOME solution (6), incubation with PHOME (7), and finally fluorescence detection (8). Reprinted from Falck *et al.* [38]. Copyright © 2012, with kind permission of Springer Science + Business Media.

LC eluent between BCD and spectrometry detection enables correlation of bioactivity with identification of individual compounds. According to Falck *et al.* [38], the detection principle of the new LC-CBD was a continuous-flow enzyme activity assay coupled online to LC with a parallel MS detection—an example of an EAD assay for search-specific enzyme inhibitors, where the substrate PHOME allowed for a sensitive and robust monitoring of bioactivity analysis of individual compounds in mixtures.

An alternative RAD assay was developed by Heus *et al.* [40] in order to diminish consumption of sample, receptor protein, and (fluorescently labeled) tracer ligand. These authors developed a nano-LC online to a light-emitting diode (LED) based capillary confocal fluorescence detection system capable of online BCD (bioassay for the acetylcholine binding protein-a structural analog of neuronal nicotinic acetylcholine receptors) with low-flow rates. This method is able to identify the bioactive compounds in small amounts of mixture samples, such as small volumes of natural extracts, in comparison to the method developed by Kool *et al.* [41] consisting of an online fluorescence enhancement assay for the acetylcholine binding protein with parallel MS identification.

10.3.3 Online MS-Based Assays for Identification and Bioassay Readout

According to Kool *et al.* [4] there are three variants of online MS-based bioassays: (1) MS is used for identification only, which is the variant already mentioned and discussed in the previous subsections of this chapter; (2) MS is used for identification and bioassay readout; and (3) MS detection is conducted in parallel with at-line microfractionation with offline bioaffinity analysis, which is beyond the scope of this chapter. In these three approaches a common objective is present: to correlate chemical information derived from MS detection with bioaffinity information [4].

One possible schematic representation of an online bioaffinity analysis setup with MS-based bioassay detection is similar to that displayed in Figure 10.7, with a MS replacing fluorescence detection, enabling the measure of different target-ligand/substrate interactions and simultaneously their identification in one bioassay such as by de Jong *et al.* [42]. MS can be used in two modes: (1) selected-ion monitoring (SIM) or selected-reaction monitoring (SRM) to be able to detect and quantify the product formed in enzymatic drug target studies or to detect and quantify free ligand concentration in ligand-binding assays; (2) full-spectrum mode to identify eluting ligand, which could involve data-dependent switching to MS^n acquisition [4]. A combination of SIM analysis of the products and full-spectrum analysis enables the detection of unknown inhibitors using an extracted-ion-chromatogram (EIC); the correlation between SIM and EIC allows for peak shape and retention time correlation between chemical information and bioaffinity (*m/z* of the inhibitor) [43]. Depending on the mode and the respective requirements in terms of sensitivity, resolution, and the possibility of analyte fragmentation, different MS specifications must be involved. It seems that until now, the Q-TOF MS has been the instrument of choice because it enables the monitoring mode (mode 1) with enough specificity and sensitivity as well as full-spectrum mode (mode 2), also with enough resolution and sensitivity with the possibility of analyte fragmentation. More information on protein-ligand affinity mass spectrometry developments can be found in the review by Jonker *et al.* [14].

Some benefits are reported when using MS instead of fluorescence readout where there is no need for fluorescent probes, tracers, or substrates. When performing the fluorescent labeling some problems may arise since ligands or substrates must be chemically modified to incorporate the fluorescent label required [4]. In addition, MS can simultaneously detect a number of compounds such as substrates, products, and/or inhibitors so that enzyme activity and/or inhibition as well as quantification of inhibition constants (IC_{50}) of active compounds from complex mixtures can be monitored [1]. However potential ESI ion contamination and associated ionization suppression due, for example, to the continuous influx of compounds such as proteins, buffers, or other bioassays compounds makes some online bioassays not truly compatible with the ESI-MS readout [4].

No studies were found on application of online bioassays with MS readout in the field of MNP. De Jong *et al.* [42] applied a HPLC-MS methodology for the screening of acetylcholinesterase (AChE) inhibitors in *Narcissus* plant extracts using MS for both readout of bioassay and acquisition of MS and MS-MS data of detected active compounds. According to the authors AChE inhibitors were detected by measuring a decrease of the conversion of substrate acetylcholine into choline by AChE using electrospray MS with ammonium bicarbonate as a compatible buffer between BCD assay and MS detector. The screening of ethanol crude *Narcissus* extract by HPLC-MS was able to detect native galanthamine with anti-AChE activity, proving to be a sensitive and robust methodology to identify the active compound in a plant extract, shortening the time to match chemical identification with bioactivity [42].

Mroczek [44] published a study comparing the screening of AChE inhibitors from solid extracted plant extracts by HPLC/ESI-TOF-MS (an HRS approach) and by a TLC-based bioautography (and HTS approach). The TLC-bioautography was shown to be able to detect more potent AChE inhibitors than known galanthamine, namely 1,2-dihydrogalanthamine in *Narcissus jonquilla* "Pipit" extract.

10.4 CONCLUSIONS AND PERSPECTIVES

The simultaneous screening and identification of bioactive compounds in complex matrices such as MNP using HTS or HRS approaches without prior purification steps are of potential interest although still difficult and challenging. In general, the methodologies described were able to detect and identify new bioactive compounds or other known compounds for which bioactivity had not yet been assigned.

Not many marine samples have been a target of HTS or HRS methods; marine sponges, however, have been the most studied samples especially by HTS methods based on postcolumn methodologies. In comparison, plant extracts have been much more studied by both methods than extracts of marine samples. The known potential of marine natural products in terms of bioactive compounds and the capacity of the simultaneous screening and identification of these compounds, especially in online combination with BCD assays, will lead to new studies in the coming years.

The online HPLC-BCD assays are seen with great interest in several fields, especially for searching for new drugs, because they have been recognized as being able to fast-track identification of novel bioactive compounds in several complex natural matrices. These approaches help to overcome a number of limitations such as those associated with preisolation when analyzing the effects of bioactive compounds after separation (postcolumn) and minimizing the number of *in vitro* bioassays.

Technological advances in HPLC coupled with MS or NMR boosted different experimental setups with great impact by delivering abundant structural information about compounds of interest in natural extracts or in their fractions and enabling efficient dereplication of known compounds. Despite the advantages of simultaneous screening and identification of bioactive compounds in complex matrices, especially by online post-column HPLC-BCD assays, only a few studies were found for MNP except some with online antioxidant assays. Some possible explanations could be related to the costs and complexity of the available experimental setups in online format, which are still limited and seems to remain in the domain of a limited number of research groups and/or scientists.

Novel and efficient strategies for searching for new compounds in extracts and fractions should be the pursuit of MNP. There is an increasing demand for biodiversity from natural resources for therapeutic drugs such as antimicrobial drugs to face the patterns of resistance of pathogens. In the future, other possible challenges for online screening approaches could involve the integration of more complex biological systems, such as cells, which will open new scientific frontiers in the search for and study of bioactive compounds.

ACKNOWLEDGMENTS

This work was supported by Portuguese Science Foundation (Fundação para a Ciência e Tecnologia), through individual research grants, references SFRH/BPD/73781/2010 and SFRH/BPD/65410/2009, under QREN-POPH funds, cofinanced by the European Social Fund and Portuguese National Funds from MCTES.

The authors wish to thank Maria C. Arau Ribeiro, PhD for editing the manuscript and for helpful suggestions in the use of the English language.

REFERENCES

[1] S. Shi, H. Zhou, Y. Zhang, X. Jiang, X. Chen, K. Huang, Coupling HPLC to on-line, post-column (bio)chemical assays for high-resolution screening of bioactive compounds from complex mixtures, Trends Anal. Chem. 28 (2009) 865–877.

[2] M.M. Wagenaar, Pre-fractionated microbial samples – The second generation natural products library at Wyeth, Molecules 13 (2008) 1406–1426.

[3] K. Duarte, T.A.P. Rocha-Santos, A.C. Freitas, A.C. Duarte, Analytical techniques for discovery compounds from marine fungi, Trends Anal. Chem. 34 (2012) 97–110.

[4] J. Kool, M. Giera, H. Irth, W.M.A. Niessen, Advances in mass spectrometry-based post column bioaffinity profiling of mixtures, Anal. Bioanal. Chem. 399 (2011) 2655–2668.

[5] M. Giera, H. Irth, Simultaneously screening and chemical characterization of bioactive compounds using LC-MS based technologies (affinity chromatography), in: B. Werner (Ed.), Effect-Directed Analysis of Complex Environmental Contamination, The Handbook of Environmental Chemistry, 15, Springer, 2011, pp. 119–141.

[6] O. Potterat, M. Hamburger, Concepts and technologies for tracking bioactive compounds in natural product extracts: generation of libraries, and hyphenation of analytical process with bioassays, Nat. Prod. Rep. 30 (2013) 546–564.

[7] J. Kool, M. Eggink, H. van Rossum, S.M. van Liempd, D.A. van Elswijk, H. Irth, J.N. Commandeur, J.H. Meerman, N.P. Vermeulen, Online biochemical detection of glutathione-S-transferase P1-specific inhibitors in complex mixtures, J. Biomol. Screen. 12 (2007) 396–405.

[8] C.J. Malherbe, D. de Beer, E. Joubert, Development of on-line high performance liquid chromatography (HPLC)-biochemical detection methods as tools in the identification of bioactives, Int. J. Mol. Sci. 13 (2012) 3101–3133.

[9] T.S. Bugni, B. Richards, L. Bhoite, D. Cimbora, M.K. Harper, C.M. Ireland, Marine natural product libraries for high-throughput screening and rapid drug discovery, J. Nat. Prod. 71 (2008) 1095–1098.

[10] P. Gribbon, A. Sewing, High-throughput Drug Discovery: What Can We Expect from HTS? Drug Discov. Today 10 (2005) 17–22.

[11] A.K. Croft, W. Groenewald, M.S. Tierney, Medicinal chemistry and ligand profiling for evaluation of promising marine bioactive molecules, In: M. Hayes (Ed.), Marine Bioactive Compounds: Sources, Characterization and Applications, Springer Science+Business Media, LLC, 2012.

[12] F.E. Koehn, High impact technologies for natural products screening, in: F. Petersen, R. Amstutz (Eds.), Progress in Drug Research, 65, Birkhäuser Verlag, Basel, 2008.

[13] T.A. Johnson, J. Sohn, W.D. Inman, S.A. Estee, S.T. Loveridge, H.C. Vervoort, K. Tenney, J. Liu, K.K. Ang, J. Ratnam, W.M. Bray, N.C. Gassner, Y.Y. Shen, R.S. Lokey, J.H. MacKerrow, K. Boundy-Mills, A. Nukanto, A. Kanti, H. Julistono, L.B.S. Kardono, L.F. Bjeldanes, P. Crews, Natural product libraries to accelerate the high-throughput discovery of therapeutics leads, J. Nat. Prod. 74 (2011) 2545–2555.

[14] N. Jonker, J. Kool, H. Irth, W.M.A. Niessen, Recent developments in protein-ligand affinity mass spectrometry, Anal. Bioanal. Chem. 399 (2011) 2669–2681.

[15] E. Calleri, C. Temporini, G. Caccialanza, G. Massolini, Target-based drug discovery: the emerging success of frontal affinity chromatography coupled to mass spectrometry, ChemMedChem 4 (2009) 905–916.

[16] H. Luo, L. Cheng, Z. Li, Z. Ding, X. Xu, Frontal immunoaffinity chromatography with mass spectrometric detection: a method for finding active compounds from traditional Chinese herbs, Anal. Chem. 75 (2003) 3994–3998.

[17] Y. Choi, R.B. van Breeman, Development of a screening assay for ligands to the estrogen receptor based on magnetic microparticles and LC-MS, Comb. Chem. High Throughput Screen. 11 (2008) 1–6.

[18] Y. Choi, K. Jermihov, S. Nam, M. Sturdy, K. Maloney, X. Qiu, L.R. Chadwick, M. Main, S. Chen, A.D. Mesecar, N.R. Farnsworth, G.F. Pauli, W. Fenical, J.M. Pezzuto, R.B. van Breeman, Screening natural products for inhibitors of quinine reductase-2 using ultrafiltration LC-MS, Anal. Chem. 83 (2011) 1048–1052.

[19] H. Vu, N.B. Pham, R.J. Quinn, Direct screening of natural product extracts using mass spectrometry, J. Biomol. Screening 13 (2008) 265–275.

[20] S. Fermas, F. Gonnet, A. Sutton, N. Charnaux, B. Mulloy, Y. Du, F. Baleux, R. Daniel, Sulfate oligosaccharides (heparin and fucoidan) binding and dimerization of stromal cell-derived factor-1 (SDF-1/CXCL 12) are coupled as evidenced by affinity CE-MS analysis, Glycobiology 18 (2008) 1054–1064.

[21] K.D. Chischester, M. Sebastian, J.W. Ammerman, C.L. Colyer, Enzymatic assay of marine bacterial phosphatases by capillary electrophoresis with laser-induced fluorescence detection, Electrophoresis 29 (2008) 3810–3816.

[22] N.C. Gassner, C.M. Tamble, J.E. Bock, N. Cotton, K.N. White, K. Tenney, R.P. St. Onge, M.J. Proctor, G. Giaever, C. Nislow, R.W. Davis, P. Crews, T.R. Holman, R.S. Lokey, Accelerating the discovery of biologically active small molecules using a high-throughput yeast halo assay, J. Nat. Prod. 70 (2007) 383–390.

[23] Y. Zhang, S. Xiao, L. Sun, Z. Ge, F. Fang, W. Zhang, Y. Wang, Y. Cheng, Rapid screening of bioactive compounds from natural products by integrating 5-channel parallel chromatography coupled with on-line mass spectrometry and microplate based assays, Anal. Chim. Acta 777 (2013) 49–56.

[24] T.S. Bugni, M.K. Harper, M.W.B. McCulloch, J. Reppart, C.M. Ireland, Fractionated marine invertebrate extract libraries for drug discovery, Molecules 13 (2008) 1372–1383.

[25] T.A. Johnson, M.V.C. Morgan, N.A. Aratow, S.A. Estee, K.V. Sashidhara, S.T. Loveridge, N.L. Segraves, P. Crews, Assessing pressurized liquid extraction for the high-throughput extraction of marine-sponge-derived natural products, J. Nat. Prod. 73 (2010) 359–364.

[26] A. Marston, Thin-layer chromatography with biological detection in phytochemistry, J. Chromatography A 1218 (2011) 2676–2683.

[27] I.M. Choma, E.M. Grzelak, Bioautography detection in thin-layer chromatography, J. Chromatogr. A 1218 (2011) 2684–2691.

[28] S. Cheng, M. Huang, J. Shiea, Thin-layer chromatography/mass spectrometry, J. Chromatogr. A 1218 (2011) 2700–2711.

[29] G. Morlock, W. Schwack, Hyphenations in planar chromatography, J. Chromatogr. A 1217 (2010) 6600–6609.

[30] A. Klöppel, W. Grasse, F. Brümmer, G.E. Morlock, HPTLC coupled with bioluminescence and mass spectrometry for bioactivity-based analysis of secondary metabolites in marine sponges, J. Planar Chromatogr. 21 (2008) 431–436.

[31] J. Chen, H. Zhao, Q. Shi, D. Zhang, H. Cheng, X. Wang, F.S. Lee, Rapid screening and identification of the antioxidants in *Hippocampus japonicus* Kaup by HPLC-ESI-TOF/MS and on-line ABTS free radical scavenging assay, J. Sep. Sci. 33 (2010) 672–677.

[32] H.A.G. Niederlander, T.A. van Beek, A. Bartasiute, I.I. Koleva, Antioxidant activity assays on-line with liquid chromatography, J. Chromatogr. A 1210 (2010) 121–134.

[33] K.H. Cha, S.W. Kang, C.Y. Kim, B.H. Um, Y.R. Na, C. Pan, Effect of pressurized liquids on extraction of antioxidants from Chlorella vulgaris, J. Agric. Food Chem. 58 (2010) 4756–4761.

[34] L.A. Marques, J. Kool, F. de Kanter, H. Lingeman, W. Niessen, H. Irth, Production and on-line acetylcholinesterase bioactivity profiling of chemical and biological degradation products of tacrine, J. Pharm. Biomed. Anal. 53 (2010) 609–616.

[35] S.P. Li, J. Zhao, B. Yang, Strategies for quality control of Chinese medicines, J. Pharma. Biomed. Anal. 55 (2011) 802–809.

[36] J. Kool, S.M. van Liempd, H. van Rossum, D.A. van Elswijk, H. Irth, J.N.M. Commandeur, N.P.E. Vermeulen, Development of Three Parallel Cytochrome P450 Enzyme Affinity Detection Systems Coupled On-line to Gradient High-Performance Liquid Chromatography, Drug Met. Disposition 35 (2007) 640–648.

[37] J. Reinen, J. Kool, N.P.E. Vermeulen, Reversed-phase liquid chromatography coupled on-line to estrogen receptor bioaffinity detection based on fluorescence polarization, Anal Bioanal. Chem. 390 (2008) 1987–1988.

[38] D. Falck, N.H. Schebb, S. Prihatiningtyas, J. Zhang, F. Heus, C. Morisseau, J. Kool, B.D. Hammock, W.M.A. Niessen, Development of on-line liquid chromatography-biochemical detection for soluble epoxide hydrolase inhibitors in mixtures, Chromatographia 76 (2013) 13–21.

[39] J.W. Newman, C. Morisseau, B.D. Hammock, Epoxide hydrolases: their roles and interactions with lipid metabolism, Prog. Lipid Res. 44 (2005) 1–51.

[40] F. Heus, M. Giera, G.E. de Kloe, D. van Iperen, J. Buijs, T.T. Nahar, A.B. Smit, H. Lingeman, I.J.P. de Esch, W.M.A. Niessen, H. Irth, J. Kool, Development of a microfluidic confocal fluorescence detection system for the hyphenation of nano-LC to on-line biochemical assays, Anal. Bioanal. Chem. 398 (2010) 3023–3032.

[41] J. Kool, G.E. de Kloe, B. Bruyneel, J.S. de Vlieger, K. Retra, M. Wijtmans, R. van Elk, A.B. Smith, R. Leurs, H. Lingeman, I.J.P. de Esch, H. Irth, Online fluorescence enhancement assay for the acetylcholine binding protein with parallel mass spectrometric identification, J. Med. Chem. 53 (2010) 4720–4730.

[42] C.F. de Jong, R.J.E. Derks, B. Bruyneel, W. Niessen, H. Irth, High-performance liquid-chromatography-mass spectrometry-based acetylcholinesterase assay for the screening of inhibitors in natural extracts, J. Chromatogr. A 1112 (2006) 303–310.

[43] A.R. de Boer, T. Letzel, D.A. van Elswijk, H. Lingeman, W.M. Niessen, H. Irth, On-line coupling of high-performance liquid chromatography to a continuous-flow enzyme assay based on electrospray ionization mass spectrometry, Anal. Chem. 76 (2004) 3155–3161.

[44] T. Mroczek, Highly efficient, selective and sensitive molecular screening of acetylcholinesterase inhibitors of natural origin by solid-phase extraction-liquid chromatography/electrospray ionization-octopole-orthogonal acceleration time-of-flight-mass spectrometry and novel thin-layer chromatography-based bioautography, J. Chromatogr. A 1216 (2009) 2519–2528.

Index

A

B

C

D

F

G

N

Q

R

S

T

U

V

W

X

Z

Kolibri 360 GmbH
Ferdinand-Jühlke-Straße 7
99095 Erfurt, Deutschland
produktsicherheit@kolibri360.de